A Problem-Solving Approach to Electric Circuits

Farzin Asadi

A Problem-Solving Approach to Electric Circuits

Volume III

Springer

Farzin Asadi
Ostim Technical University
Ankara, Türkiye

ISBN 978-3-031-95492-4   ISBN 978-3-031-95493-1   (eBook)
https://doi.org/10.1007/978-3-031-95493-1

A Problem-Solving Approach to Electric Circuits III

This Springer imprint is published by the registered company Springer Nature Switzerland AG
The registered company address is: Gewerbestrasse 11, 6330 Cham, Switzerland

If disposing of this product, please recycle the paper.

Preface

Preface

Welcome to "A Problem-Solving Approach to Electric Circuits III", a guide designed to empower you with the practical skills needed to master the fundamental principles of electrical circuits. Electrical engineering, at its core, is about understanding and manipulating the flow of electricity, and this book aims to equip you with the tools to do just that.

As you delve into the world of circuits, you'll find that theory and practice are inseparable. It's not enough to simply memorize formulas; you must learn to apply them to real-world scenarios. This book emphasizes a problem-solving methodology, guiding you through the process of analyzing circuits, identifying key concepts, and developing effective solutions.

We'll delve into the fundamental principles of circuit analysis, including Kirchhoff's Voltage and Current Laws (KVL and KCL), superposition, and the powerful Thévenin and Norton theorems. We'll also explore the behavior of first- and second-order circuits, as well as the intricacies of single-phase and three-phase systems, equipping you with the tools to analyze a wide range of electrical circuits.

This book is structured to build your confidence and proficiency step-by-step. Each chapter presents core concepts followed by a variety of solved problems, allowing you to see theory in action. You'll learn to break down complex problems into manageable parts, apply appropriate techniques, and verify your solutions.

Whether you are a student embarking on your electrical engineering journey or a professional seeking to refresh your knowledge, this book provides a comprehensive and practical approach to mastering electric circuits. We'll move beyond rote memorization and cultivate a deeper understanding of the underlying principles.

I encourage you to actively engage with the material, work through the problems, and challenge yourself to apply the concepts to new situations. By adopting a problem-solving mindset, you'll not only succeed in this course but also develop invaluable skills that will serve you throughout your engineering career.

Let's embark on this exciting journey together, and unlock the power of electric circuits!

Ankara, Türkiye   Farzin Asadi

Important Notes About This Book

Throughout this book, calculation results are rounded to four decimal places. Please be aware that this rounding may lead to slight discrepancies in expected equalities. For example, when applying Kirchhoff's Current Law (KCL) at a node, the sum of currents might appear as a small non-zero value (e.g., 0.0007) instead of precisely zero. These minor deviations are solely due to rounding and the limited number of decimal places used in the calculations, and *not* a result of errors in the underlying principles or methodologies.

Competing Interests The author has no competing interests to declare that are relevant to the content of this manuscript.

Contents

Chapter 1
The Operational Amplifier

1.1 Introduction

Operational amplifiers (Op-amps) are fundamental building blocks in analog circuits. Their versatility stems from their ability to perform a wide range of functions,
including amplification, filtering, signal conditioning, and mathematical operations.

Op-amps are crucial in applications ranging from audio equipment and instru-
mentation to industrial control systems and communication devices. Their high input
impedance, low output impedance, and high gain make them ideal for precise signal
processing and manipulation.

Op-amps require two supply terminals for proper operation. These are typically
labeled as V_{SS} and V_{DD}, or V_{EE} and V_{CC}, representing the negative and positive
supply voltages, respectively (Fig. 1.1). Additionally, they feature two input terminals: the inverting input ($-$) and the non-inverting input ($+$).

Fig. 1.1 Op-amp symbol
with power supply
connections

V_{CC}

Non-inverting input — $+$

Inverting input — $-$

output

V_{EE}

V_{DD}

Non-inverting input — $+$

Inverting input — $-$

output

V_{SS}

© The Author(s), under exclusive license to Springer Nature Switzerland AG 2025   1
F. Asadi, *A Problem-Solving Approach to Electric Circuits*,
https://doi.org/10.1007/978-3-031-95493-1_1

V_{CC}/V_{EE} terminals are typically used for op-amps that require a dual (or bipolar) power supply. This means that the op-amp needs both a positive and a negative voltage supply (for instance, +12 V and −12 V). V_{CC} is the positive supply voltage, while V_{EE} is the negative supply voltage.

V_{SS}/V_{DD} terminals are typically used for op-amps that can operate with a single power supply. This means that the op-amp only needs a positive voltage supply. V_{DD} is the positive supply voltage, while V_{SS} is typically connected to ground (0 V).

It is important to note that these are just general guidelines, and there can be exceptions. Some op-amps may be designed to operate with either a dual or a single power supply. It is always best to consult the datasheet for the specific op-amp that you are using to determine the correct power supply configuration.

The output voltage of an op-amp (V_O) can swing very close to, but not quite reach, the supply voltages. It typically falls slightly short, by a volt or two, of the positive and negative supply rails. In the most optimistic scenario:

$$V_{EE} < V_O < V_{CC}$$

$$V_{SS} < V_O < V_{DD}$$

Supply terminals are often omitted from the op-amp symbol for clarity and conciseness (Fig. 1.2).

Fig. 1.2 Op-amp symbol without power supply connections

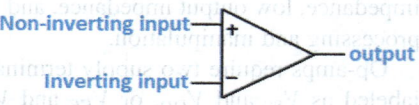

Non-inverting input

Inverting input

output

An op-amp's output stage can both source (Fig. 1.3) and sink (Fig. 1.4) current. This means it can either supply current to a load (source) or absorb current from a load (sink), depending on the circuit configuration and the load's requirements.

Fig. 1.3 Op-amp sourcing current

Fig. 1.4 Op-amp sinking current

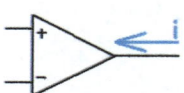

A key characteristic of op-amps is that their positive (+) and negative (−) input terminals do not draw any current (Fig. 1.5).

Fig. 1.5 Op-amp terminals draw zero current

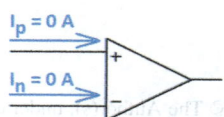

$I_p = 0\ A$

$I_n = 0\ A$

In an op-amp circuit, negative feedback means a portion of the output signal is fed back to the input in a way that opposes the original input signal. A simple way to recognize negative feedback in an op-amp circuit is to look for a direct connection, usually through one or more components (like resistors), between the op-amp's output and its inverting (−) input terminal. If you see this connection, it is a good indication that the circuit employs negative feedback.

With negative feedback in a properly configured op-amp circuit, the op-amp actively works to minimize the voltage difference between its two input terminals (Fig. 1.6). In essence, it strives to make the potential at the non-inverting input (V_A) equal to the potential at the inverting input (V_B). This "virtual short" between the inputs is a fundamental principle in op-amp circuit analysis.

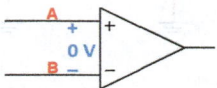

Fig. 1.6 Negative feedback causes the potential difference between the op-amp input terminals to be zero

This chapter explores op-amp circuits through a variety of illustrative examples. Calculations presented in this book are truncated to four decimal places. Consequently, minor discrepancies may arise between results obtained from different solution methodologies. These variations are solely attributable to accumulated rounding errors and do not reflect inaccuracies in the methodologies themselves.

1.2 Solved Problems

1.2.1 Example 1.1

Derive the relationship between the input voltage $V_1(t)$ and the output voltage $V_O(t)$ in Fig. 1.7.

Fig. 1.7 Circuit for Example 1.1

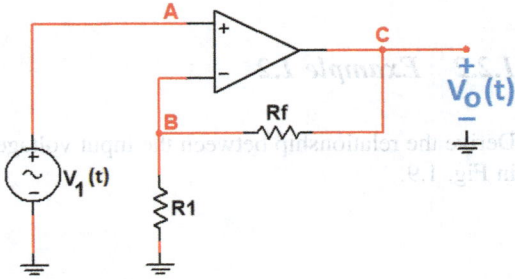

Solution:
The currents are shown in Fig. 1.8.

Fig. 1.8 Circuit currents

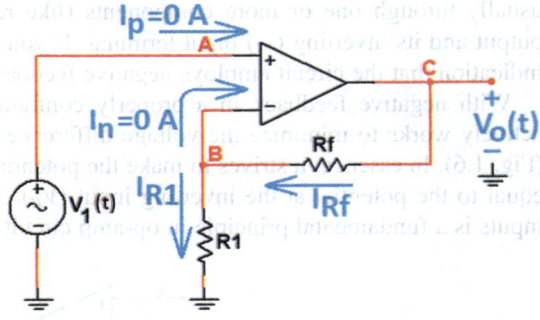

From Fig. 1.8, we have

$$KCL@B : I_{Rf} = I_n + I_{R_1} \Rightarrow I_{Rf} = 0 + I_{R_1} \Rightarrow I_{Rf} = I_{R_1} \Rightarrow \frac{v_C(t) - v_B(t)}{R_f}$$

$$= \frac{v_B(t) - 0}{R_1} \Rightarrow v_C(t) - v_B(t) = R_f \frac{v_B(t)}{R_1} \Rightarrow v_C(t)$$

$$= \frac{R_f}{R_1} v_B(t) + v_B(t) \Rightarrow v_C(t) = \left(1 + \frac{R_f}{R_1}\right) v_B(t)$$

$$v_C(t) = \left(1 + \frac{R_f}{R_1}\right) v_B(t) \Rightarrow v_C(t) = \left(1 + \frac{R_f}{R_1}\right) v_A(t)$$

$$v_A(t) = v_1(t) \Rightarrow v_C(t) = \left(1 + \frac{R_f}{R_1}\right) v_A(t) = \left(1 + \frac{R_f}{R_1}\right) v_1(t)$$

$$v_O(t) = v_C(t) = \left(1 + \frac{R_f}{R_1}\right) v_1(t)$$

$$A_V = \frac{v_O(t)}{v_1(t)} \Rightarrow A_V = \frac{\left(1 + \frac{R_f}{R_1}\right) v_1(t)}{v_1(t)} \Rightarrow A_V = \left(1 + \frac{R_f}{R_1}\right)$$

1.2.2   Example 1.2

Derive the relationship between the input voltage $V_1(t)$ and the output voltage $V_O(t)$ in Fig. 1.9.

Fig. 1.9 Circuit for Example 1.2

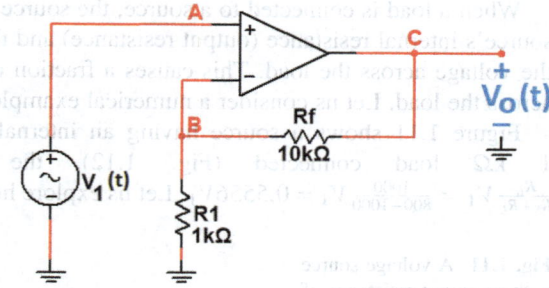

Solution:

$$v_O(t) = \left(1 + \frac{R_f}{R_1}\right)v_1(t) \Rightarrow v_O(t) = \left(1 + \frac{10k\Omega}{1k\Omega}\right)v_1(t) \Rightarrow v_O(t) = 11v_1(t)$$

1.2.3   Example 1.3

Derive the relationship between the input voltage $V_1(t)$ and the output voltage $V_O(t)$ in Fig. 1.10.

Fig. 1.10 Circuit for Example 1.3

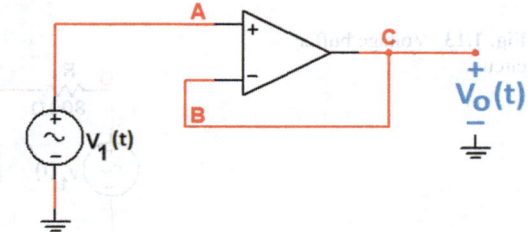

Solution:
We call the circuit shown in Fig. 1.10 a voltage buffer. For this circuit,
$$v_O(t) = \left(1 + \frac{R_f}{R_1}\right)v_1(t) \Rightarrow v_O(t) = \left(1 + \frac{0}{\infty}\right)v_1(t) \Rightarrow v_O(t) = v_1(t).$$

1.2.4   Example 1.4

Let us explore the applications of a voltage buffer and understand why it is useful despite offering no voltage amplification.

When a load is connected to a source, the source voltage is divided between the source's internal resistance (output resistance) and the load resistance, determining the voltage across the load. This causes a fraction of the source voltage to appear across the load. Let us consider a numerical example.

Figure 1.11 shows a source having an internal resistance of 800 Ω. With a 1 kΩ load connected (Fig. 1.12), the load voltage is $V_{R_L} = \frac{R_L}{R_S+R_L} V_1 = \frac{1000}{800+1000} V_1 = 0.5556 V_1$. Let us explore how to increase the load voltage.

Fig. 1.11 A voltage source with an output resistance of 800 ohms

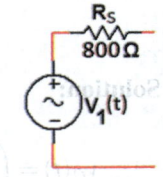

Fig. 1.12 A 1 kΩ load is connected to the source

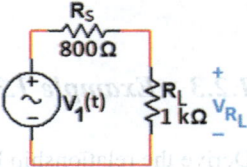

Consider the circuit shown in Fig. 1.13.

Fig. 1.13 Voltage buffer circuit

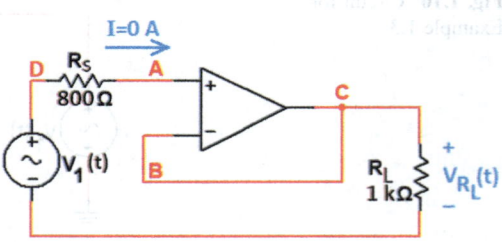

From Fig. 1.13, we have $V_D = V_1(t)$ and $V_{R_L}(t) = V_C = V_B$. No current is drawn from the source $V_1(t)$. Therefore, $V_A = V_D = V_1(t)$. Due to negative feedback, $V_B = V_A$. Therefore, $V_B = V_A = V_{R_L} = V_1(t)$, demonstrating that the input voltage is transferred entirely to the load.

1.2.5 Example 1.5

Derive the relationship between the input voltage $V_1(t)$ and the output voltage $V_O(t)$ in Fig. 1.14.

Fig. 1.14 Circuit for
Example 1.5

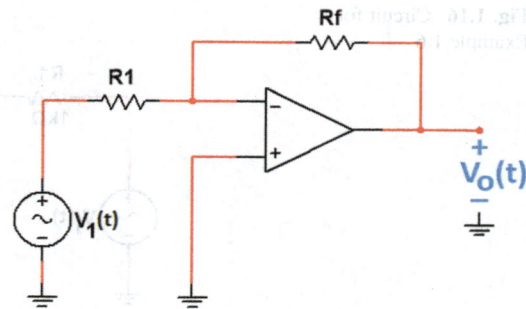

Solution:
The currents are shown in Fig. 1.15.

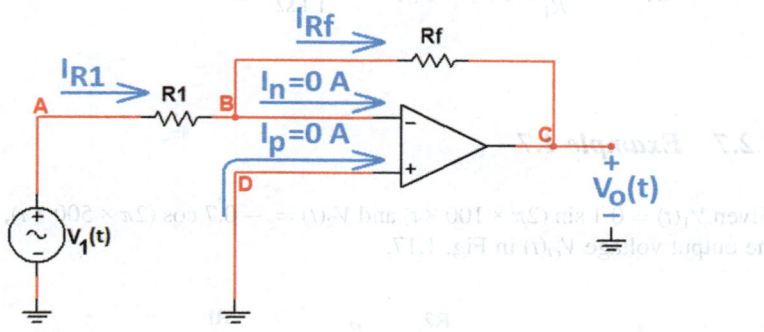

Fig. 1.15 Circuit currents

From Fig. 1.15, we have

$$v_D = 0\ \text{V} \Longrightarrow v_B = v_D \Longrightarrow v_B = 0\ \text{V}$$

$$v_A = v_1(t)$$

$$KCL@B: i_{R_1}(t) = i_{R_f}(t) + i_n(t) \Longrightarrow i_{R_1}(t) = i_{R_f}(t) + 0 \Longrightarrow i_{R_1}(t) = i_{R_f}(t) \Longrightarrow \frac{v_A(t) - v_B(t)}{R_1}$$

$$= \frac{v_B(t) - v_C(t)}{R_f} \Longrightarrow \frac{v_1(t) - 0}{R_1} = \frac{0 - v_C(t)}{R_f} \Longrightarrow v_C(t) = -\frac{R_f}{R_1} v_1(t)$$

$$v_o(t) = v_C(t) = -\frac{R_f}{R_1} v_1(t)$$

1.2.6 Example 1.6

Derive the relationship between the input voltage $V_1(t)$ and the output voltage $V_O(t)$ in Fig. 1.16.

Fig. 1.16 Circuit for Example 1.6

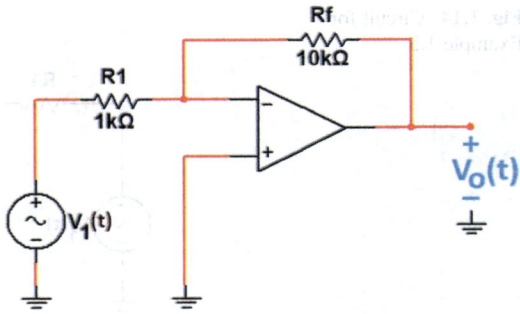

Solution:

$$v_O(t) = -\frac{R_f}{R_1}v_1(t) \Rightarrow v_O(t) = -\frac{10\text{ k}\Omega}{1\text{ k}\Omega}v_1(t) \Rightarrow v_O(t) = -10v_1(t)$$

1.2.7 Example 1.7

Given $V_1(t) = 0.1 \sin(2\pi \times 100 \times t)$ and $V_2(t) = -0.7\cos(2\pi \times 500 \times t)$, determine the output voltage $V_O(t)$ in Fig. 1.17.

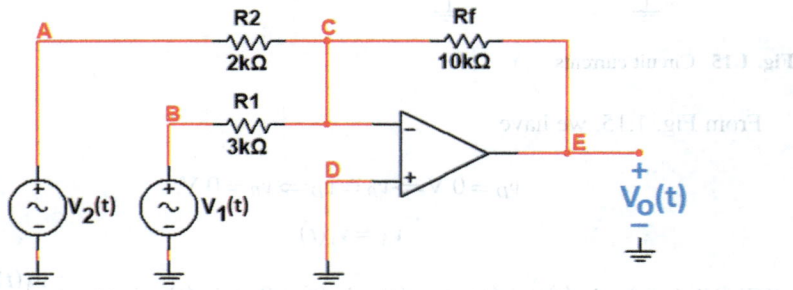

Fig. 1.17 Circuit for Example 1.7

Solution:
The currents are shown in Fig. 1.18.

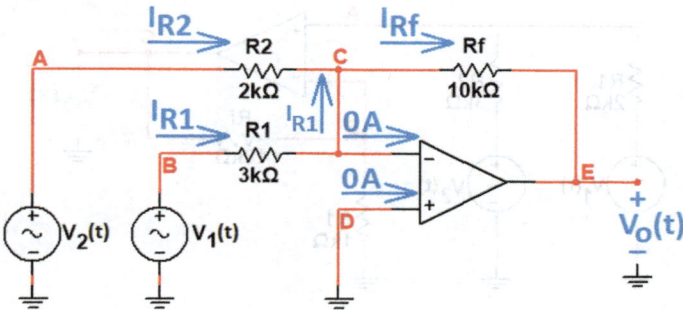

Fig. 1.18 Circuit currents

From Fig. 1.18, we have

$$v_D = 0 \text{ V}$$

$$v_C = v_D \Rightarrow v_C = 0 \text{ V}$$

$$v_A = V_2(t)$$

$$v_B = V_1(t)$$

$$KCL@C: i_{R_2}(t) + i_{R_1}(t) = i_{R_f}(t) \Rightarrow \frac{v_A - v_C}{R_2} + \frac{v_B - v_C}{R_1} = \frac{v_C - v_E}{R_f} \Rightarrow \frac{V_2(t) - 0}{2k} + \frac{V_1(t) - 0}{3k}$$

$$= \frac{0 - v_E}{10k} \Rightarrow \frac{V_2(t)}{2k} + \frac{V_1(t)}{3k} = \frac{-v_E}{10k} \Rightarrow v_E = -10k\left(\frac{V_2(t)}{2k} + \frac{V_1(t)}{3k}\right) \Rightarrow v_E$$

$$= -\frac{10k}{2k}V_2(t) - \frac{10k}{3k}V_1(t) \Rightarrow v_E = -5V_2(t) - 3.3333V_1(t) \Rightarrow v_o(t)$$

$$= v_E - 0 = v_E = -5V_2(t) - 3.3333V_1(t)$$

$$\begin{cases} V_1(t) = 0.1 \sin(2\pi \times 100 \times t) \\ V_2(t) = -0.7 \cos(2\pi \times 500 \times t) \end{cases} \Rightarrow v_o(t) = -5V_2(t) - 3.3333V_1(t) \Rightarrow v_o(t)$$

$$= -5 \times 0.1 \sin(2\pi \times 100 \times t) - 3.3333 \times -0.7 \cos(2\pi \times 500 \times t) \Rightarrow v_o(t) =$$
$$-0.5 \sin(2\pi \times 100 \times t) + 2.3333 \cos(2\pi \times 500 \times t)$$

1.2.8 Example 1.8

Given $V_1(t) = \sin(2\pi \times 100 \times t)$ and $V_2(t) = -7\cos(2\pi \times 500 \times t)$, determine the output voltage $V_O(t)$ in Fig. 1.19.

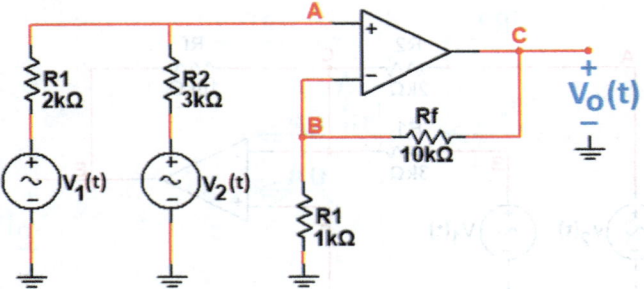

Fig. 1.19 Circuit for Example 1.8

Solution:

The currents are shown in Fig. 1.20.

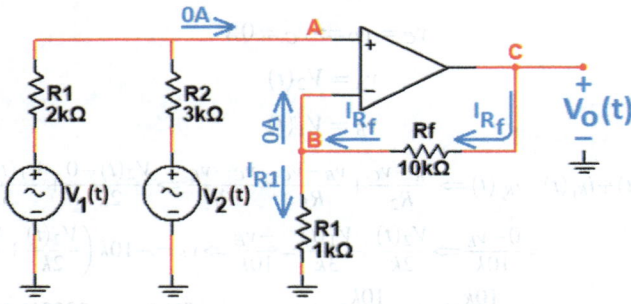

Fig. 1.20 Circuit currents

From Fig. 1.20, we have

$$v_A = v_B$$

$$KCL@B: I_{R_f} = I_{R_1} + 0 \Rightarrow I_{R_f} = I_{R_1} \Rightarrow \frac{v_C - v_B}{R_f} = \frac{v_B - 0}{R_1} \Rightarrow v_C - v_B = \frac{R_f}{R_1} v_B \Rightarrow v_C$$

$$= \left(1 + \frac{R_f}{R_1}\right) v_B \Rightarrow v_C = \left(1 + \frac{R_f}{R_1}\right) v_A \Rightarrow v_C = \left(1 + \frac{10k}{1k}\right) v_A \Rightarrow v_C = 11 v_A$$

The Op-amp does not draw any current from node A. As a result, the Op-amp can be ignored, and the voltage at node A can be determined using the circuit shown in Fig. 1.21.

Fig. 1.21 Node A voltage must be calculated

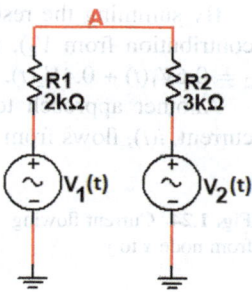

To determine the voltage at node A, we can use the principle of superposition. Figure 1.22 shows the circuit used to analyze the effect of voltage source V_1, where V_2 has been replaced with a short circuit. Applying the voltage divider formula, we find that the voltage at node A due to V_1 is $v_{A,1} = \frac{R_2}{R_2+R_1} \times V_1(t) = \frac{3}{3+2} \times V_1(t) = 0.6V_1(t)$.

Fig. 1.22 Voltage source V_2 is deactivated

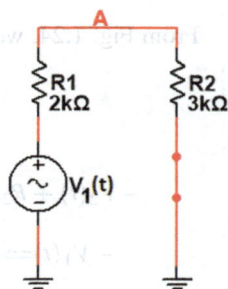

Figure 1.23 shows the circuit used to analyze the effect of voltage source V_2, where V_1 has been replaced with a short circuit. Applying the voltage divider formula, we find that the voltage at node A due to V_2 is $v_{A,2} = \frac{R_1}{R_2+R_1} \times V_2(t) = \frac{2}{3+2} \times V_2(t) = 0.4V_2(t)$.

Fig. 1.23 Voltage source V_1 is deactivated

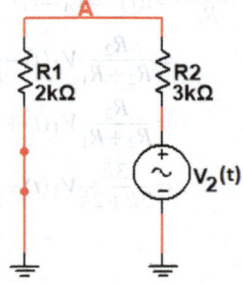

By summing the results of the two analyses (the contribution from V_1 and the contribution from V_2), we arrive at the final voltage at node A: $v_A = v_{A,\,1} + v_{A,\,2} = 0.6V_1(t) + 0.4V_2(t)$.

Another approach to calculating the voltage at node A involves assuming a current, $i(t)$, flows from node x to node y (Fig. 1.24).

Fig. 1.24 Current flowing from node x to y

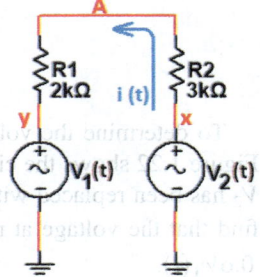

From Fig. 1.24, we have

$$v_y = V_1(t)$$

$$v_x = V_2(t)$$

$$-V_2(t) + R_2 i(t) + R_1 i(t) + V_1(t) = 0 \Longrightarrow R_2 i(t) + R_1 i(t) = V_2(t)$$

$$-V_1(t) \Longrightarrow (R_2 + R_1)i(t) = V_2(t) - V_1(t) \Longrightarrow i(t) = \frac{V_2(t) - V_1(t)}{R_2 + R_1}$$

$$\frac{v_A - v_y}{R_1} = i(t) \Longrightarrow v_A = R_1 i(t) + v_y \Longrightarrow v_A = R_1 i(t) + V_1(t) \Longrightarrow v_A = R_1 \frac{V_2(t) - V_1(t)}{R_2 + R_1} + V_1(t) \Longrightarrow v_A$$

$$= \frac{R_1}{R_2 + R_1} V_2(t) - \frac{R_1}{R_2 + R_1} V_1(t) + V_1(t) \Longrightarrow v_A = \frac{R_1}{R_2 + R_1} V_2(t) + \left(1 - \frac{R_1}{R_2 + R_1}\right) V_1(t) \Longrightarrow v_A$$

$$= \frac{R_1}{R_2 + R_1} V_2(t) + \frac{R_2}{R_2 + R_1} V_1(t) \Longrightarrow v_A = \frac{2k}{3k + 2k} V_2(t) + \frac{3k}{3k + 2k} V_1(t) \Longrightarrow v_A$$

$$= 0.6V_1(t) + 0.4V_2(t)$$

$$\frac{v_x - v_A}{R_2} = i(t) \Longrightarrow v_A = v_x - R_2 i(t) \Longrightarrow v_A = V_2(t) - R_2 i(t) \Longrightarrow v_A = V_2(t) - R_2 \frac{V_2(t) - V_1(t)}{R_2 + R_1} \Longrightarrow v_A$$

$$= \frac{R_2}{R_2 + R_1} V_1(t) - \frac{R_2}{R_2 + R_1} V_2(t) + V_2(t) \Longrightarrow v_A$$

$$= \frac{R_2}{R_2 + R_1} V_1(t) + \left(1 - \frac{R_2}{R_2 + R_1}\right) V_2(t) \Longrightarrow v_A = \frac{R_2}{R_2 + R_1} V_1(t) + \frac{R_1}{R_2 + R_1} V_2(t) \Longrightarrow v_A$$

$$= \frac{3k}{3k + 2k} V_1(t) + \frac{2k}{3k + 2k} V_2(t) \Longrightarrow v_A = 0.6V_1(t) + 0.4V_2(t)$$

The result obtained using this current analysis is identical to the result obtained using the superposition principle.

$$KCL@B : v_C = 11v_A \Rightarrow v_C = 11 \times (0.6V_1(t) + 0.4V_2(t)) \Rightarrow v_C = 0.66V_1(t) + 0.44V_2(t)$$

$$v_o(t) = v_C - 0 \Rightarrow v_o(t) = 0.66V_1(t) + 0.44V_2(t)$$

$$\begin{cases} V_1(t) = \sin(2\pi \times 100 \times t) \\ V_2(t) = -7\cos(2\pi \times 500 \times t) \end{cases} \Rightarrow v_o(t) = 0.66V_1(t) + 0.44V_2(t) \Rightarrow v_o(t)$$

$$= 0.66 \times 1 \sin(2\pi \times 100 \times t) + 0.44 \times -7\cos(2\pi \times 500 \times t) \Rightarrow v_o(t)$$

$$= 0.66 \sin(2\pi \times 100 \times t) - 3.08 \cos(2\pi \times 500 \times t)$$

1.2.9 Example 1.9

Given $V_1(t) = \sin(2\pi \times 100 \times t)$, determine the output voltage $V_O(t)$ in Fig. 1.25. The voltage across the capacitor at time $t = 0$ is 0 V.

Fig. 1.25 Circuit for Example 1.9

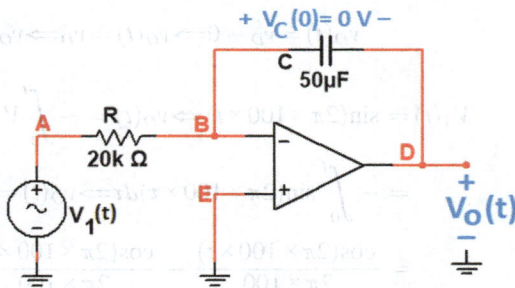

Solution:
The currents are shown in Fig. 1.26.

Fig. 1.26 Circuit currents

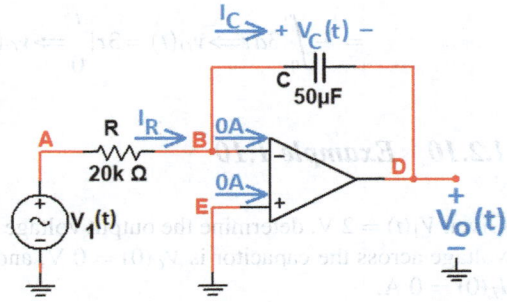

From Fig. 1.26, we have

$$v_A = V_1(t)$$

$$v_E = 0 \text{ V} \Rightarrow v_B = v_E = 0 \text{ V}$$

$$KCL@B : i_R(t) = 0 + i_C(t) \Rightarrow i_R(t) = i_C(t) \Rightarrow \frac{v_A - v_B}{R}$$

$$= i_C(t) \Rightarrow \frac{V_1(t) - 0}{R} = i_C(t) \Rightarrow i_C(t) = \frac{V_1(t)}{R}$$

$$v_C(t) = \frac{1}{C} \int_0^t i_C(\tau) d\tau + v_0(t) \Rightarrow v_C(t) = \frac{1}{C} \int_0^t \frac{V_1(\tau)}{R} d\tau + v_0(t) \Rightarrow v_C(t)$$

$$= \frac{1}{50\,\mu} \int_0^t \frac{V_1(\tau)}{20\,k} d\tau + 0 \Rightarrow \Rightarrow v_C(t) = \frac{1}{50\,\mu \times 20\,k} \int_0^t V_1(\tau) d\tau \Rightarrow v_C(t)$$

$$= \frac{1}{1} \int_0^t V_1(\tau) d\tau \Rightarrow v_C(t) = \int_0^t V_1(\tau) d\tau$$

$$v_C(t) = v_B - v_D \Rightarrow v_C(t) = v_E - v_D \Rightarrow v_C(t) = 0 - v_D \Rightarrow v_C(t)$$

$$= -v_D \Rightarrow v_D = -v_C(t) \Rightarrow v_D = -\int_0^t V_1(\tau) d\tau$$

$$v_O(t) = v_D - 0 \Rightarrow v_O(t) = v_D \Rightarrow v_O(t) = -\int_0^t V_1(\tau) d\tau$$

$$V_1(t) = \sin(2\pi \times 100 \times t) \Rightarrow v_O(t) = -\int_0^t V_1(\tau) d\tau \Rightarrow v_O(t)$$

$$= -\int_0^t \sin(2\pi \times 100 \times \tau) d\tau \Rightarrow v_O(t) = \frac{\cos(2\pi \times 100 \times \tau)}{2\pi \times 100} \Big|_0^t \Rightarrow v_O(t)$$

$$= \frac{\cos(2\pi \times 100 \times t)}{2\pi \times 100} - \frac{\cos(2\pi \times 100 \times 0)}{2\pi \times 100} \Rightarrow v_O(t)$$

$$= \frac{\cos(2\pi \times 100 \times t)}{2\pi \times 100} - \frac{1}{2\pi \times 100} \Rightarrow v_O(t) = \frac{\cos(2\pi \times 100 \times t) - 1}{2\pi \times 100}$$

$$V_1(t) = 3 \Rightarrow v_O(t) = -\int_0^t V_1(\tau) d\tau \Rightarrow v_O(t)$$

$$= -\int_0^t 3 d\tau \Rightarrow v_O(t) = 3\tau \Big|_0^t \Rightarrow v_O(t) = 3t - 3 \times 0 \Rightarrow v_O(t) = 3t$$

1.2.10   Example 1.10

Given $V_1(t) = 2$ V, determine the output voltage $V_O(t)$ in Fig. 1.27. At time $t = 0$, the voltage across the capacitor is $V_C(0) = 0$ V, and the current through the inductor is $I_L(0) = 0$ A.

Fig. 1.27 Circuit for
Example 1.10

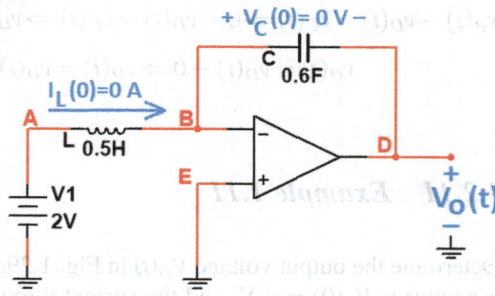

Solution:
The currents are shown in Fig. 1.28.

Fig. 1.28 Circuit currents

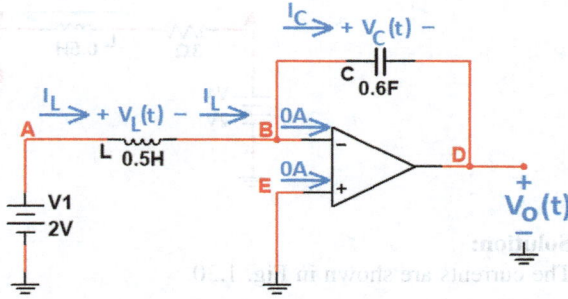

From Fig. 1.28, we have

$$v_A = 2 \text{ V}$$

$$v_E = 0 \text{ V}$$

$$v_B = v_E = 0 \text{ V}$$

$$v_L(t) = v_A - v_B \Rightarrow v_L(t) = 2 - 0 \Rightarrow v_L(t) = 2 \text{ V}$$

$$i_L(t) = \frac{1}{L} \int_0^t v_L(\tau) d\tau + i_L(0) \Rightarrow i_L(t)$$

$$= \frac{1}{0.5} \int_0^t 2 d\tau + 0 \Rightarrow i_L(t) = 2 \times 2\tau \Big|_0^t \Rightarrow i_L(t) = 2 \times (2t - 2 \times 0) \Rightarrow i_L(t) = 4t$$

$$KCL@B : i_L(t) = 0 + i_C(t) \Rightarrow i_L(t) = i_C(t) \Rightarrow i_C(t) = 4t$$

$$v_C(t) = \frac{1}{C} \int_0^t i_C(\tau) d\tau + v_C(0) \Rightarrow v_C(t) = \frac{1}{0.6} \int_0^t 4\tau d\tau + 0 \Rightarrow v_C(t) = \frac{2}{0.6} \tau^2 \Big|_0^t \Rightarrow v_C(t)$$

$$= 3.3333 (t^2 - 0^2) \Rightarrow v_C(t) = 3.3333 t^2$$

$$v_B(t) - v_D(t) = v_C(t) \Rightarrow 0 - v_D(t) = v_C(t) \Rightarrow v_D(t) = -v_C(t) \Rightarrow v_D(t) = -3.3333t^2$$

$$v_O(t) = v_D(t) - 0 \Rightarrow v_O(t) = v_D(t) \Rightarrow v_o(t) = -3.3333t^2$$

1.2.11   Example 1.11

Determine the output voltage $V_O(t)$ in Fig. 1.29. At time $t = 0$, the voltage across the capacitor is $V_C(0) = 4$ V, and the current through the inductor is $I_L(0) = 0$ A.

Fig. 1.29 Circuit for
Example 1.11

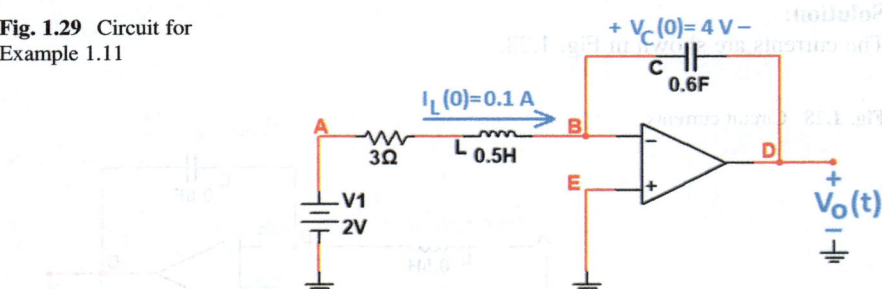

Solution:
The currents are shown in Fig. 1.30.

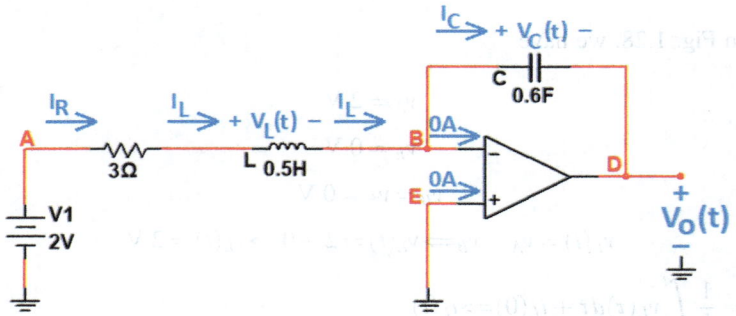

Fig. 1.30 Circuit currents

From Fig. 1.30, we have

$$v_E = 0 \text{ V}$$

$$v_B = v_E = 0 \text{ V}$$

$$v_{AB} = v_A - v_B \Rightarrow Ri_R(t) + L\frac{di_L(t)}{dt} = v_A - v_B \Rightarrow Ri_L(t) + L\frac{di_L(t)}{dt} = v_A - v_B \Rightarrow 3i_L(t)$$

$$+0.5\frac{di_L(t)}{dt} = 2 - 0 \Rightarrow 0.5\frac{di_L(t)}{dt} + 3i_L(t) = 2 \Rightarrow \frac{di_L(t)}{dt} + 6i_L(t) = 4$$

$$\frac{di_L(t)}{dt} + 6i_L(t) = 4, i_L(0) = 0.1 \text{ A}$$

$$\lambda + 6 = 0 \Rightarrow \lambda = -6$$

$$i_L(t) = C_1 e^{-6t} + \frac{4}{6} \Rightarrow i_L(t) = C_1 e^{-6t} + 0.6667$$

$$i_L(0) = 0.1 \Rightarrow C_1 e^{-6 \times 0} + 0.6667 = 0.1 \Rightarrow C_1 + 0.6667 = 0.1 \Rightarrow C_1 = -0.5667$$

$$i_L(t) = -0.5667 e^{-6t} + 0.6667$$

$$KCL@B : i_L(t) = i_C(t) + 0 \Rightarrow i_L(t) = i_C(t) \Rightarrow i_C(t) = -0.5667 e^{-6t} + 0.6667$$

$$v_C(t) = \frac{1}{C}\int_0^t i(\tau)d\tau + v_C(0) \Rightarrow v_C(t) = \frac{1}{0.6}\int_0^t \left(-0.5667 e^{-6\tau} + 0.6667\right)d\tau + 4 \Rightarrow v_C(t)$$

$$= 1.6667\left(\frac{-0.5667}{-6}e^{-6\tau} + 0.6667\tau\Big|_0^t\right) + 4 \Rightarrow v_C(t) = 0.1574 e^{-6\tau}$$

$$+1.1112\tau\Big|_0^t + 4 \Rightarrow v_C(t) = 0.1574 e^{-6t} + 1.1112t - \left(0.1574 e^{-6 \times 0} + 1.1112 \times 0\right)$$

$$+4 \Rightarrow v_C(t) = 0.1574 e^{-6t} + 1.1112t + 3.8426 \text{ V}$$

$$v_C(t) = v_B - v_D \Rightarrow v_C(t) = 0 - v_D \Rightarrow v_D = -v_C(t) \Rightarrow v_D$$

$$= -0.1574 e^{-6t} - 1.1112t - 3.8426 \text{ V}$$

$$v_O(t) = v_D(t) - 0 \Rightarrow v_O(t) = -0.1574 e^{-6t} - 1.1112t - 3.8426$$

1.2.12 Example 1.12

Determine the output voltage $V_O(t)$ in Fig. 1.31.

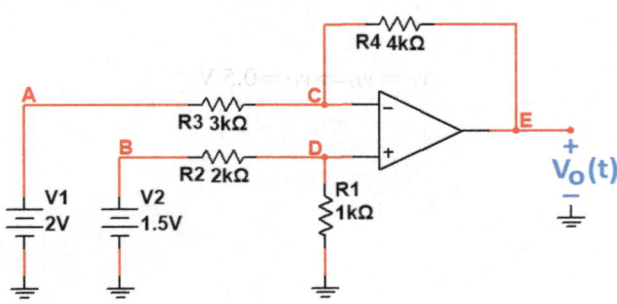

Fig. 1.31 Circuit for Example 1.12

Solution:

The currents are shown in Fig. 1.32.

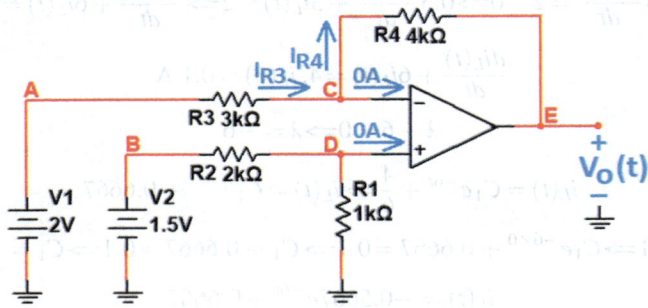

Fig. 1.32 Circuit currents

From Fig. 1.32, we have

$$v_A = V_1 = 2 \text{ V}$$

$$v_B = V_2 = 1.5 \text{ V}$$

$$v_C = v_D$$

The Op-amp does not draw any current from node D. As a result, the Op-amp can be ignored, and the voltage at node D can be determined using the circuit shown in Fig. 1.33.

Fig. 1.33 Node D voltage must be calculated

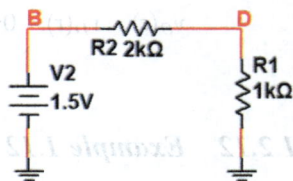

From Fig. 1.33, we have $v_D = \frac{R_1}{R_1+R_2} V_2 \Rightarrow v_D = \frac{1\,k}{1\,k+2\,k} \times 1.5 \Rightarrow v_D = 0.5$ V. Therefore,

$$v_C = v_D \Rightarrow v_C = 0.5 \text{ V}$$

$$KCL@C : I_{R_3} = I_{R_4} + 0 \Rightarrow I_{R_3} = I_{R_4} \Rightarrow \frac{v_A - v_C}{R_3} = \frac{v_C - v_E}{R_4} \Rightarrow \frac{R_4}{R_3}(v_A - v_C) = v_C - v_E \Rightarrow v_E$$

$$= v_C - \frac{R_4}{R_3}(v_A - v_C) \Rightarrow v_E = \left(1 + \frac{R_4}{R_3}\right)v_C - \frac{R_4}{R_3}v_A \Rightarrow v_E = \left(1 + \frac{4k}{3k}\right)v_C - \frac{4k}{3k}v_A \Rightarrow v_E$$

$$= 2.333v_C - 1.3333v_A \Rightarrow v_E = 2.333 \times 0.5 - 1.3333 \times 2 \Rightarrow v_E$$

$$= 2.333 \times 0.5 - 1.3333 \times 2 \Rightarrow v_E = -1.5\,\text{V}$$

$$v_O = v_E - 0 \Rightarrow v_O = v_E \Rightarrow v_O = -1.5\,\text{V}$$

1.2.13 Exercise 1.1

Determine the output voltage $V_O(t)$ in Fig. 1.34.

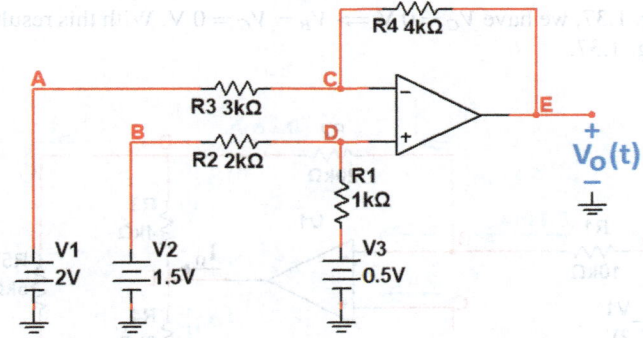

Fig. 1.34 Circuit for Exercise 1.1

1.2.14 Example 1.13

Determine the current $I_O(t)$ in Fig. 1.35.

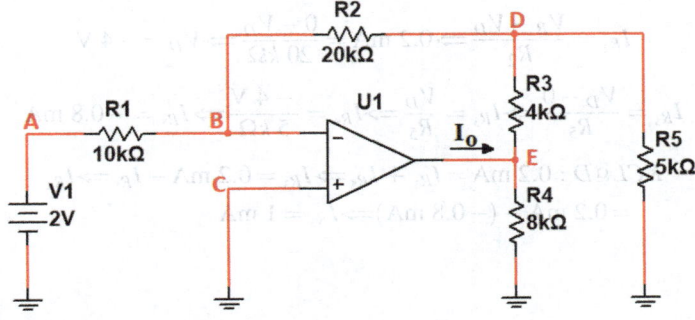

Fig. 1.35 Circuit for Example 1.13

Solution:

The currents are shown in Fig. 1.36.

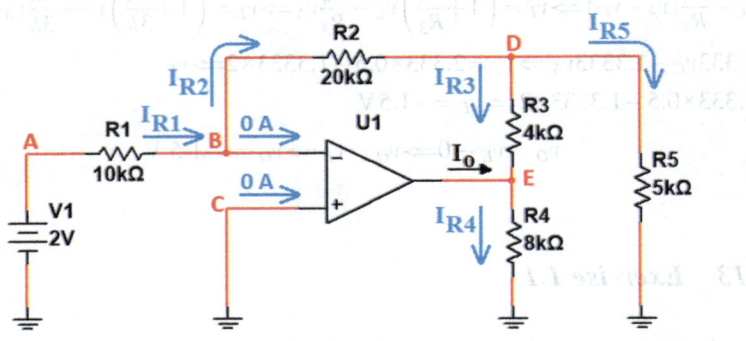

Fig. 1.36 Circuit currents

From Fig. 1.37, we have $V_C = 0$ V $\Rightarrow V_B = V_C = 0$ V. With this result, Fig. 1.36 becomes Fig. 1.37.

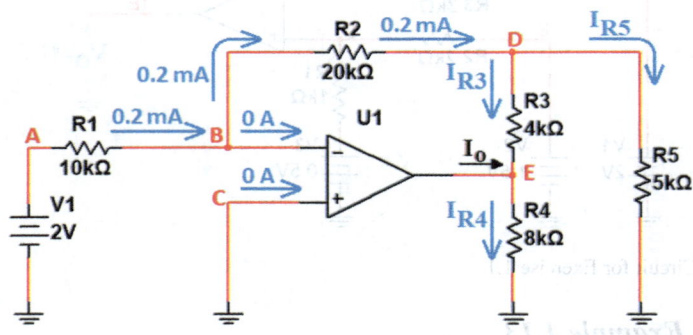

Fig. 1.37 Circuit currents

From Fig. 1.37, we have

$$I_{R_2} = \frac{V_B - V_D}{R_2} \Rightarrow 0.2 \text{ mA} = \frac{0 - V_D}{20 \text{ } k\Omega} \Rightarrow V_D = -4 \text{ V}$$

$$I_{R_5} = \frac{V_D - 0}{R_5} \Rightarrow I_{R_5} = \frac{V_D}{R_5} \Rightarrow I_{R_5} = \frac{-4 \text{ V}}{5 \text{ } k\Omega} \Rightarrow I_{R_5} = -0.8 \text{ mA}$$

$$KCL@D : 0.2 \text{ mA} = I_{R_3} + I_{R_5} \Rightarrow I_{R_3} = 0.2 \text{ mA} - I_{R_5} \Rightarrow I_{R_3}$$
$$= 0.2 \text{ mA} - (-0.8 \text{ mA}) \Rightarrow I_{R_3} = 1 \text{ mA}$$

$$I_{R_3} = \frac{V_D - V_E}{R_3} \Rightarrow 1 \text{ mA} = \frac{-4 - V_E}{4 \text{ k}\Omega} \Rightarrow 1 \text{ mA} \times 4 \text{ k}\Omega = -4 - V_E \Rightarrow$$

$$- (1 \text{ mA} \times 4 \text{ k}\Omega) = 4 + V_E \Rightarrow V_E = -(1 \text{ mA} \times 4 \text{ k}\Omega) - 4 \Rightarrow V_E$$

$$= -4 - 4 \Rightarrow V_E = -8 \text{ V}$$

$$KCL@E : I_{R_3} + I_o = I_{R_4} \Rightarrow I_{R_3} + I_o = \frac{V_E - 0}{R_4} \Rightarrow 1 \text{ mA} + I_o$$

$$= \frac{-8 - 0}{8 \text{ k}\Omega} \Rightarrow 1 \text{ mA} + I_o = -1 \text{ mA} \Rightarrow I_o = -2 \text{ mA}$$

Further Reading

1. Asadi F (2023) Applied Op Amp circuits. Springer. doi: https://doi.org/10.1007/978-981-99-3881-0
2. Asadi F (2022) Essential circuit analysis using NI Multisim™ and MATLAB®. Springer. doi: https://doi.org/10.1007/978-3-030-89850-2
3. Asadi F (2022) Essential circuit analysis using Proteus®. Springer. doi: https://doi.org/10.1007/978-981-19-4353-9
4. Asadi F (2022) Essential circuit analysis using LTspice®. Springer. doi: https://doi.org/10.1007/978-3-031-09853-6
5. Asadi F (2022). Electric circuit analysis with EasyEDA. Springer. doi: https://doi.org/10.1007/978-3-031-00292-2
6. Asadi F, Eguchi K (2022) Electric and electronic circuit simulation using TINA-TI®. River Publishers. doi: https://doi.org/10.13052/rp-9788770226851
7. Asadi F (2023) Analog electronic circuits laboratory manual. Springer. doi: https://doi.org/10.1007/978-3-031-25122-1

Chapter 2
Magnetically Coupled Circuits

2.1 Introduction

Coupled inductors are essential components in electronic circuits, particularly in applications involving transformers, power supplies, and resonant circuits. Unlike isolated inductors, coupled inductors share a magnetic field, leading to mutual inductance. This mutual inductance allows energy to be transferred between the inductors without a direct electrical connection, enabling voltage and current transformations. The degree of coupling is quantified by the coupling coefficient (k), which ranges from 0 (no coupling) to 1 (perfect coupling).

Understanding the dot convention is crucial for analyzing circuits with coupled inductors, as it defines the polarity of induced voltages based on the direction of current flow.

The essential relationships for coupled inductors are reviewed in Figs. 2.1, 2.2, 2.3, and 2.4. The sign of the mutual inductance (M) is determined by the dot convention: M is positive if currents enter or leave both dotted terminals simultaneously, and M is negative if one current enters while the other leaves a dotted terminal.

Fig. 2.1 Coupled inductor voltage-current relationships

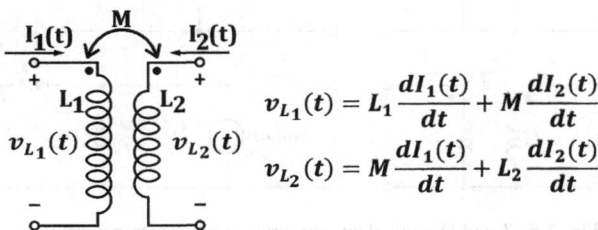

$$v_{L_1}(t) = L_1 \frac{dI_1(t)}{dt} + M \frac{dI_2(t)}{dt}$$

$$v_{L_2}(t) = M \frac{dI_1(t)}{dt} + L_2 \frac{dI_2(t)}{dt}$$

Fig. 2.2 Coupled inductor
voltage-current relationships

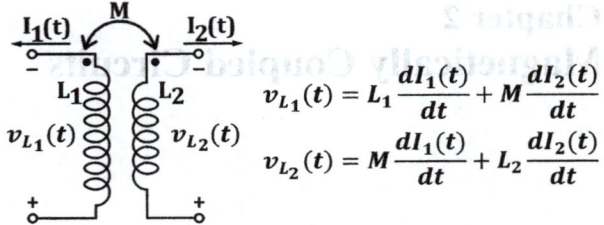

$$v_{L_1}(t) = L_1 \frac{dI_1(t)}{dt} + M \frac{dI_2(t)}{dt}$$

$$v_{L_2}(t) = M \frac{dI_1(t)}{dt} + L_2 \frac{dI_2(t)}{dt}$$

Fig. 2.3 Coupled inductor
voltage-current relationships

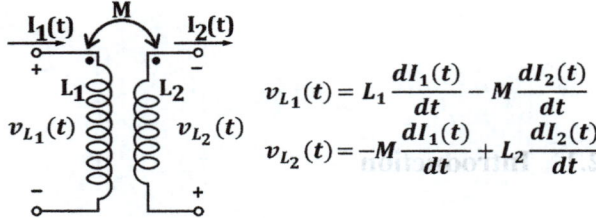

$$v_{L_1}(t) = L_1 \frac{dI_1(t)}{dt} - M \frac{dI_2(t)}{dt}$$

$$v_{L_2}(t) = -M \frac{dI_1(t)}{dt} + L_2 \frac{dI_2(t)}{dt}$$

Fig. 2.4 Coupled inductor
voltage-current relationships

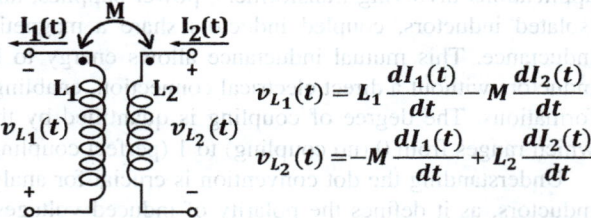

$$v_{L_1}(t) = L_1 \frac{dI_1(t)}{dt} - M \frac{dI_2(t)}{dt}$$

$$v_{L_2}(t) = -M \frac{dI_1(t)}{dt} + L_2 \frac{dI_2(t)}{dt}$$

Figure 2.5 depicts the T and Π equivalent networks of a linear transformer.

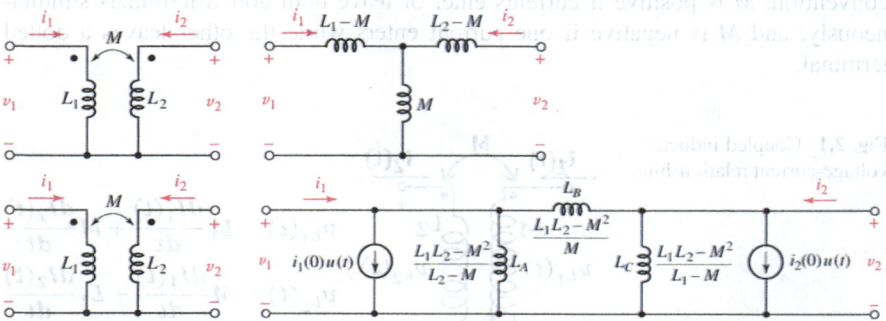

Fig. 2.5 T and Π equivalent networks of a linear transformer

The essential relationships for transformers are reviewed in Figs. 2.6, 2.7, 2.8, 2.9, 2.10, 2.11, and 2.12.

Fig. 2.6 Ideal transformer voltage and current relations

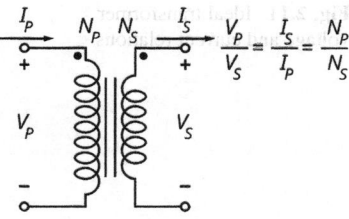

Fig. 2.7 Ideal transformer voltage and current relations

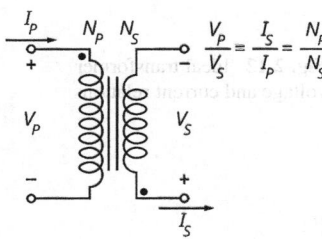

Fig. 2.8 Ideal transformer voltage and current relations

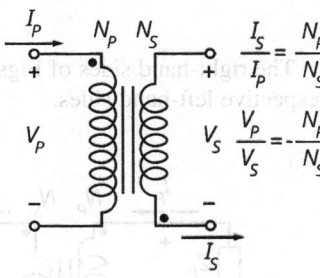

Fig. 2.9 Ideal transformer voltage and current relations

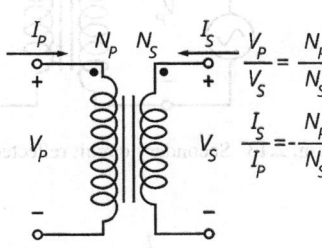

Fig. 2.10 Ideal transformer voltage and current relations

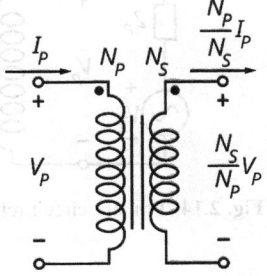

Fig. 2.11 Ideal transformer
voltage and current relations

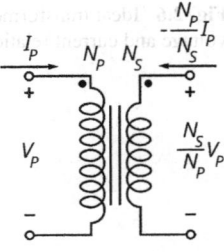

Fig. 2.12 Ideal transformer
voltage and current relations

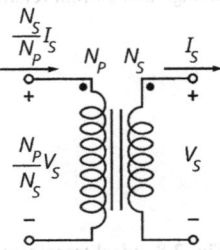

The right-hand sides of Figs. 2.13 and 2.14 show the equivalent circuits for their
respective left-hand sides.

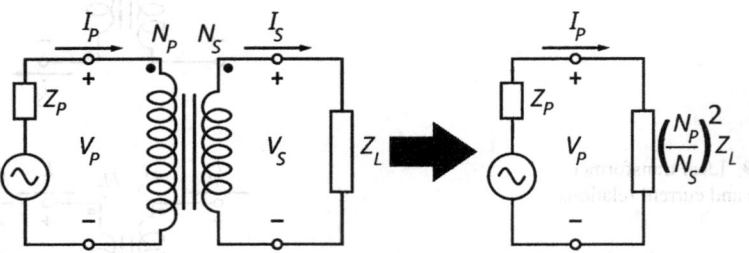

Fig. 2.13 Secondary circuit reflected to the primary side of the transformer

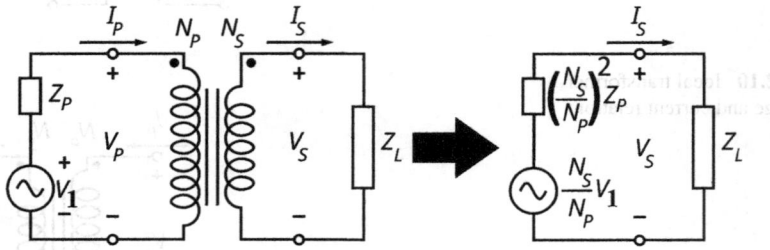

Fig. 2.14 Primary circuit reflected to the secondary side of the transformer

Calculations presented in this book are truncated to four decimal places. Consequently, minor discrepancies may arise between results obtained from different solution methodologies. These variations are solely attributable to accumulated rounding errors and do not reflect inaccuracies in the methodologies themselves.

2.2 Solved Examples

2.2.1 *Example 2.1*

Find the coupling coefficient for the circuit in Fig. 2.15.

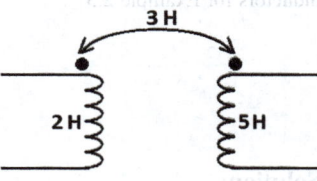

Fig. 2.15 Coupled inductors for Example 2.1

Solution:

$$k = \frac{M}{\sqrt{L_1 L_2}} \Rightarrow k = \frac{3}{\sqrt{2 \times 5}} \Rightarrow k = 0.9487$$

2.2.2 *Example 2.2*

Given the instantaneous currents in the coupled inductors as depicted in Fig. 2.16, calculate the stored energy within the system at that specific time instant.

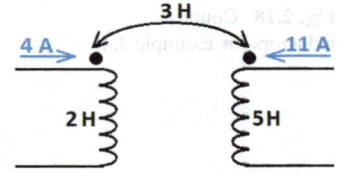

Fig. 2.16 Coupled inductors for Example 2.2

Solution:
Both currents enter the dotted terminals. Therefore,

$$W = \frac{1}{2}L_1i_1(t)^2 + \frac{1}{2}L_2i_2(t)^2 + Mi_1(t)i_2(t)$$

$$W = \frac{1}{2} \times 2 \times 4^2 + \frac{1}{2} \times 5 \times 11^2 + 3 \times 4 \times 11 \Rightarrow W = 16 + 302.5 + 132 \Rightarrow W = 450.5\,\text{J}$$

2.2.3   Example 2.3

Given the instantaneous currents in the coupled inductors as depicted in Fig. 2.17, calculate the stored energy within the system at that specific time instant.

Fig. 2.17 Coupled
inductors for Example 2.3

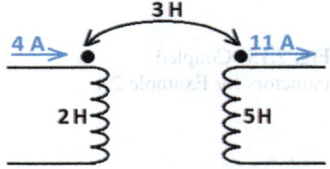

Solution:
One current enters and the other exits the dotted terminals. Therefore,

$$W = \frac{1}{2}L_1i_1(t)^2 + \frac{1}{2}L_2i_2(t)^2 - Mi_1(t)i_2(t)$$

$$W = \frac{1}{2} \times 2 \times 4^2 + \frac{1}{2} \times 5 \times 11^2 + 3 \times 4 \times 11 \Rightarrow W = 16 + 302.5 - 132 \Rightarrow W = 186.5\,\text{J}$$

2.2.4   Example 2.4

Referencing the circuit depicted in Fig. 2.18, derive the Kirchhoff's voltage law (KVL) equation for the path from node A to node B.

Fig. 2.18 Coupled
inductors for Example 2.4

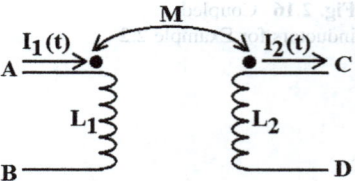

Solution:
Figure 2.19 depicts the solution.

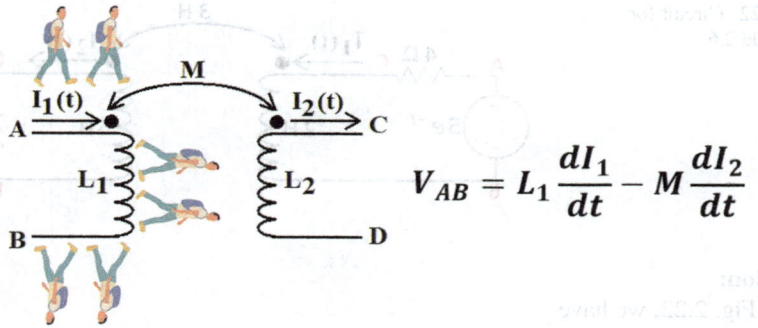

Fig. 2.19 Traveling from A to B

2.2.5 Example 2.5

Referencing the circuit depicted in Fig. 2.20, derive the Kirchhoff's voltage law (KVL) equation for the path from node B to node A.

Fig. 2.20 Coupled inductors for Example 2.5

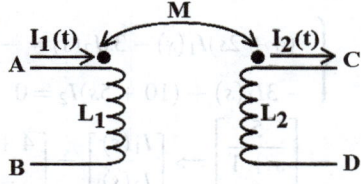

Solution:
Figure 2.21 depicts the solution. Note that the negative sign must be distributed to all terms within the parentheses.

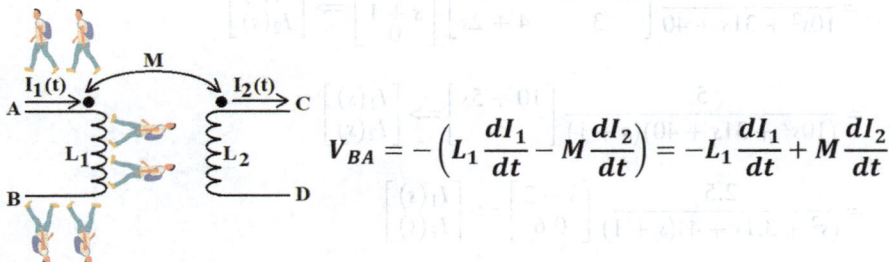

Fig. 2.21 Traveling from B to A

2.2.6 Example 2.6

Determine the currents $I_1(t)$ and $I_2(t)$ for the circuit depicted in Fig. 2.22, given that all initial conditions are zero.

Fig. 2.22 Circuit for
Example 2.6

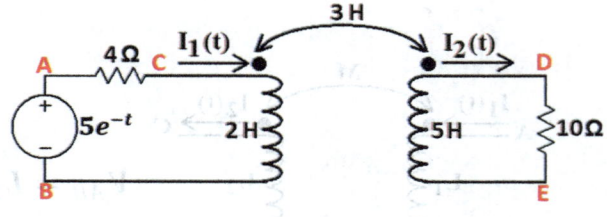

Solution:

From Fig. 2.22, we have

$$KVL@BACB: \ -5e^{-t} + 4I_1(t) + 2\frac{dI_1(t)}{dt} - 3\frac{dI_2(t)}{dt} = 0 \overset{\mathcal{L}}{\Rightarrow}$$

$$-\frac{5}{s+1} + 4I_1(s) + 2sI_1(s) - 3sI_2(s) = 0 \Longrightarrow (4+2s)I_1(s) - 3sI_2(s) = \frac{5}{s+1}$$

$$KVL@DED: \ 10I_2 + 5\frac{dI_2(t)}{dt} - 3\frac{dI_1(t)}{dt} = 0 \overset{\mathcal{L}}{\Rightarrow} 10I_2(s) + 5sI_2(s)$$

$$- 3I_1(s) = 0 \Longrightarrow -3I_1(s) + (10 + 5s)I_2 = 0$$

$$\begin{cases} (4+2s)I_1(s) - 3sI_2(s) = \dfrac{5}{s+1} \\ -3I_1(s) + (10+5s)I_2 = 0 \end{cases} \Rightarrow \begin{bmatrix} 4+2s & -3s \\ -3 & 10+5s \end{bmatrix}\begin{bmatrix} I_1(s) \\ I_2(s) \end{bmatrix}$$

$$= \begin{bmatrix} \dfrac{5}{s+1} \\ 0 \end{bmatrix} \Rightarrow \begin{bmatrix} I_1(s) \\ I_2(s) \end{bmatrix} = \begin{bmatrix} 4+2s & -3s \\ -3 & 10+5s \end{bmatrix}^{-1}\begin{bmatrix} \dfrac{5}{s+1} \\ 0 \end{bmatrix} \Rightarrow \begin{bmatrix} I_1(s) \\ I_2(s) \end{bmatrix}$$

$$= \frac{1}{(4+2s)(10+5s) - (-3)(-3s)}\begin{bmatrix} 10+5s & 3s \\ 3 & 4+2s \end{bmatrix}\begin{bmatrix} \dfrac{5}{s+1} \\ 0 \end{bmatrix} \Rightarrow \begin{bmatrix} I_1(s) \\ I_2(s) \end{bmatrix}$$

$$= \frac{1}{10s^2 + 31s + 40}\begin{bmatrix} 10+5s & 3s \\ 3 & 4+2s \end{bmatrix}\begin{bmatrix} \dfrac{5}{s+1} \\ 0 \end{bmatrix} \Rightarrow \begin{bmatrix} I_1(s) \\ I_2(s) \end{bmatrix}$$

$$= \frac{5}{(10s^2 + 31s + 40)(s+1)}\begin{bmatrix} 10+5s \\ 3 \end{bmatrix} \Rightarrow \begin{bmatrix} I_1(s) \\ I_2(s) \end{bmatrix}$$

$$= \frac{2.5}{(s^2 + 3.1s + 4)(s+1)}\begin{bmatrix} s+2 \\ 0.6 \end{bmatrix} \Rightarrow \begin{bmatrix} I_1(s) \\ I_2(s) \end{bmatrix}$$

$$= \frac{2.5}{(s^2 + 3.1s + 4)(s+1)}\begin{bmatrix} s+2 \\ 0.6 \end{bmatrix} \Rightarrow \begin{bmatrix} I_1(t) \\ I_2(t) \end{bmatrix}$$

$$= \mathcal{L}^{-1}\left\{ \frac{2.5}{(s^2 + 3.1s + 4)(s+1)}\begin{bmatrix} s+2 \\ 0.6 \end{bmatrix} \right\}$$

The inverse Laplace transform can be calculated using MATLAB. The MATLAB commands shown in Fig. 2.23 compute $I_1(t)$.

```
Command Window                                                        ⊙
>> syms s
>> vpa(ilaplace(2.5*(s+2)/(s^2+3.1*s+4)/(s+1)),4)

ans =

1.316*exp(-1.0*t) - 1.316*exp(-1.55*t)*(cos(1.264*t) - 1.068*sin(1.264*t))
fx >> |
```

Fig. 2.23 MATLAB code

According to Fig. 2.23, $I_1(t) = 1.316e^{-t} - 1.316e^{-1.55t}(\cos(1.264t) - 1.068 \sin(1.264t))$. The MATLAB commands shown in Fig. 2.24 compute $I_2(t)$.

```
Command Window                                                        ⊙
>> syms s
>> vpa(ilaplace(2.5*0.6/(s^2+3.1*s+4)/(s+1)),4)

ans =

0.7895*exp(-1.0*t) - 0.7895*exp(-1.55*t)*(cos(1.264*t) + 0.4352*sin(1.264*t))
fx >> |
```

Fig. 2.24 MATLAB code

According to Fig. 2.24, $I_2(t) = 0.7895e^{-t} - 0.7895e^{-1.55t}(\cos(1.264t) + 0.4352 \sin(1.264t))$.

2.2.7 Example 2.7

Demonstrate that the equivalent inductance (L_{eq}) between terminals A and B for the circuit shown in Fig. 2.25 is given by $L_{eq} = L_1 + L_2 + 2M$.

Fig. 2.25 Coupled inductors for Example 2.7

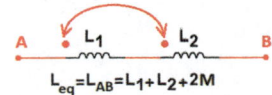

Solution:
The current $i(t)$ is defined as depicted in Fig. 2.26.

Fig. 2.26 The current $i(t)$

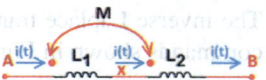

From Fig. 2.26, we have

$$V_{Ax} = L_1 \frac{di(t)}{dt} + M \frac{di(t)}{dt}$$

$$V_{xB} = L_2 \frac{di(t)}{dt} + M \frac{di(t)}{dt}$$

$$V_{AB} = V_{Ax} + V_{xB} \Rightarrow V_{AB} = L_1 \frac{di(t)}{dt} + M \frac{di(t)}{dt} + L_2 \frac{di(t)}{dt} + M \frac{di(t)}{dt} \Rightarrow V_{AB}$$

$$= (L_1 + L_2 + 2M) \frac{di(t)}{dt} \Rightarrow L_{eq} = L_1 + L_2 + 2M$$

2.2.8 Example 2.8

Calculate the equivalent inductance between terminals A and B for the circuit illustrated in Fig. 2.27.

Fig. 2.27 Coupled
inductors for Example 2.8

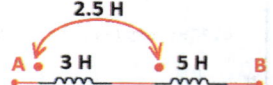

Solution

$$L_{eq} = 3 + 5 + 2 \times 2.5 = 13 \ H$$

2.2.9 Example 2.9

Demonstrate that the equivalent inductance (L_{eq}) between terminals A and B for the circuit shown in Fig. 2.28 is given by $L_{eq} = L_1 + L_2 - 2M$.

Fig. 2.28 Coupled
inductors for Example 2.9

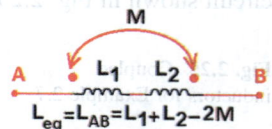

Solution
The current $i(t)$ is defined as depicted in Fig. 2.29.

Fig. 2.29 The current $i(t)$

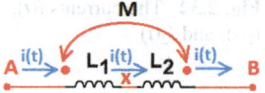

From Fig. 2.29, we have

$$V_{Ax} = L_1 \frac{di(t)}{dt} - M \frac{di(t)}{dt}$$

$$V_{xB} = L_2 \frac{di(t)}{dt} - M \frac{di(t)}{dt}$$

$$V_{AB} = V_{Ax} + V_{xB} \Rightarrow V_{AB} = L_1 \frac{di(t)}{dt} - M \frac{di(t)}{dt} + L_2 \frac{di(t)}{dt} - M \frac{di(t)}{dt} \Rightarrow V_{AB}$$

$$= (L_1 + L_2 - 2M) \frac{di(t)}{dt} \Rightarrow L_{eq} = L_1 + L_2 - 2M$$

2.2.10 Example 2.10

Calculate the equivalent inductance between terminals A and B for the circuit illustrated in Fig. 2.30.

Fig. 2.30 Coupled
inductors for Example 2.10

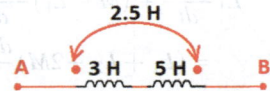

Solution:

$$L_{eq} = 3 + 5 - 2 \times 2.5 = 3\,H$$

2.2.11 Example 2.11

Demonstrate that the equivalent inductance (L_{eq}) between terminals A and B for the circuit shown in Fig. 2.31 is given by $L_{eq} = \frac{L_1 L_2 - M^2}{L_1 + L_2 - 2M}$.

Fig. 2.31 Coupled
inductors for Example 2.11

$$L_{eq} = L_{AB} = \frac{L_1 L_2 - M^2}{L_1 + L_2 - 2M}$$

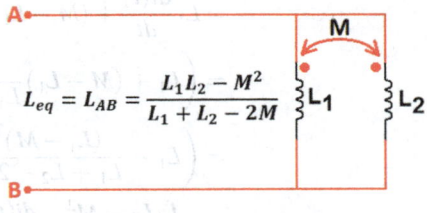

Solution:
The currents $i(t)$, $i_1(t)$, and $i_2(t)$ are defined as illustrated in Fig. 2.32.

Fig. 2.32 The currents $i(t)$, $i_1(t)$, and $i_2(t)$

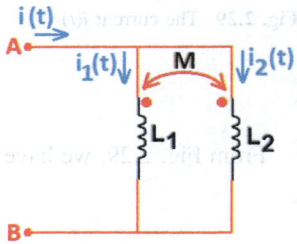

From Fig. 2.32, we have

$$KCL@A : i(t) = i_1(t) + i_2(t) \Rightarrow i_1(t) = i(t) - i_2(t)$$

$$V_{AB} = L_1 \frac{di_1(t)}{dt} + M \frac{di_2(t)}{dt} \Rightarrow V_{AB}$$

$$= L_1 \frac{d}{dt}(i(t) - i_2(t)) + M \frac{di_2(t)}{dt} \Rightarrow V_{AB} = L_1 \frac{di(t)}{dt} + (M - L_1)\frac{di_2(t)}{dt}$$

$$V_{AB} = L_2 \frac{di_2(t)}{dt} + M \frac{di_1(t)}{dt} \Rightarrow V_{AB} = L_2 \frac{di_2(t)}{dt} + M \frac{d}{dt}(i(t) - i_2(t)) \Rightarrow V_{AB}$$

$$= (L_2 - M)\frac{di_2(t)}{dt} + M \frac{di(t)}{dt}$$

$$L_1 \frac{di(t)}{dt} + (M - L_1)\frac{di_2(t)}{dt} = (L_2 - M)\frac{di_2(t)}{dt} + M \frac{di(t)}{dt} \Rightarrow (L_1 - M)\frac{di(t)}{dt}$$

$$= (L_1 + L_2 - 2M)\frac{di_2(t)}{dt} \Rightarrow \int (L_1 - M)\frac{di(t)}{dt}$$

$$= \int (L_1 + L_2 - 2M)\frac{di_2(t)}{dt} \Rightarrow (L_1 - M)\int \frac{di(t)}{dt}$$

$$= (L_1 + L_2 - 2M)\int \frac{di_2(t)}{dt} \Rightarrow (L_1 - M)i(t) = (L_1 + L_2 - 2M)i_2(t) \Rightarrow i_2(t)$$

$$= \frac{L_1 - M}{L_1 + L_2 - 2M}i(t)$$

$$V_{AB} = L_1 \frac{di(t)}{dt} + (M - L_1)\frac{di_2(t)}{dt} \Rightarrow V_{AB} = L_1 \frac{di(t)}{dt}$$

$$+ (M - L_1)\frac{d}{dt}\left(\frac{L_1 - M}{L_1 + L_2 - 2M}i(t)\right) \Rightarrow V_{AB}$$

$$= L_1 \frac{di(t)}{dt} + (M - L_1)\frac{L_1 - M}{L_1 + L_2 - 2M}\frac{di(t)}{dt} \Rightarrow V_{AB}$$

$$= \left(L_1 + (M - L_1)\frac{L_1 - M}{L_1 + L_2 - 2M}\right)\frac{di(t)}{dt} \Rightarrow V_{AB}$$

$$= \left(L_1 - \frac{(L_1 - M)^2}{L_1 + L_2 - 2M}\right)\frac{di(t)}{dt} \Rightarrow V_{AB}$$

$$= \frac{L_1 L_2 - M^2}{L_1 + L_2 - 2M}\frac{di(t)}{dt} \Rightarrow L_{eq} = \frac{L_1 L_2 - M^2}{L_1 + L_2 - 2M}$$

2.2.12 Example 2.12

Calculate the equivalent inductance between terminals A and B for the circuit illustrated in Fig. 2.33.

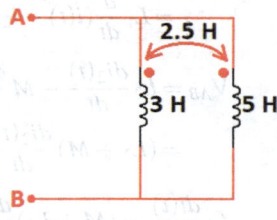

Fig. 2.33 Coupled inductors for Example 2.12

Solution:

$$L_{eq} = \frac{3 \times 5 - 2.5^2}{3 + 5 - 2 \times 2.5} \Rightarrow L_{eq} = 2.9167\ H$$

2.2.13 Example 2.13

Demonstrate that the equivalent inductance (L_{eq}) between terminals A and B for the circuit shown in Fig. 2.34 is given by $L_{eq} = \frac{L_1 L_2 - M^2}{L_1 + L_2 + 2M}$.

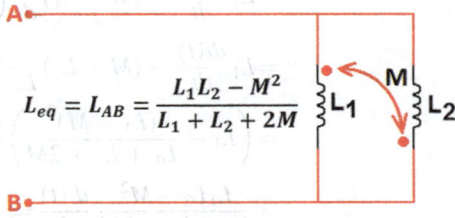

Fig. 2.34 Coupled inductors for Example 2.13

$$L_{eq} = L_{AB} = \frac{L_1 L_2 - M^2}{L_1 + L_2 + 2M}$$

Solution:
The currents $i(t)$, $i_1(t)$, and $i_2(t)$ are defined as illustrated in Fig. 2.35.

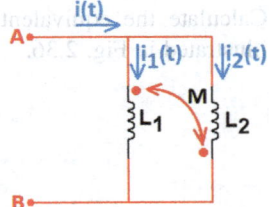

Fig. 2.35 The currents $i(t)$, $i_1(t)$, and $i_2(t)$

From Fig. 2.35, we have

$$KCL@A : i(t) = i_1(t) + i_2(t) \Rightarrow i_1(t) = i(t) - i_2(t)$$

$$V_{AB} = L_1 \frac{di_1(t)}{dt} - M \frac{di_2(t)}{dt} \Rightarrow V_{AB}$$

$$= L_1 \frac{d}{dt}(i(t) - i_2(t)) - M \frac{di_2(t)}{dt} \Rightarrow V_{AB} = L_1 \frac{di(t)}{dt} - (M + L_1) \frac{di_2(t)}{dt}$$

$$V_{AB} = L_2 \frac{di_2(t)}{dt} - M \frac{di_1(t)}{dt} \Rightarrow V_{AB} = L_2 \frac{di_2(t)}{dt} - M \frac{d}{dt}(i(t) - i_2(t)) \Rightarrow V_{AB}$$

$$= (L_2 + M) \frac{di_2(t)}{dt} - M \frac{di(t)}{dt}$$

$$L_1 \frac{di(t)}{dt} - (M + L_1) \frac{di_2(t)}{dt} = (L_2 + M) \frac{di_2(t)}{dt} - M \frac{di(t)}{dt} \Rightarrow (L_1 + M) \frac{di(t)}{dt}$$

$$= (L_1 + L_2 + 2M) \frac{di_2(t)}{dt} \Rightarrow \int (L_1 + M) \frac{di(t)}{dt}$$

$$= \int (L_1 + L_2 + 2M) \frac{di_2(t)}{dt} \Rightarrow (L_1 + M) \int \frac{di(t)}{dt}$$

$$= (L_1 + L_2 + 2M) \int \frac{di_2(t)}{dt} \Rightarrow (L_1 + M) i(t)$$

$$= (L_1 + L_2 + 2M) i_2(t) \Rightarrow i_2(t) = \frac{L_1 + M}{L_1 + L_2 + 2M} i(t)$$

$$V_{AB} = L_1 \frac{di(t)}{dt} - (M + L_1) \frac{di_2(t)}{dt} \Rightarrow V_{AB}$$

$$= L_1 \frac{di(t)}{dt} - (M + L_1) \frac{d}{dt}\left(\frac{L_1 + M}{L_1 + L_2 + 2M} i(t)\right) \Rightarrow V_{AB}$$

$$= L_1 \frac{di(t)}{dt} - (M + L_1) \frac{L_1 + M}{L_1 + L_2 + 2M} \frac{di(t)}{dt} \Rightarrow V_{AB}$$

$$= \left(L_1 - \frac{(L_1 + M)^2}{L_1 + L_2 + 2M}\right) \frac{di(t)}{dt} \Rightarrow V_{AB}$$

$$= \frac{L_1 L_2 - M^2}{L_1 + L_2 + 2M} \frac{di(t)}{dt} \Rightarrow L_{eq} = \frac{L_1 L_2 - M^2}{L_1 + L_2 + 2M}$$

2.2.14 Example 2.14

Calculate the equivalent inductance between terminals A and B for the circuit illustrated in Fig. 2.36.

Fig. 2.36 Coupled
inductors for Example 2.14

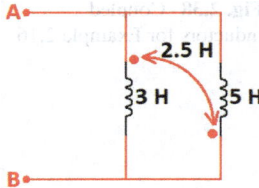

Solution:

$$L_{eq} = \frac{3 \times 5 - 2.5^2}{3 + 5 + 2 \times 2.5} \Rightarrow L_{eq} = 0.6731\ H$$

2.2.15   *Example 2.15*

A summary of equivalent inductance values for different connection configurations
is presented in Fig. 2.37.

Fig. 2.37 Equivalent
inductance formulas for
coupled inductors

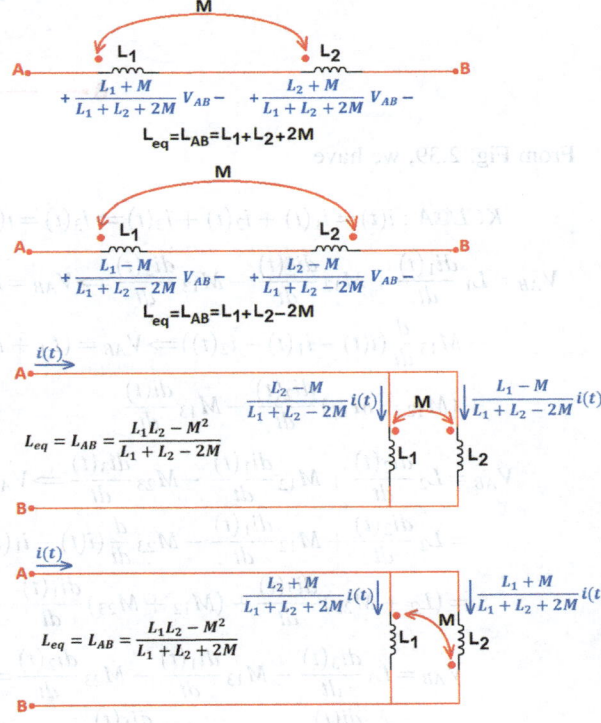

2.2.16   *Example 2.16*

Calculate the equivalent inductance between terminals *A* and *B* for the circuit
illustrated in Fig. 2.38.

Fig. 2.38 Coupled
inductors for Example 2.16

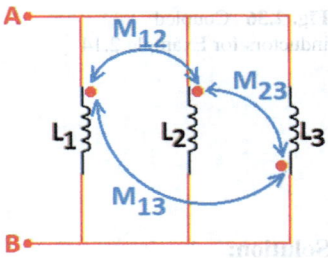

Solution:
The currents $i(t)$, $i_1(t)$, and $i_2(t)$ are defined as illustrated in Fig. 2.39.

Fig. 2.39 The currents $i(t)$,
$i_1(t)$, and $i_2(t)$

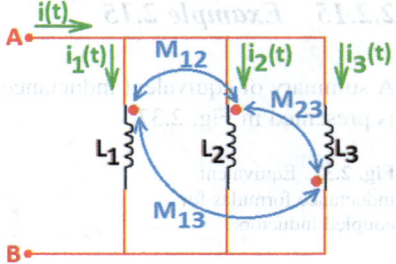

From Fig. 2.39, we have

$$KCL@A : i(t) = i_1(t) + i_2(t) + i_3(t) \Longrightarrow i_3(t) = i(t) - i_1(t) - i_2(t)$$

$$V_{AB} = L_1 \frac{di_1(t)}{dt} + M_{12} \frac{di_2(t)}{dt} - M_{13} \frac{di_3(t)}{dt} \Longrightarrow V_{AB} = L_1 \frac{di_1(t)}{dt} + M_{12} \frac{di_2(t)}{dt}$$

$$- M_{13} \frac{d}{dt}(i(t) - i_1(t) - i_2(t)) \Longrightarrow V_{AB} = (L_1 + M_{13}) \frac{di_1(t)}{dt}$$

$$+ (M_{12} + M_{13}) \frac{di_2(t)}{dt} - M_{13} \frac{di(t)}{dt}$$

$$V_{AB} = L_2 \frac{di_2(t)}{dt} + M_{12} \frac{di_1(t)}{dt} - M_{23} \frac{di_3(t)}{dt} \Longrightarrow V_{AB}$$

$$= L_2 \frac{di_2(t)}{dt} + M_{12} \frac{di_1(t)}{dt} - M_{23} \frac{d}{dt}(i(t) - i_1(t) - i_2(t)) \Longrightarrow V_{AB}$$

$$= (L_2 + M_{23}) \frac{di_2(t)}{dt} + (M_{12} + M_{23}) \frac{di_1(t)}{dt} - M_{23} \frac{di(t)}{dt}$$

$$V_{AB} = L_3 \frac{di_3(t)}{dt} - M_{13} \frac{di_1(t)}{dt} - M_{23} \frac{di_2(t)}{dt} \Longrightarrow V_{AB}$$

$$= L_3 \frac{di(t)}{dt} - (M_{13} + L_3) \frac{di_1(t)}{dt} - (M_{23} + L_3) \frac{di_2(t)}{dt}$$

$$(L_1 + M_{13})\frac{di_1(t)}{dt} + (M_{12} + M_{13})\frac{di_2(t)}{dt} - M_{13}\frac{di(t)}{dt} = (L_2 + M_{23})\frac{di_2(t)}{dt} + (M_{12} + M_{23})\frac{di_1(t)}{dt} - M_{23}\frac{di_2(t)}{dt} \Rightarrow$$

$$(L_1 + M_{13})\frac{di_1(t)}{dt} + (M_{12} + M_{13})\frac{di_2(t)}{dt} - M_{13}\frac{di(t)}{dt} = L_3\frac{di(t)}{dt} - (M_{13} + L_3)\frac{di_1(t)}{dt} - (M_{23} + L_3)\frac{di_2(t)}{dt}$$

$$(L_1 + M_{13} - M_{12} - M_{23})\frac{di_1(t)}{dt} - M_{13}\frac{di(t)}{dt} = (M_{13} - M_{23})\frac{di(t)}{dt}$$

$$(L_1 + 2M_{13} + L_3)\frac{di_1(t)}{dt} + (M_{12} + M_{13} + L_3 + M_{23})\frac{di_2(t)}{dt} = (L_3 + M_{13})\frac{di(t)}{dt}$$

$$\begin{bmatrix} L_1 + M_{13} - M_{12} - M_{23} & M_{12} + M_{13} - M_{23} - L_2 \\ L_1 + 2M_{13} + L_3 & M_{12} + M_{13} + L_3 + M_{23} \end{bmatrix}\begin{bmatrix}\frac{di_1(t)}{dt} \\ \frac{di_2(t)}{dt}\end{bmatrix} = \begin{bmatrix} M_{13} - M_{23} \\ L_3 + M_{13}\end{bmatrix}\frac{di(t)}{dt} \Rightarrow$$

$$\begin{bmatrix}\frac{di_1(t)}{dt} \\ \frac{di_2(t)}{dt}\end{bmatrix} = \begin{bmatrix} L_1 + M_{13} - M_{12} - M_{23} & M_{12} + M_{13} - M_{23} - L_2 \\ L_1 + 2M_{13} + L_3 & M_{12} + M_{13} + L_3 + M_{23} \end{bmatrix}^{-1}\begin{bmatrix} M_{13} - M_{23} \\ L_3 + M_{13}\end{bmatrix}\frac{di(t)}{dt}$$

$$\begin{bmatrix}\int\frac{di_1(t)}{dt}dt \\ \int\frac{di_2(t)}{dt}dt\end{bmatrix} = \begin{bmatrix} L_1 + M_{13} - M_{12} - M_{23} & M_{12} + M_{13} - M_{23} - L_2 \\ L_1 + 2M_{13} + L_3 & M_{12} + M_{13} + L_3 + M_{23} \end{bmatrix}^{-1}\begin{bmatrix} M_{13} - M_{23} \\ L_3 + M_{13}\end{bmatrix}\int\frac{di(t)}{dt}dt \Rightarrow$$

$$\begin{bmatrix}i_1(t) \\ i_2(t)\end{bmatrix} = \begin{bmatrix} L_1 + M_{13} - M_{12} - M_{23} & M_{12} + M_{13} - M_{23} - L_2 \\ L_1 + 2M_{13} + L_3 & M_{12} + M_{13} + L_3 + M_{23} \end{bmatrix}^{-1}\begin{bmatrix} M_{13} - M_{23} \\ L_3 + M_{13}\end{bmatrix}i(t)$$

MATLAB can be employed to perform the calculations (Fig. 2.40).

```
Command Window
>> syms L1 L2 L3 M12 M13 M23
>> I=simplify(inv([L1+M13-M12-M23 M12+M13-M23-L2;L1+2*M13+L3 M12+M13+L3+M23])*[M13-M23;L3+M13]);
>> pretty(I)
/      2                                                        \
| M23   - L2 L3 - L2 M13 + L3 M12 + M12 M23 - M13 M23           |
| ----------------------------------------------------          |
|                         #1                                    |
|                                                               |
|      2                                                        |
| M13   - L1 L3 + L3 M12 - L1 M23 + M12 M13 - M13 M23           |
| ----------------------------------------------------          |
\                         #1                                    /

where

           2                                              2                         2
 #1 == M12   + 2 M12 M13 + 2 M12 M23 + 2 L3 M12 + M13   - 2 M13 M23 - 2 L2 M13 + M23   - 2 L1 M23 - L1 L2 - L1 L3 - L2 L3

fx >> |
```

Fig. 2.40 MATLAB code

According to the results shown in Fig. 2.40,

$$
\begin{bmatrix} i_1(t) \\ i_2(t) \end{bmatrix} = \begin{bmatrix} \dfrac{M_{23}^2 - L_2L_3 - L_2M_{13} + L_3M_{12} + M_{12}M_{23} - M_{13}M_{23}}{den} \\ \dfrac{M_{13}^2 - L_1L_3 + L_3M_{12} - L_1M_{23} + M_{12}M_{13} - M_{13}M_{23}}{den} \end{bmatrix} i(t)
$$

where, $den = M_{12}^2 + 2M_{12}M_{13} + 2M_{12}M_{23} + 2L_3M_{12} + M_{13}^2 - 2M_{13}M_{23} - 2L_2M_{13} + M_{23}^2 - 2L_1M_{23} - L_1L_2 - L_1L_3 - L_2L_3$. Therefore,

$$
V_{AB} = (L_1 + M_{13})\frac{di_1(t)}{dt} + (M_{12} + M_{13})\frac{di_2(t)}{dt} - M_{13}\frac{di(t)}{dt} \Rightarrow V_{AB}
$$

$$
= (L_1 + M_{13})\frac{M_{23}^2 - L_2L_3 - L_2M_{13} + L_3M_{12} + M_{12}M_{23} - M_{13}M_{23}}{den}\frac{di(t)}{dt}
$$

$$
+ (M_{12} + M_{13})\frac{M_{13}^2 - L_1L_3 + L_3M_{12} - L_1M_{23} + M_{12}M_{13} - M_{13}M_{23}}{den}\frac{di(t)}{dt} - M_{13}\frac{di(t)}{dt} \Rightarrow V_{AB}
$$

$$
= \left((L_1 + M_{13})\frac{M_{23}^2 - L_2L_3 - L_2M_{13} + L_3M_{12} + M_{12}M_{23} - M_{13}M_{23}}{den} \right.
$$

$$
\left. + (M_{12} + M_{13})\frac{M_{13}^2 - L_1L_3 + L_3M_{12} - L_1M_{23} + M_{12}M_{13} - M_{13}M_{23}}{den} - M_{13} \right)\frac{di(t)}{dt} \Rightarrow L_{eq}
$$

$$
= \left((L_1 + M_{13})\frac{M_{23}^2 - L_2L_3 - L_2M_{13} + L_3M_{12} + M_{12}M_{23} - M_{13}M_{23}}{den} \right.
$$

$$
\left. + (M_{12} + M_{13})\frac{M_{13}^2 - L_1L_3 + L_3M_{12} - L_1M_{23} + M_{12}M_{13} - M_{13}M_{23}}{den} - M_{13} \right)
$$

Manual simplification of this expression presents a challenge. Therefore, we will utilize MATLAB for the simplification process. The MATLAB code employed for this simplification is presented in Fig. 2.41.

```
Command Window
>> syms L1 L2 L3 M12 M13 M23
>> I=simplify(inv([L1+M13-M12-M23 M12+M13-M23-L2;L1+2*M13+L3 M12+M13+L3+M23])*[M13-M23;L3+M13]);
>> pretty(simplify((L1+M13)*I(1)+(M12+M13)*I(2)-M13))
                    2                    2                2
          L3 M12  - 2 M12 M13 M23 + L2 M13  + L1 M23  - L1 L2 L3
  ------------------------------------------------------------------------
    2                                   2                          2
  M12  + 2 M12 M13 + 2 M12 M23 + 2 L3 M12 + M13  - 2 M13 M23 - 2 L2 M13 + M23  - 2 L1 M23 - L1 L2 - L1 L3 - L2 L3
fx >> |
```

Fig. 2.41 MATLAB code

According to the result shown in Fig. 2.41,

$$V_{AB} = \frac{L_3 M_{12}^2 - 2M_{12}M_{13}M_{23} + L_2 M_{13}^2 + L_1 M_{23}^2 - L_1 L_2 L_3}{den} \frac{di(t)}{dt} \Rightarrow L_{eq}$$

$$= \frac{L_3 M_{12}^2 - 2M_{12}M_{13}M_{23} + L_2 M_{13}^2 + L_1 M_{23}^2 - L_1 L_2 L_3}{den}$$

2.2.17 Example 2.17

According to the formula obtained in Example 2.16, the equivalent inductance between terminals A and B in Fig. 2.42 is 6.3426 H, with the current division between the inductors as displayed in Fig. 2.43. It is recommended to solve this circuit step by step to make sure you arrive at the given results.

Fig. 2.42 Coupled inductors for Example 2.17

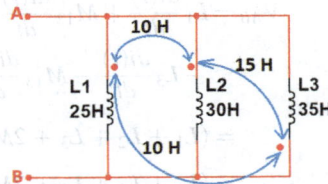

Fig. 2.43 Current division between the inductors

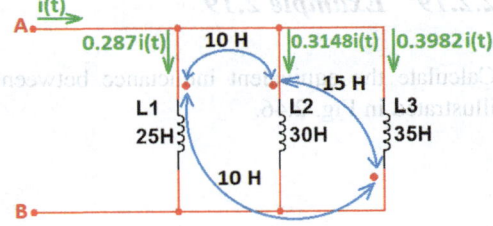

2.2.18   Example 2.18

Calculate the equivalent inductance between terminals A and B for the circuit illustrated in Fig. 2.44.

Fig. 2.44 Coupled
inductors for Example 2.18

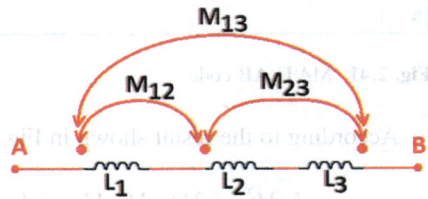

Solution:
The current $i(t)$ is defined as depicted in Fig. 2.45.

Fig. 2.45 The current $i(t)$

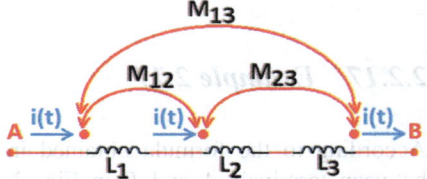

From Fig. 2.45, we have

$$V_{AB} = L_1 \frac{di(t)}{dt} + M_{12} \frac{di(t)}{dt} - M_{13} \frac{di(t)}{dt} + L_2 \frac{di(t)}{dt} + M_{12} \frac{di(t)}{dt} - M_{23} \frac{di(t)}{dt}$$
$$+ L_3 \frac{di(t)}{dt} - M_{13} \frac{di(t)}{dt} - M_{23} \frac{di(t)}{dt} \Rightarrow V_{AB}$$
$$= (L_1 + L_2 + L_3 + 2M_{12} - 2M_{13} - 2M_{23}) \frac{di(t)}{dt} \Rightarrow L_{eq}$$
$$= (L_1 + L_2 + L_3 + 2M_{12} - 2M_{13} - 2M_{23})$$

2.2.19   Example 2.19

Calculate the equivalent inductance between terminals A and B for the circuit illustrated in Fig. 2.46.

Fig. 2.46 Coupled inductors for Example 2.19

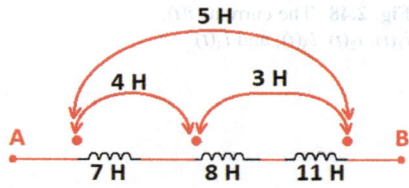

Solution:
According to the results from Example 2.18, we have $L_{eq} = 7 + 8 + 11 + 2 \times 4 - 2 \times 5 - 2 \times 3 = 18\ H$.

2.2.20 Example 2.20

Calculate the equivalent inductance between terminals A and B for the circuit illustrated in Fig. 2.47.

Fig. 2.47 Circuit for Example 2.20

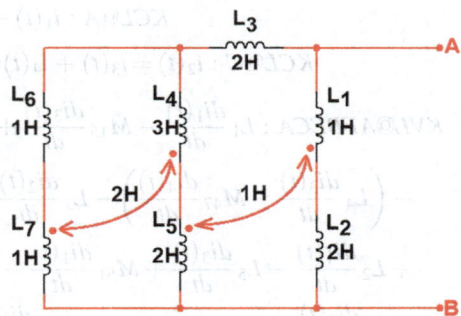

Solution:
The currents $i(t)$, $i_1(t)$, $i_2(t)$, $i_3(t)$, and $i_4(t)$ are defined as illustrated in Fig. 2.48.

Fig. 2.48 The currents $i(t)$, $i_1(t)$, $i_2(t)$, $i_3(t)$, and $i_4(t)$

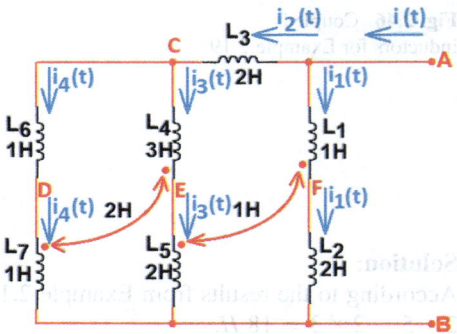

From Fig. 2.48, we have

$$M_{15} = M_{51} = 1\ H$$

$$M_{47} = M_{74} = 2\ H$$

$$V_{AB} = L_1 \frac{di_1(t)}{dt} - M_{15} \frac{di_3(t)}{dt} + L_2 \frac{di_1(t)}{dt} \Longrightarrow V_{AB} = (L_1 + L_2) \frac{di_1(t)}{dt} - M_{15} \frac{di_3(t)}{dt}$$

$$KCL@A : i_1(t) + i_2(t) = i(t)$$

$$KCL@C : i_2(t) = i_3(t) + i_4(t) \Longrightarrow i_2(t) - i_3(t) - i_4(t) = 0$$

$$KVL@AFBECA : L_1 \frac{di_1(t)}{dt} - M_{15} \frac{di_3(t)}{dt} + L_2 \frac{di_1(t)}{dt} - \left(L_5 \frac{di_3(t)}{dt} - M_{51} \frac{di_1(t)}{dt} \right)$$

$$- \left(L_4 \frac{di_3(t)}{dt} - M_{47} \frac{di_4(t)}{dt} \right) - L_3 \frac{di_2(t)}{dt} = 0 \Longrightarrow L_1 \frac{di_1(t)}{dt} - M_{15} \frac{di_3(t)}{dt}$$

$$+ L_2 \frac{di_1(t)}{dt} - L_5 \frac{di_3(t)}{dt} + M_{51} \frac{di_1(t)}{dt} - L_4 \frac{di_3(t)}{dt} + M_{47} \frac{di_4(t)}{dt}$$

$$- L_3 \frac{di_2(t)}{dt} = 0 \Longrightarrow (L_1 + L_2 + M_{51}) \frac{di_1(t)}{dt}$$

$$- L_3 \frac{di_2(t)}{dt} - (L_4 + L_5 + M_{15}) \frac{di_3(t)}{dt} + M_{47} \frac{di_4(t)}{dt} = 0$$

$$KVL@CEBDC : L_4 \frac{di_3(t)}{dt} - M_{47} \frac{di_4(t)}{dt} + L_5 \frac{di_3(t)}{dt} - M_{51} \frac{di_1(t)}{dt}$$

$$- \left(L_7 \frac{di_4(t)}{dt} - M_{74} \frac{di_3(t)}{dt} \right) - L_6 \frac{di_4(t)}{dt} = 0 \Longrightarrow - M_{51} \frac{di_1(t)}{dt} + L_4 \frac{di_3(t)}{dt}$$

$$- M_{47} \frac{di_4(t)}{dt} + L_5 \frac{di_3(t)}{dt} - L_7 \frac{di_4(t)}{dt} + M_{74} \frac{di_3(t)}{dt} - L_6 \frac{di_4(t)}{dt} = 0 \Longrightarrow$$

$$- M_{51} \frac{di_1(t)}{dt} + (L_4 + L_5 + M_{74}) \frac{di_3(t)}{dt} - (L_6 + L_7 + M_{47}) \frac{di_4(t)}{dt} = 0$$

$$
\begin{cases}
i_1(t) + i_2(t) = i(t) \\
i_2(t) - i_3(t) - i_4(t) = 0 \\
(L_1 + L_2 + M_{51})\dfrac{di_1(t)}{dt} - L_3\dfrac{di_2(t)}{dt} - (L_4 + L_5 + M_{15})\dfrac{di_3(t)}{dt} + M_{47}\dfrac{di_4(t)}{dt} = 0 \\
- M_{51}\dfrac{di_1(t)}{dt} + (L_4 + L_5 + M_{74})\dfrac{di_3(t)}{dt} - (L_6 + L_7 + M_{47})\dfrac{di_4(t)}{dt} = 0
\end{cases}
$$

$$
\Rightarrow
\begin{cases}
\dfrac{di_1(t)}{dt} + \dfrac{di_2(t)}{dt} = \dfrac{di(t)}{dt} \\
\dfrac{di_2(t)}{dt} - \dfrac{di_3(t)}{dt} - \dfrac{di_4(t)}{dt} = 0 \\
(L_1 + L_2 + M_{51})\dfrac{di_1(t)}{dt} - L_3\dfrac{di_2(t)}{dt} - (L_4 + L_5 + M_{15})\dfrac{di_3(t)}{dt} + M_{47}\dfrac{di_4(t)}{dt} = 0 \\
- M_{51}\dfrac{di_1(t)}{dt} + (\dot{L}_4 + L_5 + M_{74})\dfrac{di_3(t)}{dt} - (L_6 + L_7 + M_{47})\dfrac{di_4(t)}{dt} = 0
\end{cases}
$$

$$
\Rightarrow
\begin{bmatrix}
1 & 1 & 0 & 0 \\
0 & 1 & -1 & -1 \\
L_1 + L_2 + M_{51} & -L_3 & -L_4 - L_5 - M_{15} & M_{47} \\
-M_{51} & 0 & L_4 + L_5 + M_{74} & -L_6 - L_7 - M_{47}
\end{bmatrix}
\begin{bmatrix}
\dfrac{di_1(t)}{dt} \\[2mm]
\dfrac{di_2(t)}{dt} \\[2mm]
\dfrac{di_3(t)}{dt} \\[2mm]
\dfrac{di_4(t)}{dt}
\end{bmatrix}
$$

$$
=
\begin{bmatrix}
\dfrac{di(t)}{dt} \\[2mm]
0 \\[2mm]
0 \\[2mm]
0
\end{bmatrix}
\Rightarrow
\begin{bmatrix}
1 & 1 & 0 & 0 \\
0 & 1 & -1 & -1 \\
4 & -2 & -6 & 2 \\
-1 & 0 & 7 & -4
\end{bmatrix}
\begin{bmatrix}
\dfrac{di_1(t)}{dt} \\[2mm]
\dfrac{di_2(t)}{dt} \\[2mm]
\dfrac{di_3(t)}{dt} \\[2mm]
\dfrac{di_4(t)}{dt}
\end{bmatrix}
=
\begin{bmatrix}
\dfrac{di(t)}{dt} \\[2mm]
0 \\[2mm]
0 \\[2mm]
0
\end{bmatrix}
\Rightarrow
\begin{bmatrix}
\dfrac{di_1(t)}{dt} \\[2mm]
\dfrac{di_2(t)}{dt} \\[2mm]
\dfrac{di_3(t)}{dt} \\[2mm]
\dfrac{di_4(t)}{dt}
\end{bmatrix}
$$

$$
=
\begin{bmatrix}
1 & 1 & 0 & 0 \\
0 & 1 & -1 & -1 \\
4 & -2 & -6 & 2 \\
-1 & 0 & 7 & -4
\end{bmatrix}^{-1}
\begin{bmatrix}
\dfrac{di(t)}{dt} \\[2mm]
0 \\[2mm]
0 \\[2mm]
0
\end{bmatrix}
$$

MATLAB can be employed to perform the calculations (Fig. 2.49).

```
Command Window                                                        ⊙
  >> inv([1 1 0 0;0 1 -1 -1;4 -2 -6 2;-1 0 7 -4])*[di_dt;0;0;0]

ans =

  (8*di_dt)/17
  (9*di_dt)/17
  (4*di_dt)/17
  (5*di_dt)/17

  >> pretty(ans)
  / 8 di_dt \
  | ------- |
  |   17    |
  |         |
  | 9 di_dt |
  | ------- |
  |   17    |
  |         |
  | 4 di_dt |
  | ------- |
  |   17    |
  |         |
  | 5 di_dt |
  | ------- |
  \   17    /

fx >>
```

Fig. 2.49 MATLAB code

According to the result shown in Fig. 2.49, $\begin{bmatrix} \dfrac{di_1(t)}{dt} \\ \dfrac{di_2(t)}{dt} \\ \dfrac{di_3(t)}{dt} \\ \dfrac{di_4(t)}{dt} \end{bmatrix} = \begin{bmatrix} \dfrac{8}{17}\dfrac{di(t)}{dt} \\ \dfrac{9}{17}\dfrac{di(t)}{dt} \\ \dfrac{4}{17}\dfrac{di(t)}{dt} \\ \dfrac{5}{17}\dfrac{di(t)}{dt} \end{bmatrix}$. The

equivalent inductance between terminals A and B is

$$V_{AB}(t) = (L_1 + L_2)\frac{di_1(t)}{dt} - M_{15}\frac{di_3(t)}{dt} \Rightarrow V_{AB}(t) = 3\frac{di_1(t)}{dt} - \frac{di_3(t)}{dt} \Rightarrow V_{AB}(t)$$

$$= 3 \times \frac{8}{17}\frac{di(t)}{dt} - \frac{4}{17}\frac{di(t)}{dt} \Rightarrow V_{AB}(t) = \frac{24}{17}\frac{di(t)}{dt} - \frac{4}{17}\frac{di(t)}{dt} \Rightarrow V_{AB}(t)$$

$$= \frac{20}{17}\frac{di(t)}{dt} \Rightarrow L_{eq} = \frac{20}{17}$$

2.2.21 Example 2.21

Determine the impedance observed between terminals A and B for the circuit depicted in Fig. 2.50, given an angular frequency (ω) of 100 $\frac{Rad}{s}$.

Fig. 2.50 Circuit for Example 2.21

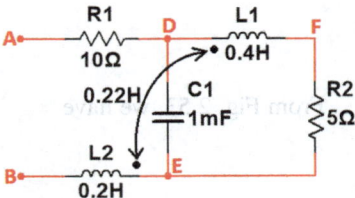

Solution:
A sinusoidal source with an angular frequency of 100 $\frac{Rad}{s}$ will be connected to terminals A and B, and the corresponding current will be measured (Fig. 2.51).

Fig. 2.51 $I_1(t)$ needs to be measured

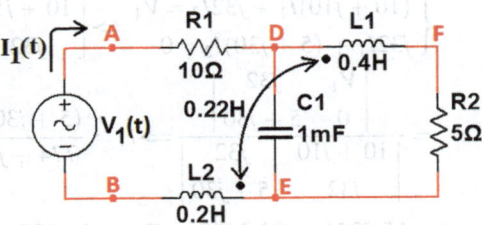

The sinusoidal steady-state equivalent circuit for the circuit depicted in Fig. 2.51 is illustrated in Fig. 2.52 ($\omega = 100 \ \frac{Rad}{s}$).

Fig. 2.52 The sinusoidal steady-state equivalent for Fig. 2.51

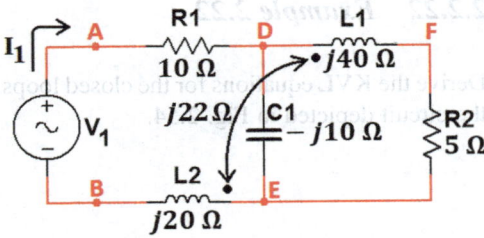

The currents I_1 and I_2 are defined as illustrated in Fig. 2.53.

Fig. 2.53 The currents I_1 and I_2

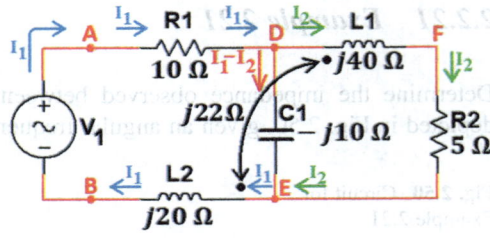

From Fig. 2.53, we have

$$Z_{AB} = \frac{V_1}{I_1}$$

$$KVL@BADEB: \; -V_1 + 10I_1 - j10(I_1 - I_2) + j20I_1 + j22I_2$$

$$= 0 \Rightarrow (10 + j10)I_1 + j32I_2 = V_1$$

$$KVL@DFED: j40I_2 + j22I_1 + 5I_2 - (-j10(I_1 - I_2)) = 0 \Rightarrow j32I_1 + (5 + j30)I_2 = 0$$

$$\begin{cases} (10 + j10)I_1 + j32I_2 = V_1 \\ j32I_1 + (5 + j30)I_2 = 0 \end{cases} \Rightarrow \begin{bmatrix} 10 + j10 & j32 \\ j32 & 5 + j30 \end{bmatrix} \begin{bmatrix} I_1 \\ I_2 \end{bmatrix} = \begin{bmatrix} V_1 \\ 0 \end{bmatrix} \Rightarrow I_1$$

$$= \frac{\begin{vmatrix} V_1 & j32 \\ 0 & 5 + j30 \end{vmatrix}}{\begin{vmatrix} 10 + j10 & j32 \\ j32 & 5 + j30 \end{vmatrix}} \Rightarrow I_1 = \frac{(5 + j30)V_1}{774 + j350} \Rightarrow \frac{V_1}{I_1} = \frac{774 + j350}{5 + j30} \Rightarrow \frac{V_1}{I_1}$$

$$= 15.5351 - j23.2108 \Rightarrow Z_{AB} = 15.5351 - j23.2108 \Rightarrow Z_{AB}$$

$$= 27.9299e^{-j0.9810} \Omega \Rightarrow Z_{AB} = 27.9299e^{-j56.2054°} \Omega$$

2.2.22 Example 2.22

Derive the KVL equations for the closed loops *ABDEFA*, *DFED*, and *ABDFA* within the circuit depicted in Fig. 2.54.

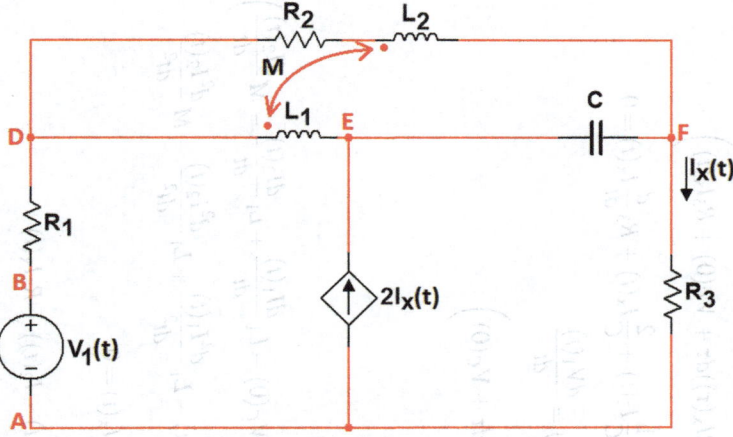

Fig. 2.54 Circuit for Example 2.22

Solution:

The currents $I_1(t)$ and $I_2(t)$ are defined as illustrated in Fig. 2.55.

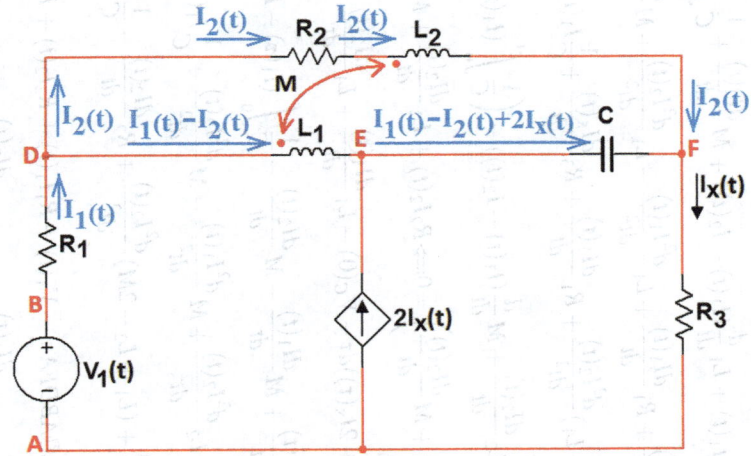

Fig. 2.55 The currents $I_1(t)$ and $I_2(t)$

From Fig. 2.55, we have

$$KVL@ABDEFA : \quad -V_1(t) + R_1I_1(t) + L_1\frac{d}{dt}(I_1(t)-I_2(t)) + M\frac{dI_2(t)}{dt} + \frac{1}{C}\int_0^t(I_1(\tau)-I_2(\tau)+2I_x(\tau))d\tau + V_C(0) + R_3I_x(t) = 0$$

$$\Rightarrow \frac{d}{dt}\left(-V_1(t) + R_1I_1(t) + L_1\frac{d}{dt}(I_1(t)-I_2(t)) + M\frac{dI_2(t)}{dt} + \frac{1}{C}\int_0^t(I_1(\tau)-I_2(\tau)+2I_x(\tau))d\tau + V_C(0) + R_3I_x(t)\right)$$

$$= \frac{d}{dt}(0) \Rightarrow -\frac{dV_1(t)}{dt} + R_1\frac{dI_1(t)}{dt} + L_1\frac{d^2I_1(t)}{dt^2} - L_1\frac{d^2I_2(t)}{dt^2} + M\frac{d^2I_2(t)}{dt^2} + \frac{1}{C}I_1(t) - \frac{1}{C}I_2(t) + \frac{2}{C}I_x(t) = \frac{dV_1(t)}{dt}$$

$$\Rightarrow L_1\frac{d^2I_1(t)}{dt^2} + (M-L_1)\frac{d^2I_2(t)}{dt^2} + R_1\frac{dI_1(t)}{dt} + R_3\frac{dI_x(t)}{dt} + \frac{1}{C}I_1(t) - \frac{1}{C}I_2(t) + \frac{2}{C}I_x(t) = \frac{dV_1(t)}{dt}$$

$$KVL@DFED : \quad R_2I_2(t) + L_2\frac{dI_2(t)}{dt} + M\frac{d}{dt}(I_1(t)-I_2(t)) - \left(L_1\frac{d}{dt}(I_1(t)-I_2(t)) + M\frac{dI_2(t)}{dt}\right)$$

$$-\left(\frac{1}{C}\int_0^t(I_1(\tau)-I_2(\tau)+2I_x(\tau))d\tau + V_C(0)\right) = 0 \Rightarrow R_2I_2(t) + L_2\frac{dI_2(t)}{dt} + M\frac{dI_1(t)}{dt} - M\frac{dI_2(t)}{dt}$$

$$-\frac{1}{C}\int_0^t(I_1(\tau)-I_2(\tau)+2I_x(\tau))d\tau - V_C(0) - L_1\frac{dI_1(t)}{dt} + L_1\frac{dI_2(t)}{dt} - M\frac{dI_2(t)}{dt} = 0$$

$$\Rightarrow \frac{d}{d}\left(R_2I_2(t) + L_2\frac{dI_2(t)}{dt} + M\frac{dI_2(t)}{dt} + M\frac{dI_1(t)}{dt} - M\frac{dI_2(t)}{dt} - \frac{1}{C}\int_0^t(I_1(\tau)-I_2(\tau)+2I_x(\tau))d\tau - V_C(0) - L_1\frac{dI_1(t)}{dt} + L_1\frac{dI_2(t)}{dt} - M\frac{dI_2(t)}{dt}\right)$$

$$= \frac{d}{dt}(0) \Rightarrow R_2\frac{dI_2(t)}{dt} + L_2\frac{d^2I_2(t)}{dt^2} + M\frac{d^2I_2(t)}{dt^2} + M\frac{d^2I_1(t)}{dt^2} - M\frac{d^2I_2(t)}{dt^2} - \frac{1}{C}I_1(t) + \frac{1}{C}I_2(t) - \frac{2}{C}I_x(t) - L_1\frac{d^2I_1(t)}{dt^2} - L_1\frac{d^2I_2(t)}{dt^2} - M\frac{d^2I_2(t)}{dt^2} = 0$$

$$= 0 \Rightarrow R_2\frac{dI_2(t)}{dt} + L_2\frac{d^2I_2(t)}{dt^2} + M\frac{d^2I_2(t)}{dt^2} + R_2\frac{dI_2(t)}{dt} - \frac{1}{C}I_1(t) + \frac{1}{C}I_2(t) - \frac{2}{C}I_x(t) = 0$$

$$= 0 \Rightarrow (M-L_1)\frac{d^2I_1(t)}{dt^2} + (L_1+L_2-2M)\frac{d^2I_2(t)}{dt^2} + R_2\frac{dI_2(t)}{dt} - \frac{1}{C}I_1(t) + \frac{1}{C}I_2(t) - \frac{2}{C}I_x(t) = 0$$

$$KVL@ABDFA : \quad -V_1(t) + R_1I_1(t) + R_2I_2(t) + L_2\frac{dI_2(t)}{dt} + M\frac{d}{dt}(I_1(t)-I_2(t)) + R_3I_x(t)$$

$$= 0 \Rightarrow M\frac{dI_1(t)}{dt} + (L_2-M)\frac{dI_2(t)}{dt} + R_1I_1(t) + R_2I_2(t) + R_3I_x(t) - V_1(t) = 0$$

2.2.23   *Example 2.23*

Calculate the steady-state capacitor voltage, $V_C(t)$, for the circuit illustrated in Fig. 2.56.

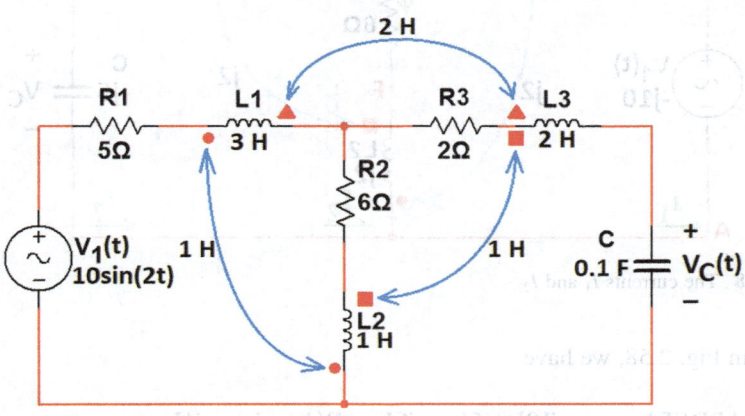

Fig. 2.56 Circuit for Example 2.23

Solution:
The sinusoidal steady-state equivalent circuit for the circuit depicted in Fig. 2.56 is illustrated in Fig. 2.57.

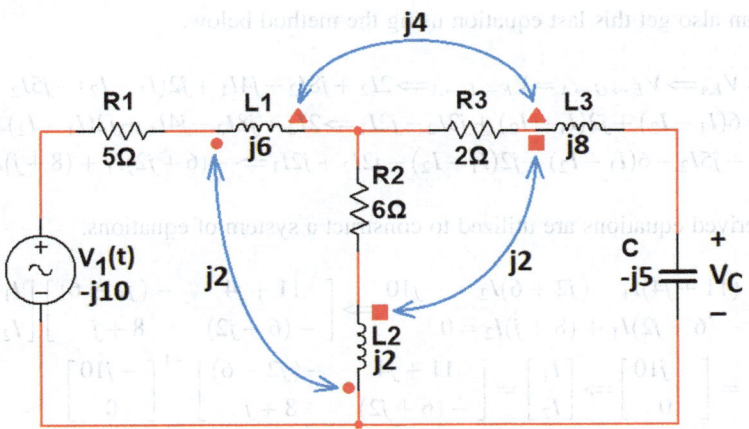

Fig. 2.57 Sinusoidal steady-state equivalent for Fig. 2.56

The currents I_1 and I_2 are defined as illustrated in Fig. 2.58.

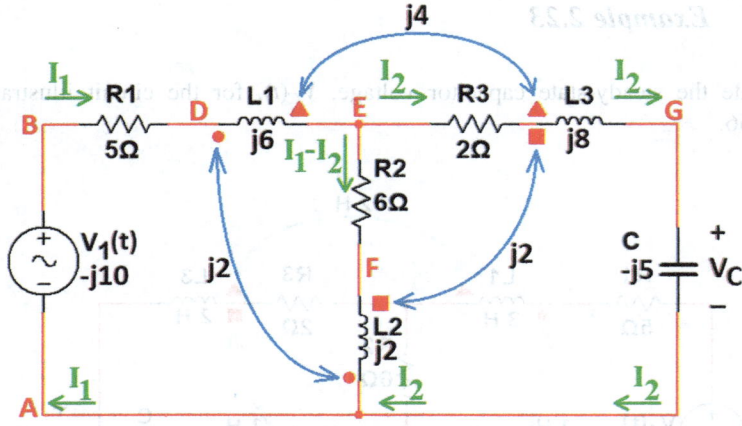

Fig. 2.58 The currents I_1 and I_2

From Fig. 2.58, we have

$KVL@ABDEFA : -(-j10) + 5I_1 + j6I_1 - j2(I_1 - I_2) - j4I_2$
$\quad + 6(I_1 - I_2) + j2(I_1 - I_2) + j2I_2 - j2I_1 = 0 \Rightarrow (11 + j4)I_1 - (j2 + 6)I_2 = -j10$

$KVL@EGAFE : 2I_2 + j8I_2 - j4I_1 + j2(I_1 - I_2) - j5I_2 - (j2(I_1 - I_2) + j2I_2 - j2I_1)$
$\quad - 6(I_1 - I_2) = 0 \Rightarrow 2I_2 + j8I_2 - j4I_1 + j2(I_1 - I_2) - j5I_2 - j2(I_1 - I_2)$
$\quad - j2I_2 + j2I_1 - 6(I_1 - I_2) = 0 \Rightarrow -(6 + j2)I_1 + (8 + j)I_2 = 0$

You can also get this last equation using the method below.

$V_{EA} = V_{EA} \Rightarrow V_{E \to G \to A} = V_{E \to F \to A} \Rightarrow 2I_2 + j8I_2 - j4I_1 + j2(I_1 - I_2) - j5I_2$
$\quad = 6(I_1 - I_2) + j2(I_1 - I_2) + j2I_2 - j2I_1 \Rightarrow 2I_2 + j8I_2 - j4I_1 + j2(I_1 - I_2)$
$\quad - j5I_2 - 6(I_1 - I_2) - j2(I_1 - I_2) - j2I_2 + j2I_1 \Rightarrow -(6 + j2)I_1 + (8 + j)I_2 = 0$

The derived equations are utilized to construct a system of equations.

$$\begin{cases} (11 + j4)I_1 - (j2 + 6)I_2 = -j10 \\ -(6 + j2)I_1 + (8 + j)I_2 = 0 \end{cases} \Rightarrow \begin{bmatrix} 11 + j4 & -(j2 + 6) \\ -(6 + j2) & 8 + j \end{bmatrix} \begin{bmatrix} I_1 \\ I_2 \end{bmatrix}$$

$$= \begin{bmatrix} -j10 \\ 0 \end{bmatrix} \Rightarrow \begin{bmatrix} I_1 \\ I_2 \end{bmatrix} = \begin{bmatrix} 11 + j4 & -(j2 + 6) \\ -(6 + j2) & 8 + j \end{bmatrix}^{-1} \begin{bmatrix} -j10 \\ 0 \end{bmatrix}$$

MATLAB can be employed to perform the calculations (Fig. 2.59).

Fig. 2.59 MATLAB code

```
Command Window                                                    ⊙
>> I=inv([11+4j  -(2j+6);-(2j+6)  8+j])*[-10j;0]

I =

  -0.3263 - 1.4192i
  -0.0326 - 1.1419i

>> abs(I)

ans =

   1.4563
   1.1424

>> angle(I)

ans =

  -1.7968
  -1.5994

fx >>
```

According to the result shown in Fig. 2.59, $\begin{bmatrix} I_1 \\ I_2 \end{bmatrix} = \begin{bmatrix} 1.4563e^{-j1.7968} \\ 1.1424e^{-j1.5994} \end{bmatrix}$.
Therefore,

$$V_C = -j5 \times I_2 \Rightarrow V_C = -j5 \times 1.1424e^{-j1.5994}$$
$$= 5.712e^{-j3.1702} \rightarrow V_C(t) = 5.712\cos(2t - 3.1702)$$

2.2.24 Example 2.24

Calculate the Thévenin equivalent as viewed from terminals A and B of the circuit depicted in Fig. 2.60.

Fig. 2.60 Circuit for
Example 2.24

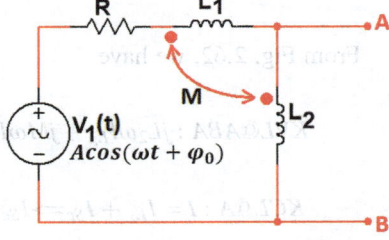

Solution:

The sinusoidal steady-state equivalent circuit for the circuit depicted in Fig. 2.60 is illustrated in Fig. 2.61.

Fig. 2.61 Sinusoidal steady-state equivalent for Fig. 2.60

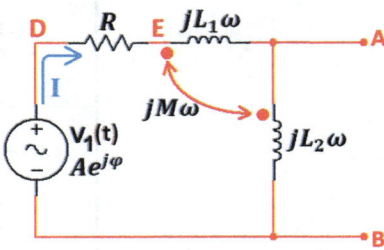

From Fig. 2.61, we have

$$KVL@BDEAB: \ -Ae^{j\varphi} + RI + jL_1\omega I + jM\omega I + jL_2\omega I + jM\omega I$$

$$= 0 \Rightarrow RI + j(L_1 + L_2 + 2M)\omega = Ae^{j\varphi} \Rightarrow I = \frac{Ae^{j\varphi}}{R + j(L_1 + L_2 + 2M)\omega}$$

$$V_{Th} = V_{AB} \Rightarrow V_{Th} = jL_2\omega I + jM\omega I \Rightarrow V_{Th} = j(L_2 + M)\omega I \Rightarrow V_{Th}$$

$$= j(L_2 + M)\omega \frac{Ae^{j\varphi}}{R + j(L_1 + L_2 + 2M)\omega}$$

The determination of the Thévenin equivalent voltage is complete. We now proceed to calculate the Thévenin equivalent impedance. To this end, let us calculate the short-circuit current (Fig. 2.62).

Fig. 2.62 I_{Sc} needs to be calculated

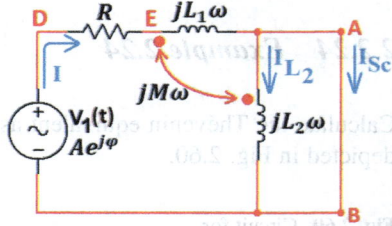

From Fig. 2.62, we have

$$KVL@ABA: jL_2\omega I_{L_2} + jM\omega I = 0 \Rightarrow L_2 I_{L_2} + MI = 0 \Rightarrow I_{L_2} = -\frac{M}{L_2}I$$

$$KCL@A: I = I_{L_2} + I_{Sc} \Rightarrow I_{Sc} = I - I_{L_2} \Rightarrow I_{Sc} = I - \left(-\frac{M}{L_2}I\right) \Rightarrow I_{Sc}$$

$$= \left(1 + \frac{M}{L_2}\right)I \Rightarrow I_{Sc} = \left(\frac{L_2 + M}{L_2}\right)I$$

$$KVL@BDEAB : \ -Ae^{j\varphi} + RI + jL_1\omega I + jM\omega I_{L_2} = 0 \Rightarrow (R + jL_1\omega)I$$

$$+ jM\omega I_{L_2} = Ae^{j\varphi} \Rightarrow (R + jL_1\omega)I - jM\omega\frac{M}{L_2}I = Ae^{j\varphi}$$

$$\Rightarrow \left(R + j\left(L_1 - \frac{M^2}{L_2}\right)\omega\right)I = Ae^{j\varphi} \Rightarrow I = \frac{Ae^{j\varphi}}{R + j\left(L_1 - \frac{M^2}{L_2}\right)\omega}$$

$$I_{Sc} = \left(\frac{L_2 + M}{L_2}\right)I \Rightarrow I_{Sc} = \left(\frac{L_2 + M}{L_2}\right)\frac{Ae^{j\varphi}}{R + j\left(L_1 - \frac{M^2}{L_2}\right)\omega}$$

The Thévenin equivalent impedance is

$$Z_{Th} = \frac{V_{AB}}{I_{Sc}} \Rightarrow Z_{Th} = \frac{j(L_2 + M)\omega\frac{Ae^{j\varphi}}{R + j(L_1 + L_2 + 2M)\omega}}{\left(\frac{L_2 + M}{L_2}\right)\frac{Ae^{j\varphi}}{R + j\left(L_1 - \frac{M^2}{L_2}\right)\omega}} \Rightarrow Z_{Th} = \omega\frac{(M^2 - L_1L_2)\omega + jRL_2}{R + j(L_1 + L_2 + 2M)\omega}$$

The Thévenin equivalent circuit, as observed from terminals A and B, is illustrated in Fig. 2.63.

Fig. 2.63 Thevenin equivalent of the circuit in Fig. 2.60

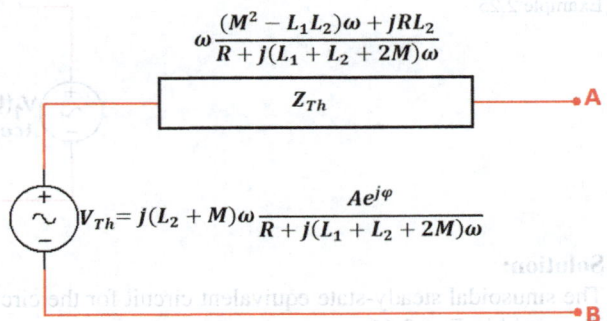

It is important to note that assuming zero inductor current (Fig. 2.64) will lead to an erroneous result. It is crucial to understand that zero inductor voltage is not a sufficient condition for zero inductor current.

Fig. 2.64 A wrong approach

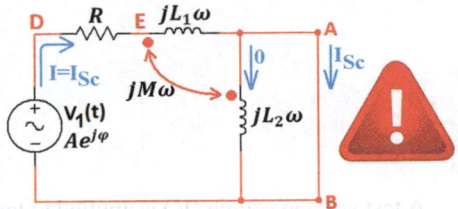

Assuming zero inductor current (Fig. 2.64) will lead to the following erroneous result.

$$KVL@BDEAB: -Ae^{j\varphi} + RI_{Sc} + jL_1\omega I_{Sc} + jM\omega \times 0$$

$$= 0 \Rightarrow RI_{Sc} + jL_1\omega I_{Sc} = Ae^{j\varphi} \Rightarrow (R + jL_1\omega)I_{Sc} = Ae^{j\varphi} \Rightarrow I_{Sc} = \frac{Ae^{j\varphi}}{R + jL_1\omega} \ (!!!)$$

$$Z_{Th} = \frac{V_{AB}}{I_{Sc}} \Rightarrow Z_{Th} = \frac{j(L_2 + M)\omega \frac{Ae^{j\varphi}}{R + j(L_1 + L_2 + 2M)\omega}}{\frac{Ae^{j\varphi}}{R + jL_1\omega}} \Rightarrow Z_{Th}$$

$$= j(L_2 + M)\omega \frac{R + jL_1\omega}{R + j(L_1 + L_2 + 2M)\omega} \Rightarrow Z_{Th}$$

$$= (L_2 + M)\omega \frac{-L_1\omega + jR}{R + j(L_1 + L_2 + 2M)\omega} \ (!!!)$$

2.2.25 Example 2.25

Determine the Thévenin equivalent impedance with respect to terminals A and B for the circuit depicted in Fig. 2.65.

Fig. 2.65 Circuit for Example 2.25

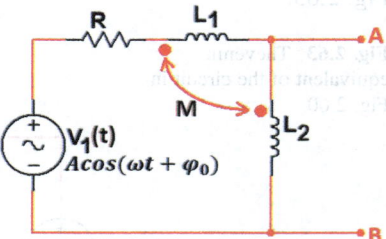

Solution:
The sinusoidal steady-state equivalent circuit for the circuit depicted in Fig. 2.65 is illustrated in Fig. 2.66.

Fig. 2.66 Sinusoidal steady-state equivalent for Fig. 2.65

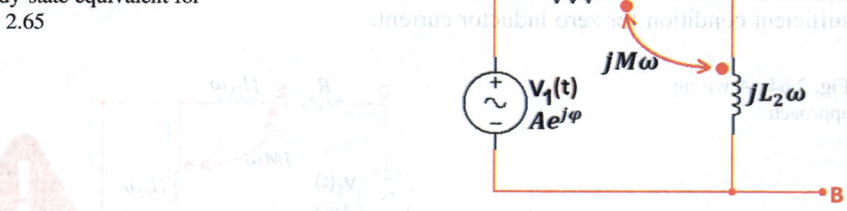

A test voltage source (V_t) is applied to terminals A and B, and the resulting current is measured (Fig. 2.67). It is important to note that the independent voltage source V_1 is deactivated.

Fig. 2.67 I_t needs to be measured

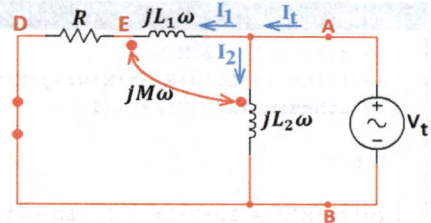

From Fig. 2.67, we have

$$KCL@A : I_t = I_1 + I_2$$

$$KVL@BAEDB : -V_t + jL_1\omega I_1 - jM\omega I_2 + RI_1 = 0 \Rightarrow (jL_1\omega + R)I_1 - jM\omega I_2 = V_t$$

$$V_{AB} = V_t \Rightarrow jL_2\omega I_2 - jM\omega I_1 = V_t$$

$$\begin{cases} (jL_1\omega + R)I_1 - jM\omega I_2 = V_t \\ jL_2\omega I_2 - jM\omega I_1 = V_t \end{cases} \Rightarrow \begin{cases} (jL_1\omega + R)I_1 - jM\omega I_2 = V_t \\ -jM\omega I_1 + jL_2\omega I_2 = V_t \end{cases}$$

$$\Rightarrow \begin{bmatrix} jL_1\omega + R & -jM \\ -jM\omega & jL_2\omega \end{bmatrix}\begin{bmatrix} I_1 \\ I_2 \end{bmatrix} = \begin{bmatrix} V_t \\ V_t \end{bmatrix} \Rightarrow \begin{bmatrix} I_1 \\ I_2 \end{bmatrix} = \begin{bmatrix} jL_1\omega + R & -jM\omega \\ -jM\omega & jL_2\omega \end{bmatrix}^{-1}\begin{bmatrix} V_t \\ V_t \end{bmatrix}$$

$$\Rightarrow \begin{bmatrix} I_1 \\ I_2 \end{bmatrix} = \frac{1}{(jL_1\omega + R)(jL_2\omega) + M^2\omega^2}\begin{bmatrix} jL_2\omega & jM\omega \\ jM\omega & jL_1\omega + R \end{bmatrix}\begin{bmatrix} V_t \\ V_t \end{bmatrix} \Rightarrow \begin{bmatrix} I_1 \\ I_2 \end{bmatrix}$$

$$= \frac{1}{(M^2 - L_1L_2)\omega^2 + jRL_2\omega}\begin{bmatrix} j(L_2 + M)\omega V_t \\ (j(M + L_1)\omega + R)V_t \end{bmatrix}$$

$$I_t = I_1 + I_2 \Rightarrow I_t = \frac{j(L_2 + M)\omega V_t}{(M^2 - L_1L_2)\omega^2 + jRL_2\omega} + \frac{(j(M + L_1)\omega + R)V_t}{(M^2 - L_1L_2)\omega^2 + jRL_2\omega} \Rightarrow I_t$$

$$= \frac{j(L_1 + L_2 + 2M)\omega + R}{(M^2 - L_1L_2)\omega^2 + jRL_2\omega}V_t \Rightarrow \frac{V_t}{I_t} = \frac{(M^2 - L_1L_2)\omega^2 + jRL_2\omega}{j(L_1 + L_2 + 2M)\omega + R} \Rightarrow Z_{Th}$$

$$= \frac{(M^2 - L_1L_2)\omega^2 + jRL_2\omega}{R + j(L_1 + L_2 + 2M)\omega} \Rightarrow Z_{Th} = \omega\frac{(M^2 - L_1L_2)\omega + jRL_2}{R + j(L_1 + L_2 + 2M)\omega}$$

2.2.26 Example 2.26

In the preceding example, the calculations were performed manually. The MATLAB code presented in Fig. 2.68 computes the values of I_1, I_2, and $\frac{V_t}{I_1 + I_2}$, where I_1 and I_2 satisfy the system of equations: $\begin{bmatrix} I_1 \\ I_2 \end{bmatrix} = \begin{bmatrix} jL_1\omega + R & -jM\omega \\ -jM\omega & jL_2\omega \end{bmatrix}^{-1}\begin{bmatrix} V_t \\ V_t \end{bmatrix}$. The result obtained is identical to that derived in Example 2.25.

```
Command Window                                                        ⊙

>> syms R L1 L2 M w Vt
>> I=inv([j*L1*w+R -j*M*w;-j*M*w j*L2*w])*[Vt;Vt];
>> Zth=simplify(Vt/sum(I))


Zth =


(w*(w*M^2 + L2*R*1i - L1*L2*w))/(R + L1*w*1i + L2*w*1i + M*w*2i)

>> pretty(Zth)
          2
 w (w M  + L2 R 1i - L1 L2 w)
 ------------------------------
 R + L1 w 1i + L2 w 1i + M w 2i

fx >>
```

Fig. 2.68 MATLAB code

2.2.27 *Example 2.27*

(A) Determine the value of the load impedance (Z_L) that maximizes power transfer to the load in the circuit illustrated in Fig. 2.69. (B) Calculate the maximum power transferable to the load impedance Z_L.

Fig. 2.69 Circuit for Example 2.27

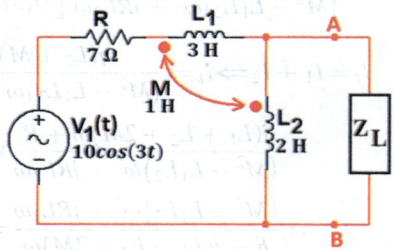

Solution:
We will utilize the results derived in Example 2.25 to determine the Thévenin equivalent circuit as observed from terminals A and B.

$$V_{Th} = j(L_2 + M)\omega \frac{Ae^{j\varphi}}{R + j(L_1 + L_2 + 2M)\omega} \Rightarrow V_{Th}$$

$$= j(2+1) \times 3 \times \frac{10e^{j0}}{7 + j(3 + 2 + 2 \times 1)3} \Rightarrow V_{Th} = \frac{j90}{7 + j21} \Rightarrow V_{Th}$$

$$= 3.8571 + j1.2857 \Rightarrow V_{Th} = 4.0658e^{j0.3218}$$

$$Z_{Th} = \omega \frac{(M^2 - L_1 L_2)\omega + jRL_2}{R + j(L_1 + L_2 + 2M)\omega} \Rightarrow Z_{Th} = 3 \frac{(1^2 - 3 \times 2)3 + j7 \times 2}{7 + j(3 + 2 + 2 \times 1)3} \Rightarrow Z_{Th}$$

$$= \frac{-45 + j42}{7 + j21} \Rightarrow Z_{Th} = 1.1571 + j2.5286$$

The Thévenin equivalent circuit is illustrated in Fig. 2.70.

Fig. 2.70 Thevenin equivalent circuit

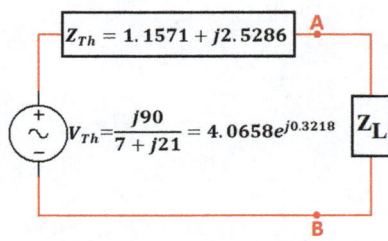

The load impedance (Z_L) for maximum power transfer and the corresponding maximum transferable power are as follows:

$$Z_L = Z_{Th}{}^* \Rightarrow Z_L = (1.1571 + j2.5286)^* \Rightarrow Z_L = 1.1571 - j2.5286$$

$$Z_L = Z_{Th}{}^* \Rightarrow P_{L,max} = \frac{|V_{Th}|^2}{4 Re\{Z_{Th}\}} = \frac{4.0658^2}{4 \times 1.1571} = 3.5716 \, W$$

2.2.28 Example 2.28

Formulate the differential equations describing the relationship between currents $I_1(t)$ and $I_2(t)$ and the sources $I_s(t)$ and $V_s(t)$ for the circuit depicted in Fig. 2.71.

Fig. 2.71 Circuit for Example 2.28

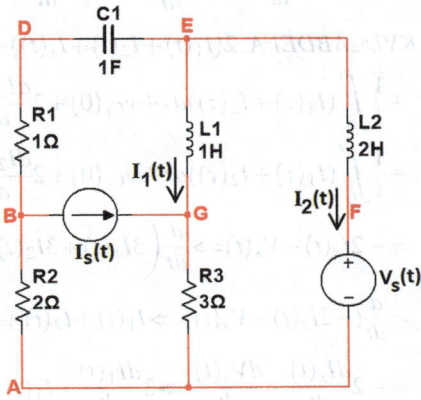

Solution:
This example serves as a preparatory exercise for the subsequent example. The current flow in the distinct branches is depicted in Fig. 2.72.

Fig. 2.72 The current flow in the distinct branches

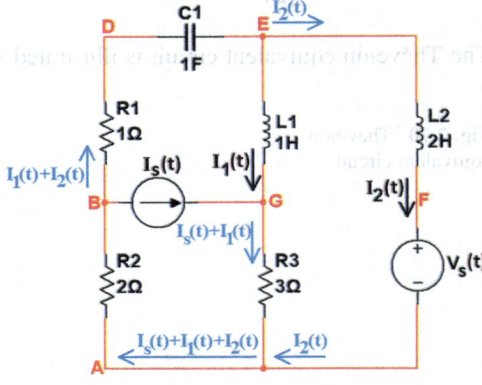

From Fig. 2.72, we have

$$KVL@ABDEGA: 2(I_1(t) + I_2(t) + I_s(t)) + 1(I_1(t) + I_2(t))$$

$$+ \frac{1}{1}\int_0^t (I_1(\tau) + I_2(\tau))d\tau + v_{C_1}(0) + 1\frac{dI_1(t)}{dt} + 3(I_s(t) + I_1(t))$$

$$= 0 \Rightarrow 6I_1(t) + 3I_2(t) + \int_0^t (I_1(\tau) + I_2(\tau))d\tau + v_{C_1}(0) + \frac{dI_1(t)}{dt}$$

$$= -5I_s(t) \Rightarrow \frac{d}{dt}\left(6I_1(t) + 3I_2(t) + \int_0^t (I_1(\tau) + I_2(\tau))d\tau + v_{C_1}(0) + \frac{dI_1(t)}{dt}\right)$$

$$= \frac{d}{dt}(-5I_s(t)) \Rightarrow I_1(t) + I_2(t) + 6\frac{dI_1(t)}{dt} + 3\frac{dI_2(t)}{dt} + \frac{d^2I_1(t)}{dt^2}$$

$$= -5\frac{dI_s(t)}{dt} \Rightarrow \frac{d^2I_1(t)}{dt^2} + 6\frac{dI_1(t)}{dt} + I_1(t) + 3\frac{dI_2(t)}{dt} + I_2(t) = -5\frac{dI_s(t)}{dt}$$

$$KVL@ABDEFA: 2(I_1(t) + I_2(t) + I_s(t)) + 1(I_1(t) + I_2(t))$$

$$+ \frac{1}{1}\int_0^t (I_1(\tau) + I_2(\tau))d\tau + v_{C_1}(0) + 2\frac{dI_2(t)}{dt} + V_s(t) = 0 \Rightarrow 3I_1(t) + 3I_2(t)$$

$$+ \frac{1}{1}\int_0^t (I_1(\tau) + I_2(\tau))d\tau + v_{C_1}(0) + 2\frac{dI_2(t)}{dt}$$

$$= -2I_s(t) - V_s(t) \Rightarrow \frac{d}{dt}\left(3I_1(t) + 3I_2(t) + \frac{1}{1}\int_0^t (I_1(\tau) + I_2(\tau))d\tau + v_{C_1}(0) + 2\frac{dI_2(t)}{dt}\right)$$

$$= \frac{d}{dt}(-2I_s(t) - V_s(t)) \Rightarrow I_1(t) + I_2(t) + 3\frac{dI_1(t)}{dt} + 3\frac{dI_2(t)}{dt} + 2\frac{d^2I_2(t)}{dt^2}$$

$$= -2\frac{dI_s(t)}{dt} - \frac{dV_s(t)}{dt} \Rightarrow 3\frac{dI_1(t)}{dt} + I_1(t) + 2\frac{d^2I_2(t)}{dt^2} + 3\frac{dI_2(t)}{dt} + I_2(t)$$

$$= -2\frac{dI_s(t)}{dt} - \frac{dV_s(t)}{dt}$$

$$\begin{cases} \dfrac{d^2I_1(t)}{dt^2}+6\dfrac{dI_1(t)}{dt}+I_1(t)+3\dfrac{dI_2(t)}{dt}+I_2(t)=-5\dfrac{dI_s(t)}{dt} \\ 3\dfrac{dI_1(t)}{dt}+I_1(t)+2\dfrac{d^2I_2(t)}{dt^2}+3\dfrac{dI_2(t)}{dt}+I_2(t)=-2\dfrac{dI_s(t)}{dt}-\dfrac{dV_s(t)}{dt} \end{cases}$$

$$\Rightarrow \begin{cases} (D^2+6D+1)I_1(t)+(3D+1)I_2(t)=-5DI_s(t) \\ (3D+1)I_1(t)+(2D^2+3D+1)I_2(t)=-2DI_s(t)-DV_s(t) \end{cases}$$

$$\Rightarrow \begin{bmatrix} D^2+6D+1 & 3D+1 \\ 3D+1 & 2D^2+3D+1 \end{bmatrix}\begin{bmatrix} I_1(t) \\ I_2(t) \end{bmatrix}=\begin{bmatrix} -5D & 0 \\ -2D & -D \end{bmatrix}\begin{bmatrix} I_s(t) \\ V_s(t) \end{bmatrix}$$

$$\Rightarrow \begin{bmatrix} I_1(t) \\ I_2(t) \end{bmatrix}=\begin{bmatrix} D^2+6D+1 & 3D+1 \\ 3D+1 & 2D^2+3D+1 \end{bmatrix}^{-1}\begin{bmatrix} -5D & 0 \\ -2D & -D \end{bmatrix}\begin{bmatrix} I_s(t) \\ V_s(t) \end{bmatrix}$$

MATLAB can be employed to perform the calculations (Fig. 2.73).

```
Command Window

>> syms D
>> I=simplify(inv([D^2+6*D+1 3*D+1;3*D+1 2*D^2+3*D+1])*[-5*D 0;-2*D -D]);
>> pretty(I)
/           2                          \
|    10 D  + 9 D + 3        3 D + 1     |
|  - ----------------,      -------     |
|         #1                  #1        |
|                                       |
|        2                    2         |
|   - 2 D  + 3 D + 3        D  + 6 D + 1 |
|   ----------------,  -   ------------- |
\         #1                  #1         /

where

             3       2
    #1 == 2 D  + 15 D  + 12 D + 3

fx >> |
```

Fig. 2.73 MATLAB code

The differential equations describing the relationships between currents $I_1(t)$ and $I_2(t)$ and the sources $I_s(t)$ and $V_s(t)$ are as follows:

$$\begin{bmatrix} I_1(t) \\ I_2(t) \end{bmatrix} = \begin{bmatrix} \dfrac{10D^2+9D+3}{2D^3+15D^2+12D+3} & \dfrac{3D+1}{2D^3+15D^2+12D+3} \\[2ex] \dfrac{-2D^2+3D+3}{2D^3+15D^2+12D+3} & \dfrac{D^2+6D+1}{-2D^3+15D^2+12D+3} \end{bmatrix} \begin{bmatrix} I_s(t) \\ V_s(t) \end{bmatrix} \Rightarrow \begin{bmatrix} I_1(t) \\ I_2(t) \end{bmatrix}$$

$$= \frac{1}{2D^3+15D^2+12D+3}\begin{bmatrix} -(10D^2+9D+3) & 3D+1 \\ -2D^2+3D+3 & -(D^2+6D+1) \end{bmatrix}\begin{bmatrix} I_s(t) \\ V_s(t) \end{bmatrix} \Rightarrow \begin{bmatrix} I_s(t) \\ V_s(t) \end{bmatrix}$$

$$= \begin{bmatrix} 3D+1 \\ -(D^2+6D+1) \end{bmatrix}\begin{bmatrix} I_s(t) \\ V_s(t) \end{bmatrix} \Rightarrow \begin{bmatrix} (2D^3+15D^2+12D+3)I_1(t) \\ (2D^3+15D^2+12D+3)I_2(t) \end{bmatrix}$$

$$= \begin{bmatrix} -(10D^2+9D+3)I_s(t)+(3D+1)V_s(t) \\ (-2D^2+3D+3)I_s(t)-(D^2+6D+1)V_s(t) \end{bmatrix}$$

$$\Rightarrow \begin{bmatrix} 2\dfrac{d^3I_1(t)}{dt^3}+15\dfrac{d^2I_1(t)}{dt^2}+12\dfrac{dI_1(t)}{dt}+3I_1(t) = -10\dfrac{d^2I_s(t)}{dt^2}-9\dfrac{dI_s(t)}{dt}-3I_s(t)+3\dfrac{dV_s(t)}{dt}+V_s(t) \\[2ex] 2\dfrac{d^3I_2(t)}{dt^3}+15\dfrac{d^2I_2(t)}{dt^2}+12\dfrac{dI_2(t)}{dt}+3I_2(t) = -2\dfrac{d^2I_s(t)}{dt^2}+3\dfrac{dI_s(t)}{dt}+3I_s(t)-6\dfrac{d^2V_s(t)}{dt^2}-\dfrac{dV_s(t)}{dt}-V_s(t) \end{bmatrix}$$

2.2.29 Example 2.29

Formulate the differential equations describing the relationship between currents $I_1(t)$ and $I_2(t)$ and the sources $I_s(t)$ and $V_s(t)$ for the circuit depicted in Fig. 2.74.

Fig. 2.74 Circuit for
Example 2.29

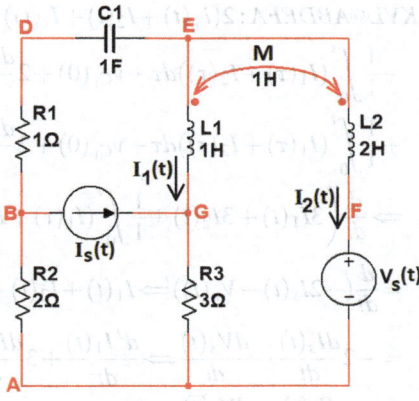

Solution:
The current flow in the distinct branches is depicted in Fig. 2.75.

Fig. 2.75 The current flow
in the distinct branches

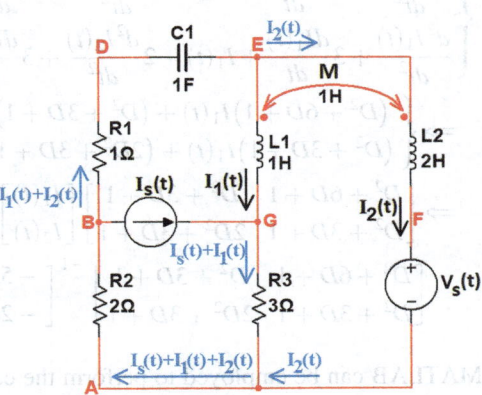

From Fig. 2.75, we have

$$KVL@ABDEGA : 2(I_1(t) + I_2(t) + I_s(t)) + 1(I_1(t) + I_2(t))$$

$$+ \frac{1}{1}\int_0^t (I_1(\tau) + I_2(\tau))d\tau + v_{C_1}(0) + 1\frac{dI_1(t)}{dt} + 1\frac{dI_2(t)}{dt} + 3(I_s(t) + I_1(t)) = 0 \Rightarrow 6I_1(t)$$

$$+ 3I_2(t) + \int_0^t (I_1(\tau) + I_2(\tau))d\tau + v_{C_1}(0) + \frac{dI_1(t)}{dt} + \frac{dI_2(t)}{dt}$$

$$= -5I_s(t) \Rightarrow \frac{d}{dt}\left(6I_1(t) + 3I_2(t) + \int_0^t (I_1(\tau) + I_2(\tau))d\tau + v_{C_1}(0) + \frac{dI_1(t)}{dt} + \frac{dI_2(t)}{dt}\right)$$

$$= \frac{d}{dt}(-5I_s(t)) \Rightarrow I_1(t) + I_2(t) + 6\frac{dI_1(t)}{dt} + 3\frac{dI_2(t)}{dt} + \frac{d^2I_1(t)}{dt^2} + \frac{d^2I_2(t)}{dt^2}$$

$$= -5\frac{dI_s(t)}{dt} \Rightarrow \frac{d^2I_1(t)}{dt^2} + 6\frac{dI_1(t)}{dt} + I_1(t) + \frac{d^2I_2(t)}{dt^2} + 3\frac{dI_2(t)}{dt} + I_2(t) = -5\frac{dI_s(t)}{dt}$$

$KVL@ABDEFA : 2(I_1(t) + I_2(t) + I_s(t)) + 1(I_1(t) + I_2(t))$

$$+ \frac{1}{1}\int_0^t (I_1(\tau) + I_2(\tau))d\tau + v_{C_1}(0) + 2\frac{dI_2(t)}{dt} + 1\frac{dI_1(t)}{dt} + V_s(t) = 0 \Rightarrow 3I_1(t) + 3I_2(t)$$

$$+ \frac{1}{1}\int_0^t (I_1(\tau) + I_2(\tau))d\tau + v_{C_1}(0) + 2\frac{dI_2(t)}{dt} + \frac{dI_1(t)}{dt} = -2I_s(t) - V_s(t)$$

$$\Rightarrow \frac{d}{dt}\left(3I_1(t) + 3I_2(t) + \frac{1}{1}\int_0^t (I_1(\tau) + I_2(\tau))d\tau + v_{C_1}(0) + 2\frac{dI_2(t)}{dt} + \frac{dI_1(t)}{dt}\right)$$

$$= \frac{d}{dt}(-2I_s(t) - V_s(t)) \Rightarrow I_1(t) + I_2(t) + 3\frac{dI_1(t)}{dt} + 3\frac{dI_2(t)}{dt} + 2\frac{d^2I_2(t)}{dt^2} + \frac{d^2I_1(t)}{dt^2}$$

$$= -2\frac{dI_s(t)}{dt} - \frac{dV_s(t)}{dt} \Rightarrow \frac{d^2I_1(t)}{dt^2} + 3\frac{dI_1(t)}{dt} + I_1(t) + 2\frac{d^2I_2(t)}{dt^2} + 3\frac{dI_2(t)}{dt} + I_2(t)$$

$$= -2\frac{dI_s(t)}{dt} - \frac{dV_s(t)}{dt}$$

$$\begin{cases} \dfrac{d^2I_1(t)}{dt^2} + 6\dfrac{dI_1(t)}{dt} + I_1(t) + \dfrac{d^2I_2(t)}{dt^2} + 3\dfrac{dI_2(t)}{dt} + I_2(t) = -5\dfrac{dI_s(t)}{dt} \\ \dfrac{d^2I_1(t)}{dt^2} + 3\dfrac{dI_1(t)}{dt} + I_1(t) + 2\dfrac{d^2I_2(t)}{dt^2} + 3\dfrac{dI_2(t)}{dt} + I_2(t) = -2\dfrac{dI_s(t)}{dt} - \dfrac{dV_s(t)}{dt} \end{cases}$$

$$\Rightarrow \begin{cases} (D^2 + 6D + 1)I_1(t) + (D^2 + 3D + 1)I_2(t) = -5DI_s(t) \\ (D^2 + 3D + 1)I_1(t) + (2D^2 + 3D + 1)I_2(t) = -2DI_s(t) - DV_s(t) \end{cases}$$

$$\Rightarrow \begin{bmatrix} D^2 + 6D + 1 & D^2 + 3D + 1 \\ D^2 + 3D + 1 & 2D^2 + 3D + 1 \end{bmatrix}\begin{bmatrix} I_1(t) \\ I_2(t) \end{bmatrix} = \begin{bmatrix} -5D & 0 \\ -2D & -D \end{bmatrix}\begin{bmatrix} I_s(t) \\ V_s(t) \end{bmatrix} \Rightarrow \begin{bmatrix} I_1(t) \\ I_2(t) \end{bmatrix}$$

$$= \begin{bmatrix} D^2 + 6D + 1 & D^2 + 3D + 1 \\ D^2 + 3D + 1 & 2D^2 + 3D + 1 \end{bmatrix}^{-1}\begin{bmatrix} -5D & 0 \\ -2D & -D \end{bmatrix}\begin{bmatrix} I_s(t) \\ V_s(t) \end{bmatrix}$$

MATLAB can be employed to perform the calculations (Fig. 2.76).

```
Command Window
>> syms D
>> I=simplify(inv([D^2+6*D+1 D^2+3*D+1;D^2+3*D+1 2*D^2+3*D+1])*[-5*D 0;-2*D -D]);
>> pretty(I)
 /       2                    2              \
 |   8 D  + 9 D + 3        D  + 3 D + 1      |
 | - --------------,       -----------      |
 |        #1                   #1           |
 |                                          |
 |      2                      2            |
 |   (D  + D + 1) 3        D  + 6 D + 1      |
 |   -------------,    -  -------------      |
 \        #1                   #1           /

where

              3       2
    #1 == D  + 9 D  + 10 D + 3
fx >>
```

Fig. 2.76 MATLAB code

The differential equations describing the relationships between currents $I_1(t)$ and $I_2(t)$ and the sources $I_s(t)$ and $V_s(t)$ are as follows:

$$\begin{bmatrix} I_1(t) \\ I_2(t) \end{bmatrix} = \begin{bmatrix} -\dfrac{8D^2+9D+3}{D^3+9D^2+10D+3} & \dfrac{D^2+3D+1}{D^3+9D^2+10D+3} \\ 3\dfrac{D^2+D+1}{D^3+9D^2+10D+3} & -\dfrac{D^2+6D+1}{D^3+9D^2+10D+3} \end{bmatrix} \begin{bmatrix} I_s(t) \\ V_s(t) \end{bmatrix} \Rightarrow \begin{bmatrix} I_1(t) \\ I_2(t) \end{bmatrix}$$

$$= \frac{1}{D^3+9D^2+10D+3} \begin{bmatrix} -(8D^2+9D+3) & D^2+3D+1 \\ -3D^2+3D+3 & -(D^2+6D+1) \end{bmatrix} \begin{bmatrix} I_s(t) \\ V_s(t) \end{bmatrix}$$

$$\Rightarrow \begin{bmatrix} (D^3+9D^2+10D+3)I_1(t) \\ (D^3+9D^2+10D+3)I_2(t) \end{bmatrix} = \begin{bmatrix} -(8D^2+9D+3) & D^2+3D+1 \\ -3D^2+3D+3 & -(D^2+6D+1) \end{bmatrix} \begin{bmatrix} I_s(t) \\ V_s(t) \end{bmatrix}$$

$$\Rightarrow \begin{cases} \dfrac{d^3I_1(t)}{dt^3}+9\dfrac{d^2I_1(t)}{dt^2}+10\dfrac{dI_1(t)}{dt}+3I_1(t) \\ \quad = -8\dfrac{d^2I_s(t)}{dt^2}-9\dfrac{dI_s(t)}{dt}-3I_s(t)+\dfrac{dV_s^2(t)}{dt^2}+3\dfrac{dV_s(t)}{dt}+V_s(t) \\ \dfrac{d^3I_2(t)}{dt^3}+9\dfrac{d^2I_2(t)}{dt^2}+10\dfrac{dI_2(t)}{dt}+3I_2(t) \\ \quad = -3\dfrac{d^2I_s(t)}{dt^2}+3\dfrac{dI_s(t)}{dt}+3I_s(t)-\dfrac{d^2V_s(t)}{dt^2}-6\dfrac{dV_s(t)}{dt}-V_s(t) \end{cases}$$

As an exercise, employ MATLAB to solve these differential equations with $I_s(t)=1\,A$ and $V_s(t)=2\,V$. The initial conditions are taken to be zero.

## 2.2.30	Example 2.30

Calculate the steady-state capacitor voltage, $V_O(t)$, for the circuit illustrated in Fig. 2.77.

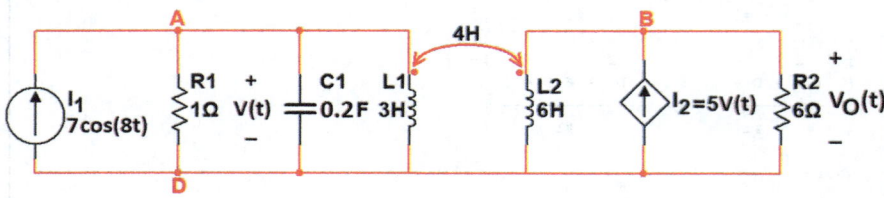

Fig. 2.77	Circuit for Example 2.30

Solution:
The sinusoidal steady-state equivalent circuit for the circuit depicted in Fig. 2.77 is illustrated in Fig. 2.78.

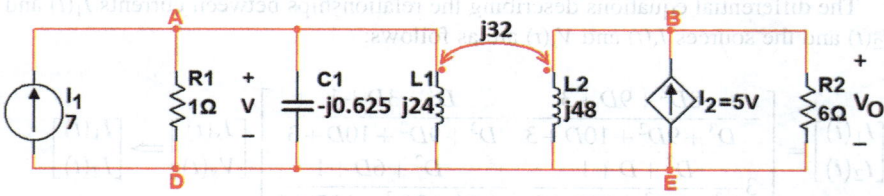

Fig. 2.78	Sinusoidal steady-state equivalent for Fig. 2.77

The currents $I_1, I_{R1},\ I_{C1}, I_{L1},\ I_{L2}, I_2,$ and I_{R_2} are defined as illustrated in Fig. 2.79.

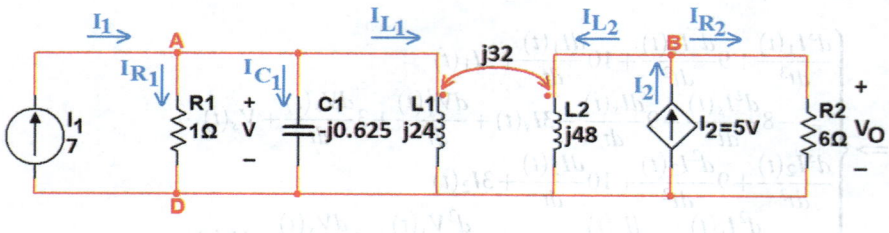

Fig. 2.79	The currents and

From Fig. 2.79, we have

$$V_D = 0\ V$$

$$KCL@A : I_{R1} + I_{C1} + I_{L1} = I_1 \Rightarrow \frac{V}{1} + \frac{V}{-j0.625} + I_{L1} = 7 \Rightarrow \frac{V}{1}$$

$$+ \frac{V}{-j0.625} + I_{L1} = 7 \Rightarrow (1 + 1.6j)V + I_{L1} = 7$$

$$KCL@B : I_{R2} + I_{L2} = I_2 \Rightarrow \frac{V_O}{6} + I_{L2} = 5V \Rightarrow 0.1667V_O + I_{L2} = 5V$$

$$V = j24I_{L1} + j32I_{L2} \Rightarrow j24I_{L1} + j32I_{L2} - V = 0$$

$$V_O = j48I_{L2} + j32I_{L1} \Rightarrow j48I_{L2} + j32I_{L1} - V_O = 0$$

$$\begin{cases} (1 + 1.6j)V + I_{L1} = 7 \\ 0.1667V_O + I_{L2} = 5V \\ j24I_{L1} + j32I_{L2} - V = 0 \\ j48I_{L2} + j32I_{L1} - V_O = 0 \end{cases} \Rightarrow \begin{bmatrix} 1 & 0 & 1 + 1.6j & 0 \\ 0 & 1 & -5 & 0.1667 \\ j24 & j32 & -1 & 0 \\ j32 & j48 & 0 & -1 \end{bmatrix} \begin{bmatrix} I_{L1} \\ I_{L2} \\ V \\ V_O \end{bmatrix}$$

$$= \begin{bmatrix} 7 \\ 0 \\ 0 \\ 0 \end{bmatrix} \Rightarrow \begin{bmatrix} I_{L1} \\ I_{L2} \\ V \\ V_O \end{bmatrix} = \begin{bmatrix} 1 & 0 & 1 + 1.6j & 0 \\ 0 & 1 & -5 & 0.1667 \\ j24 & j32 & -1 & 0 \\ j32 & j48 & 0 & -1 \end{bmatrix}^{-1} \begin{bmatrix} 7 \\ 0 \\ 0 \\ 0 \end{bmatrix}$$

MATLAB can be employed to perform the calculations (Fig. 2.80).

```
Command Window
>> x=inv([1 0 1+1.6*j 0;0 1 -5 0.1667;j*24 j*32 -1 0;j*32 j*48 0 -1])*[7;0;0;0]

x =

   5.7871 + 2.1397i
  -4.3761 - 1.5853i
  -0.6209 - 1.1462i
   7.6272 -24.8676i

>> abs(x)

ans =

   6.1700
   4.6544
   1.3035
  26.0110

>> angle(x)

ans =

   0.3541
  -2.7940
  -2.0673
  -1.2732

fx >>
```

Fig. 2.80 MATLAB code

According to the information provided in Fig. 2.80,

$$\begin{bmatrix} I_{L1} \\ I_{L2} \\ V \\ V_O \end{bmatrix} = \begin{bmatrix} 6.1700e^{j0.3541} \\ 4.6544e^{-j2.7940} \\ 1.3035e^{-j2.0673} \\ 26.0110e^{-j1.2732} \end{bmatrix} \Rightarrow \begin{bmatrix} I_{L1}(t) \\ I_{L2}(t) \\ V(t) \\ V_O(t) \end{bmatrix} = \begin{bmatrix} 6.1700\cos{(8t+0.3541)} \\ 4.6544\cos{(8t-2.794)} \\ 1.3035\cos{(8t-2.0673)} \\ 26.0110\ \cos{(8t-1.2732)} \end{bmatrix}$$

2.2.31 Example 2.31

Calculate the steady-state capacitor voltage, $V_O(t)$, for the circuit illustrated in Fig. 2.81.

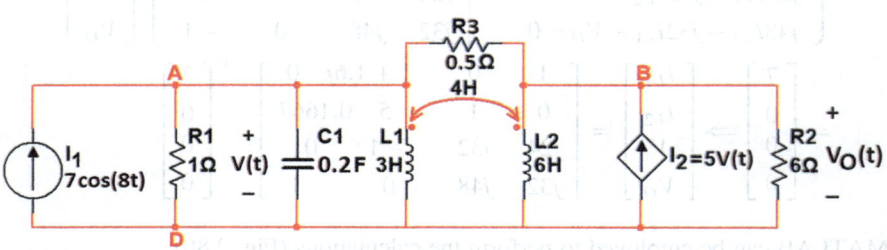

Fig. 2.81 Circuit for Example 2.31

Solution:
The sinusoidal steady-state equivalent circuit for the circuit depicted in Fig. 2.81 is illustrated in Fig. 2.82.

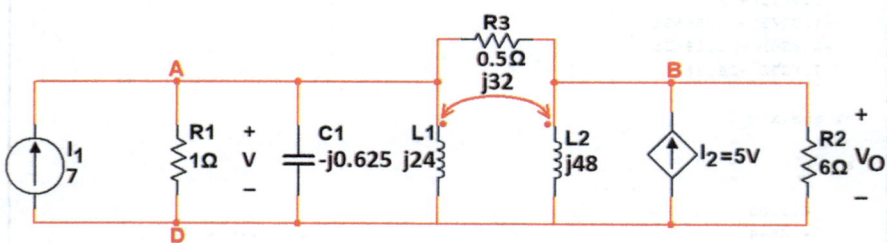

Fig. 2.82 The sinusoidal steady-state equivalent for Fig. 2.81

The currents I_1, I_{R1}, I_{C1}, I_{L1}, I_{R3}, I_{L2}, I_2, and I_{R_2} are defined as illustrated in Fig. 2.83.

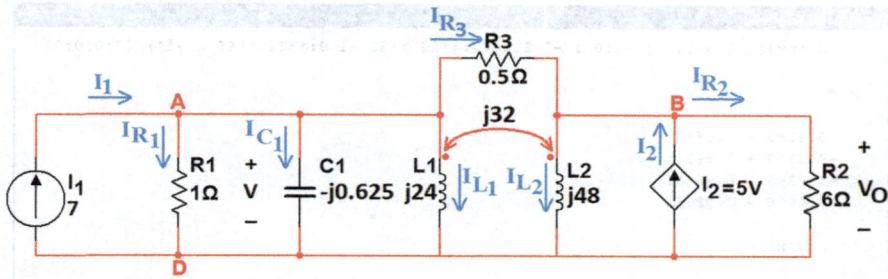

Fig. 2.83 The currents $I_1, I_{R1}, I_{C1}, I_{L1}, I_{R3}, I_{L2}, I_2$, and I_{R_2}

From Fig. 2.83, we have

$$V_D = 0 \ V$$

$$KCL@A : I_{R1} + I_{R3} + I_{C1} + I_{L1} = I_1 \Rightarrow \frac{V}{1} + \frac{V}{-j0.625} + I_{L1} + \frac{V_A - V_B}{0.5}$$

$$= 7 \Rightarrow \frac{V}{1} + \frac{V}{-j0.625} + I_{L1} + \frac{V - V_O}{0.5} = 7 \Rightarrow (3 + 1.6j)V - 2V_O + I_{L_1} = 7$$

$$KCL@B : I_{R2} + I_{L2} = I_2 + I_{R3} \Rightarrow \frac{V_O}{6} + I_{L2} = 5V + \frac{V - V_O}{0.5} \Rightarrow 0.1667 V_O$$

$$+ I_{L2} = 7V - 2V_O \Rightarrow I_{L2} - 7V + 2.1667 V_O = 0$$

$$V = j24 I_{L1} + j32 I_{L2} \Rightarrow j24 I_{L1} + j32 I_{L2} - V = 0$$

$$V_O = j48 I_{L2} + j32 I_{L1} \Rightarrow j48 I_{L2} + j32 I_{L1} - V_O = 0$$

$$\begin{cases} (3 + 1.6j)V - 2V_O + I_{L_1} = 7 \\ I_{L2} - 7V + 2.1667 V_O = 0 \\ j24 I_{L1} + j32 I_{L2} - V = 0 \\ j48 I_{L2} + j32 I_{L1} - V_O = 0 \end{cases} \Rightarrow \begin{bmatrix} 1 & 0 & 3 + 1.6j & -2 \\ 0 & 1 & -7 & 2.1667 \\ j24 & j32 & -1 & 0 \\ j32 & j48 & 0 & -1 \end{bmatrix} \begin{bmatrix} I_{L1} \\ I_{L2} \\ V \\ V_O \end{bmatrix}$$

$$= \begin{bmatrix} 7 \\ 0 \\ 0 \\ 0 \end{bmatrix} \Rightarrow \begin{bmatrix} I_{L1} \\ I_{L2} \\ V \\ V_O \end{bmatrix} = \begin{bmatrix} 1 & 0 & 3 + 1.6j & -2 \\ 0 & 1 & -7 & 2.1667 \\ j24 & j32 & -1 & 0 \\ j32 & j48 & 0 & -1 \end{bmatrix}^{-1} \begin{bmatrix} 7 \\ 0 \\ 0 \\ 0 \end{bmatrix}$$

MATLAB can be employed to perform the calculations (Fig. 2.84).

```
Command Window                                                              ⊙
 >> x=inv([1 0 3+1.6*j -2;0 1 -7 2.1667;j*24 j*32 -1 0;j*32 j*48 0 -1])*[7;0;0;0]

 x =

    0.4080 - 0.6647i
   -0.3309 + 0.5493i
   -1.6245 - 0.7965i
   -5.0956 - 2.8266i

 >> abs(x)

 ans =

    0.7799
    0.6413
    1.8092
    5.8271

 >> angle(x)

 ans =

   -1.0203
    2.1130
   -2.6858
   -2.6351

fx >> |
```

Fig. 2.84 MATLAB code

According to the information provided in Fig. 2.84,

$$
\begin{bmatrix} I_{L1} \\ I_{L2} \\ V \\ V_O \end{bmatrix} = \begin{bmatrix} 0.7799e^{-j1.0203} \\ 0.6413e^{j2.1130} \\ 1.8092e^{-j2.6858} \\ 5.8271e^{-j2.6351} \end{bmatrix} \rightarrow \begin{bmatrix} I_{L1}(t) \\ I_{L2}(t) \\ V(t) \\ V_O(t) \end{bmatrix} = \begin{bmatrix} 0.7799\cos(8t - 1.0203) \\ 0.6413\cos(8t - 2.1130) \\ 1.8092\cos(8t - 2.6858) \\ 5.8271\ \cos(8t - 2.6351) \end{bmatrix}
$$

2.2.32   Example 2.32

Calculate the steady-state resistor R_2 voltage, $V_{R_2}(t)$, for the circuit illustrated in Fig. 2.85.

Fig. 2.85 Circuit for Example 2.32

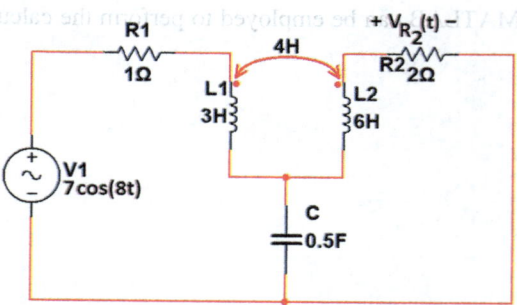

Solution:
The currents $I_1(t)$ and $I_2(t)$ are defined as illustrated in Fig. 2.86.

Fig. 2.86 The currents $I_1(t)$ and $I_2(t)$

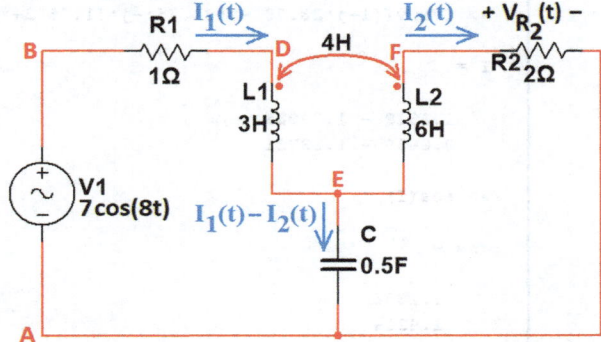

The sinusoidal steady-state equivalent circuit for the circuit depicted in Fig. 2.86 is illustrated in Fig. 2.87.

Fig. 2.87 The sinusoidal steady-state equivalent for Fig. 2.86

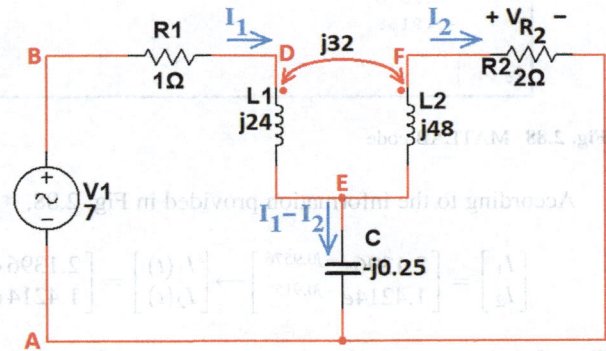

From Fig. 2.87, we have

$$KVL@ABDEA: \ -7 + I_1 + j24I_1 - j32I_2 - j0.25(I_1 - I_2)$$
$$= 0 \Rightarrow (1 + j23.75)I_1 - j31.75I_2 = 7$$

$$KVL@FAEF: 2I_2 - (-j0.25(I_1 - I_2)) + j48I_2 - j32I_1$$
$$= 0 \Rightarrow -j31.75I_1 + (2 + j47.75)I_2 = 0$$

$$\begin{cases} (1 + j23.75)I_1 - j31.75I_2 = 7 \\ -j31.75I_1 + (2 + j47.75)I_2 = 0 \end{cases} \Rightarrow \begin{bmatrix} 1 + j23.75 & -j31.75 \\ -j31.75 & 2 + j47.75 \end{bmatrix} \begin{bmatrix} I_1 \\ I_2 \end{bmatrix}$$

$$= \begin{bmatrix} 7 \\ 0 \end{bmatrix} \Rightarrow \begin{bmatrix} I_1 \\ I_2 \end{bmatrix} = \begin{bmatrix} 1 + j23.75 & -j31.75 \\ -j31.75 & 2 + j47.75 \end{bmatrix}^{-1} \begin{bmatrix} 7 \\ 0 \end{bmatrix}$$

MATLAB can be employed to perform the calculations (Fig. 2.88).

```
Command Window                                                    ⊙

>> I=inv([1+j*23.75 -j*31.75;-j*31.75 2+j*47.75])*[7;0]

I =

      1.2312 - 1.7498i
      0.8659 - 1.1272i

>> abs(I)

ans =

      2.1396
      1.4214

>> angle(I)

ans =

     -0.9576
     -0.9158

fx >> |
```

Fig. 2.88 MATLAB code

According to the information provided in Fig. 2.88,

$$\begin{bmatrix} I_1 \\ I_2 \end{bmatrix} = \begin{bmatrix} 2.1396e^{-j0.9576} \\ 1.4214e^{-j0.9158} \end{bmatrix} \rightarrow \begin{bmatrix} I_1(t) \\ I_2(t) \end{bmatrix} = \begin{bmatrix} 2.1396\cos(8t - 0.9576) \\ 1.4214\cos(8t - 0.9158) \end{bmatrix}$$

Therefore,

$$V_{R2}(t) = R_2 I_2(t) \Rightarrow V_{R2}(t) = 2 \times 1.4214\cos(8t - 0.9158) \Rightarrow V_{R2}(t)$$
$$= 2.8428\cos(8t - 0.9158)$$

2.2.33 Example 2.33

Calculate the steady-state resistor R_2 voltage, $V_{R_2}(t)$, for the circuit illustrated in Fig. 2.89.

Fig. 2.89 Circuit for
Example 2.33

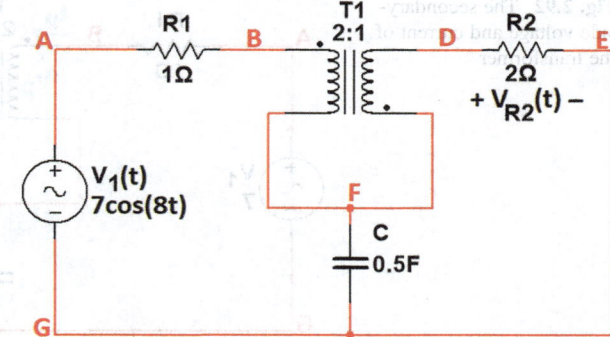

Solution:
The sinusoidal steady-state equivalent circuit for the circuit depicted in Fig. 2.89 is
illustrated in Fig. 2.90.

Fig. 2.90 The sinusoidal
steady-state equivalent for
Fig. 2.89

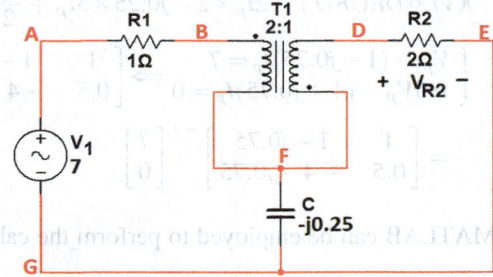

The primary-side voltage and current of the transformer are illustrated in
Fig. 2.91.

Fig. 2.91 The primary-side
voltage and current of the
transformer

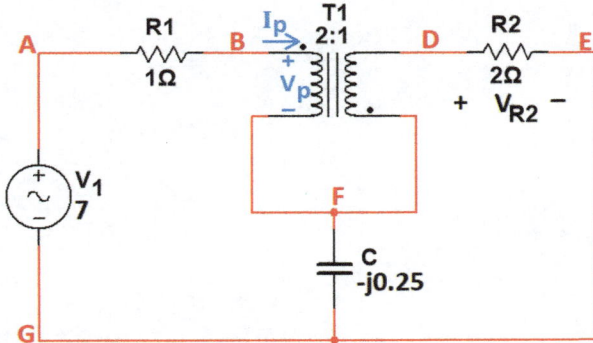

The secondary-side voltage and current of the transformer are illustrated in
Fig. 2.92.

Fig. 2.92 The secondary-side voltage and current of the transformer

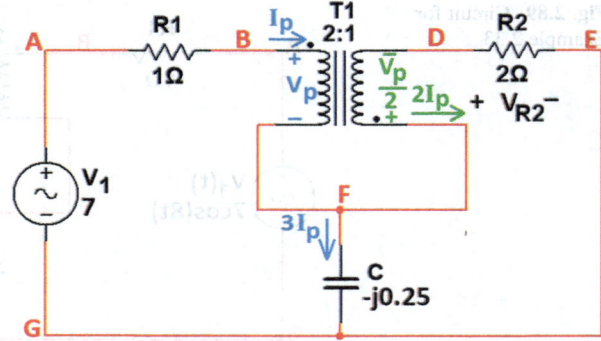

From Fig. 2.92, we have

$$KVL@GABFG: \quad -7 + 1I_p + V_p - j0.25 \times 3I_p = 0 \Rightarrow V_p + (1 - j0.75)I_p = 7$$

$$KVL@DEGFD: \quad -2I_p \times 2 - j0.25 \times 3I_p + \frac{V_p}{2} = 0 \Rightarrow 0.5V_p - (4 + j0.75)I_p = 0$$

$$\begin{cases} V_p + (1 - j0.75)I_p = 7 \\ 0.5V_p - (4 + j0.75)I_p = 0 \end{cases} \Rightarrow \begin{bmatrix} 1 & 1 - j0.75 \\ 0.5 & -4 - j0.75 \end{bmatrix} \begin{bmatrix} V_p \\ I_p \end{bmatrix} = \begin{bmatrix} 7 \\ 0 \end{bmatrix} \Rightarrow \begin{bmatrix} V_p \\ I_p \end{bmatrix}$$

$$= \begin{bmatrix} 1 & 1 - j0.75 \\ 0.5 & -4 - j0.75 \end{bmatrix}^{-1} \begin{bmatrix} 7 \\ 0 \end{bmatrix}$$

MATLAB can be employed to perform the calculations (Fig. 2.93).

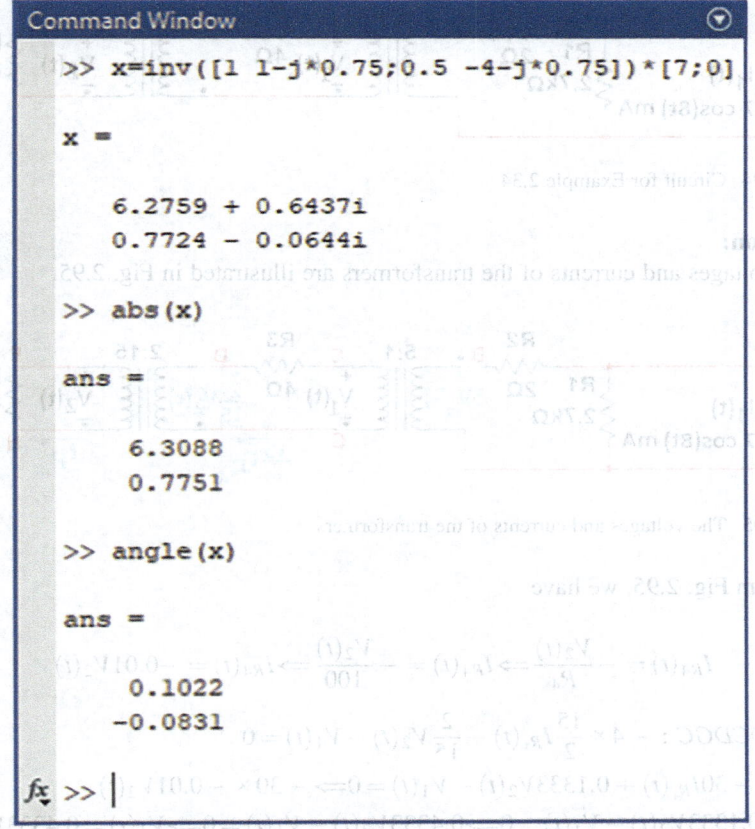

Fig. 2.93 MATLAB code

According to the information provided in Fig. 2.93,

$$\begin{bmatrix} V_p \\ I_p \end{bmatrix} = \begin{bmatrix} 6.3088e^{j0.1022} \\ 0.7751e^{-j0.0831} \end{bmatrix} \rightarrow \begin{bmatrix} V_p(t) \\ I_p(t) \end{bmatrix} = \begin{bmatrix} 6.3088\cos(8t + 0.1022) \\ 0.7751\cos(8t - 0.0831) \end{bmatrix}$$

Therefore,

$$V_{R2} = -2I_p \times R_2 \Rightarrow V_{R2} = -2 \times 0.7751e^{-j0.0831} \times 2 \Rightarrow V_{R2}$$
$$= -3.1004e^{-j0.0831} \rightarrow V_{R2}(t) = -3.1004\cos(8t - 0.0831)$$

2.2.34 Example 2.34

Calculate the voltages $V_1(t)$ and $V_2(t)$ with respect to the circuit depicted in Fig. 2.94.

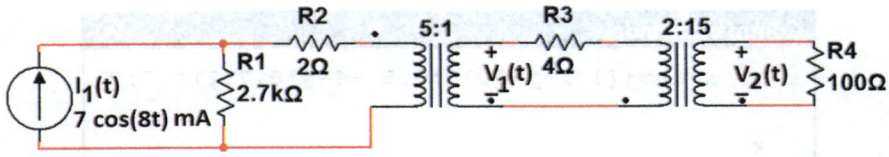

Fig. 2.94 Circuit for Example 2.34

Solution:
The voltages and currents of the transformers are illustrated in Fig. 2.95.

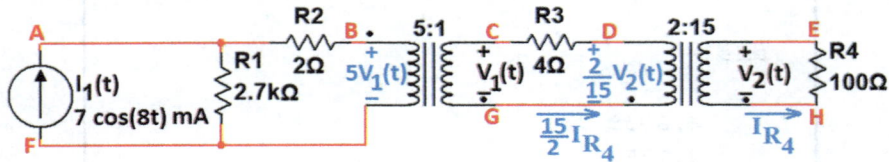

Fig. 2.95 The voltages and currents of the transformers

From Fig. 2.95, we have

$$I_{R4}(t) = -\frac{V_2(t)}{R_4} \Rightarrow I_{R4}(t) = -\frac{V_2(t)}{100} \Rightarrow I_{R4}(t) = -0.01V_2(t)$$

$$KVL@CDGC : -4 \times \frac{15}{2} I_{R_4}(t) + \frac{2}{15} V_2(t) - V_1(t) = 0$$

$$\Rightarrow -30 I_{R_4}(t) + 0.1333 V_2(t) - V_1(t) = 0 \Rightarrow -30 \times -0.01 V_2(t)$$

$$+0.1333 V_2(t) - V_1(t) = 0 \Rightarrow 0.4333 V_2(t) - V_1(t) = 0 \Rightarrow V_1(t) = 0.4333 V_2(t)$$

Figure 2.96 represents the equivalent circuit of Fig. 2.95. Resistor R_5 constitutes the equivalent resistance observed from terminals D and G.

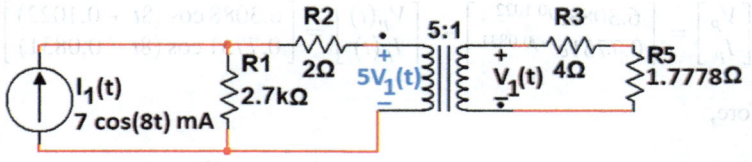

Fig. 2.96 Equivalent circuit for Fig. 2.95

Figure 2.97 represents the equivalent circuit of Fig. 2.96.

Fig. 2.97 Equivalent circuit
for Fig. 2.96

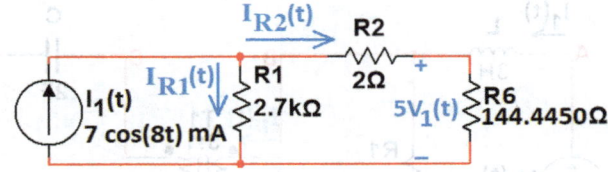

From Fig. 2.97, we have

$$I_{R2}(t) = \frac{R_1}{R_1 + (R_2 + R_6)} I_1(t) \Rightarrow I_{R2}(t)$$

$$= \frac{2700}{2700 + 2 + 144.4450} 0.007 \cos(8t) \Rightarrow I_{R2}(t) = 0.0066 \cos(8t)$$

$$V_{R6}(t) = 5V_1(t) \Rightarrow R_6 I_{R2}(t) = 5V_1(t) \Rightarrow V_1(t) = \frac{R_6 I_{R2}(t)}{5} \Rightarrow V_1(t)$$

$$= \frac{144.4450 \times 0.0066 \cos(8t)}{5} \Rightarrow V_1(t) = 0.1907 \cos(8t)$$

$$V_1(t) = 0.4333 V_2(t) \Rightarrow V_2(t) = \frac{1}{0.4333} V_1(t) \Rightarrow V_2(t) = 2.3079 V_1(t) \Rightarrow V_2(t)$$

$$= 2.3079 \times 0.1907 \cos(8t) \Rightarrow V_2(t) = 0.4401 \cos(8t)$$

2.2.35 Example 2.35

Calculate the steady-state resistor R_2 voltage, $V_O(t)$, for the circuit illustrated in
Fig. 2.98.

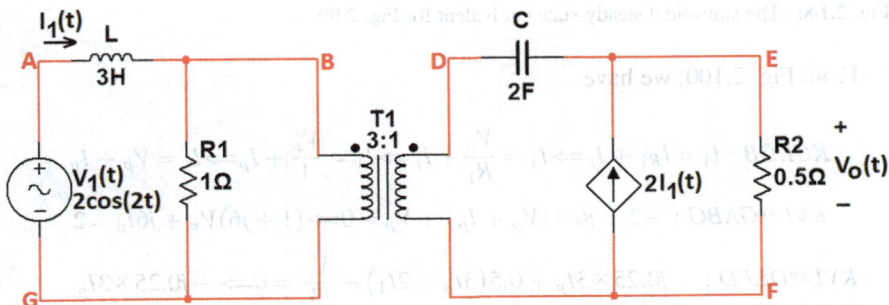

Fig. 2.98 Circuit for Example 2.35

Solution:
The primary-side voltage and current of the transformer are illustrated in Fig. 2.99.

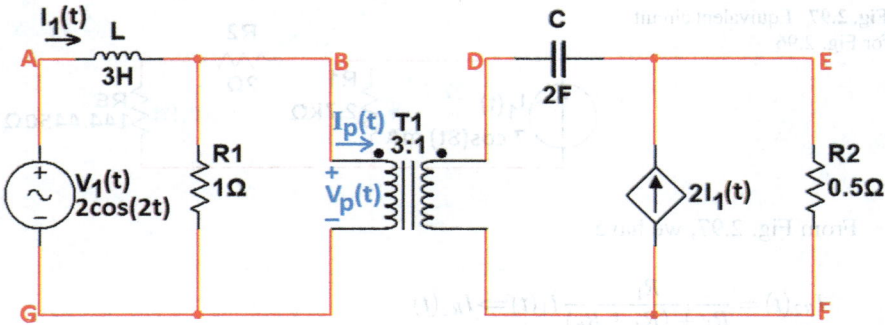

Fig. 2.99 The primary-side voltage and current of the transformer

The sinusoidal steady-state equivalent circuit for the circuit depicted in Fig. 2.99 is illustrated in Fig. 2.100.

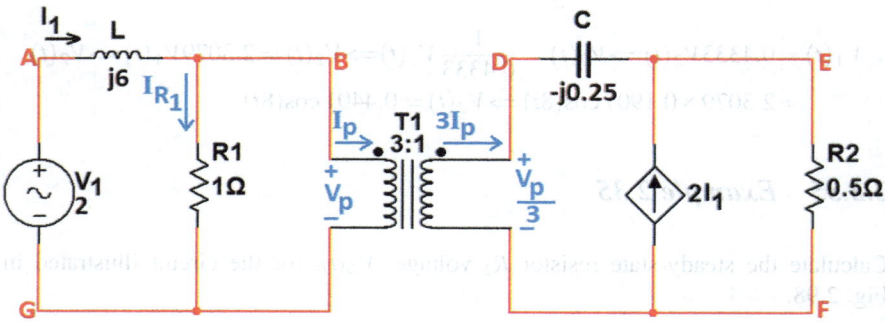

Fig. 2.100 The sinusoidal steady-state equivalent for Fig. 2.99

From Fig. 2.100, we have

$$KCL@B : I_1 = I_{R1} + I_p \Rightarrow I_1 = \frac{V_p}{R_1} + I_p \Rightarrow I_1 = \frac{V_p}{1} + I_p \Rightarrow I_1 = V_p + I_p$$

$$KVL@GABG : -2 + j6 \times (V_p + I_p) + V_p = 0 \Rightarrow (1 + j6)V_p + j6I_p = 2$$

$$KVL@DEFD : -j0.25 \times 3I_p + 0.5(3I_p + 2I_1) - \frac{V_p}{3} = 0 \Rightarrow -j0.25 \times 3I_p$$

$$+ 0.5(3I_p + 2(V_p + I_p)) - \frac{V_p}{3} = 0 \Rightarrow (2.5 - j0.75)I_p + \frac{2}{3}V_p$$

$$= 0 \Rightarrow (2.5 - j0.75)I_p + 0.6667V_p = 0 \Rightarrow 0.6667V_p + (2.5 - j0.75)I_p = 0$$

$$\begin{cases} (1+j6)V_p + j6I_p = 2 \\ 0.6667V_p + (2.5 - j0.75)I_p = 0 \end{cases} \Rightarrow \begin{bmatrix} 1+j6 & j6 \\ 0.6667 & 2.5 - j0.75 \end{bmatrix} \begin{bmatrix} V_p \\ I_p \end{bmatrix}$$

$$= \begin{bmatrix} 2 \\ 0 \end{bmatrix} \Rightarrow \begin{bmatrix} V_p \\ I_p \end{bmatrix} = \begin{bmatrix} 1+j6 & j6 \\ 0.6667 & 2.5 - j0.75 \end{bmatrix}^{-1} \begin{bmatrix} 2 \\ 0 \end{bmatrix}$$

MATLAB can be employed to perform the calculations (Fig. 2.101).

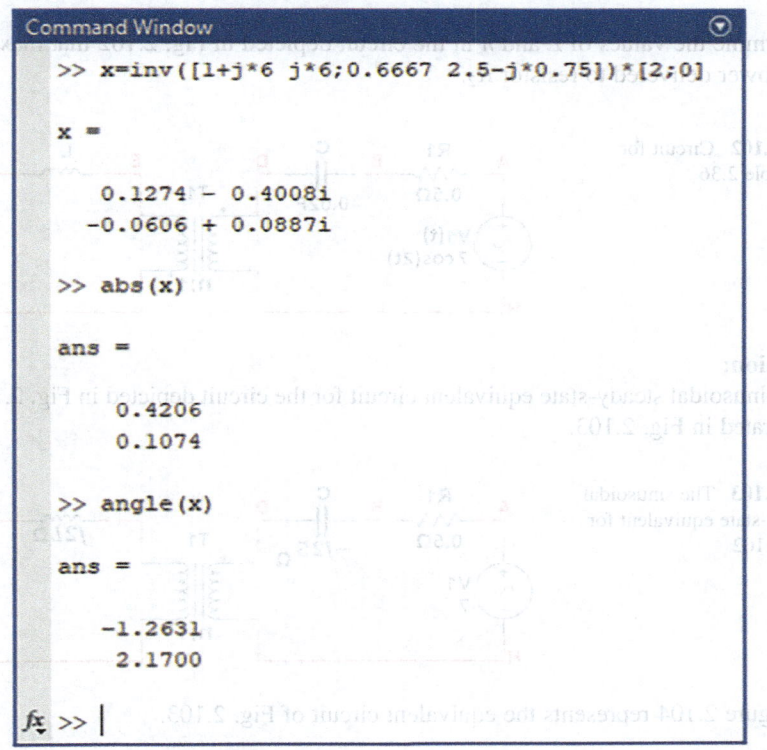

Command Window

```
>> x=inv([1+j*6 j*6;0.6667 2.5-j*0.75])*[2;0]

x =

    0.1274 - 0.4008i
   -0.0606 + 0.0887i

>> abs(x)

ans =

    0.4206
    0.1074

>> angle(x)

ans =

   -1.2631
    2.1700

fx >> |
```

Fig. 2.101 MATLAB code

According to the information provided in Fig. 2.101,

$$\begin{bmatrix} V_p \\ I_p \end{bmatrix} = \begin{bmatrix} 0.4206e^{-j1.2631} \\ 0.1074e^{j2.1700} \end{bmatrix} \rightarrow \begin{bmatrix} V_p(t) \\ I_p(t) \end{bmatrix} = \begin{bmatrix} 0.4206\cos(8t - 1.2631) \\ 0.1074\cos(8t - 2.1700) \end{bmatrix}$$

Therefore,

$$V_o = V_{R2} \Rightarrow V_o = (2I_1 + 3I_p)R_2 \Rightarrow V_o = (2(V_p + I_p) + 3I_p)R_2 \Rightarrow V_o$$
$$= (2V_p + 5I_p) \times 0.5 \Rightarrow V_o = V_p + 2.5I_p \Rightarrow V_o = 0.4206e^{-j1.2631}$$
$$+ 2.5 \times 0.1074e^{j2.1700} \Rightarrow V_o = 0.4206e^{-j1.2631} + 0.2685e^{j2.1700} \Rightarrow V_o$$
$$= 0.1807e^{-j1.7042} \rightarrow V_o(t) = 0.1807\cos(8t - 1.7042) = 0.1807\sin(8t - 0.1334)$$

2.2.36 Example 2.36

Determine the values of L and n in the circuit depicted in Fig. 2.102 that maximize the power delivered to resistor R_2.

Fig. 2.102 Circuit for Example 2.36

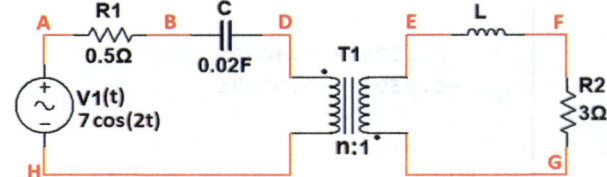

Solution:
The sinusoidal steady-state equivalent circuit for the circuit depicted in Fig. 2.102 is illustrated in Fig. 2.103.

Fig. 2.103 The sinusoidal steady-state equivalent for Fig. 2.102

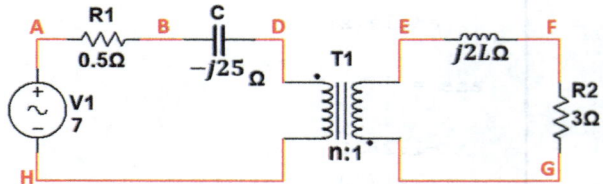

Figure 2.104 represents the equivalent circuit of Fig. 2.103.

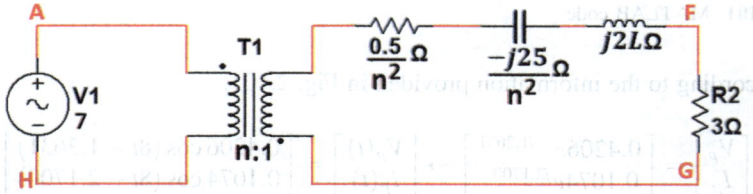

Fig. 2.104 Equivalent circuit for Fig. 2.103

The circuit of Fig. 2.104 can be simplified to that illustrated in Fig. 2.105.

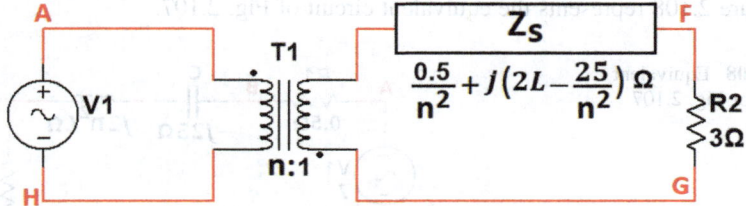

Fig. 2.105 Equivalent circuit for Fig. 2.104

From Fig. 2.105, we have

$$R_2 = Z_s^* \Rightarrow \begin{cases} \dfrac{0.5}{n^2} = 3 \\ 2L - \dfrac{25}{n^2} = 0 \end{cases} \Rightarrow \begin{cases} n^2 = \dfrac{0.5}{3} = \dfrac{1}{6} \\ 2L - \dfrac{25}{n^2} = 0 \end{cases} \Rightarrow \begin{cases} n = 0.4082 \\ 2L - \dfrac{25}{\frac{1}{6}} = 0 \end{cases}$$

$$\Rightarrow \begin{cases} n = 0.4082 \\ 2L - 150 = 0 \end{cases} \Rightarrow \begin{cases} n = 0.408 \\ L = 75 \ H \end{cases}$$

2.2.37 Example 2.37

Determine the values of L and n in the circuit depicted in Fig. 2.106 that maximize the power delivered to resistor R_2.

Fig. 2.106 Circuit for Example 2.37

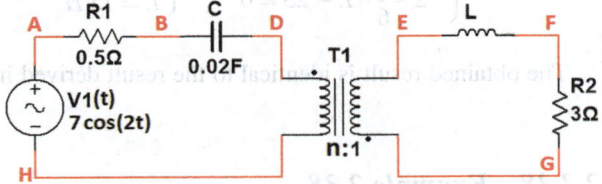

Solution:
This problem is addressed in Example 2.36. However, a distinct methodology is employed herein. The sinusoidal steady-state equivalent circuit for the circuit depicted in Fig. 2.106 is illustrated in Fig. 2.107.

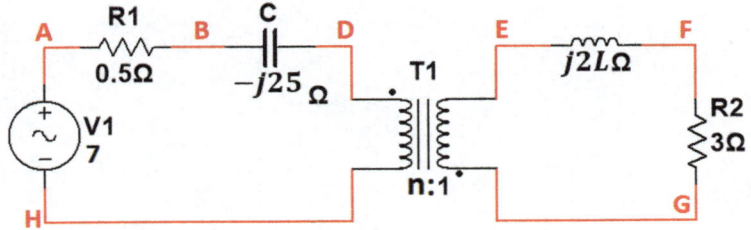

Fig. 2.107 Sinusoidal steady-state equivalent for Fig. 2.106

Figure 2.108 represents the equivalent circuit of Fig. 2.107.

Fig. 2.108 Equivalent circuit for Fig. 2.107

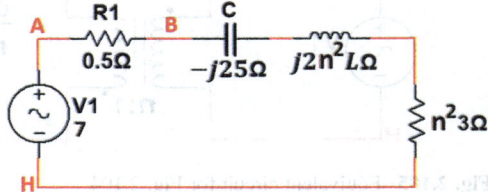

The circuit of Fig. 2.108 can be simplified to that illustrated in Fig. 2.109.

Fig. 2.109 Equivalent circuit for Fig. 2.108

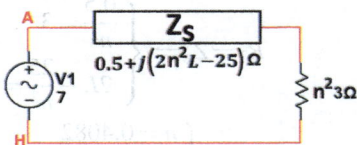

From Fig. 2.109, we have

$$3n^2 = \left(0.5 + j(2n^2L - 25)\right)^* \Rightarrow \begin{cases} 3n^2 = 0.5 \\ 2n^2L - 25 = 0 \end{cases} \Rightarrow \begin{cases} n^2 = \dfrac{1}{6} \\ 2n^2L - 25 = 0 \end{cases}$$

$$\Rightarrow \begin{cases} n = 0.4082 \\ 2 \times \dfrac{1}{6} \times L - 25 = 0 \end{cases} \Rightarrow \begin{cases} n = 0.4082 \\ L = 75\ H \end{cases}$$

The obtained result is identical to the result derived in Example 2.36.

2.2.38 Example 2.38

Calculate the power dissipated by resistor R_3 in the circuit of Fig. 2.110.

Fig. 2.110 Circuit for
Example 2.38

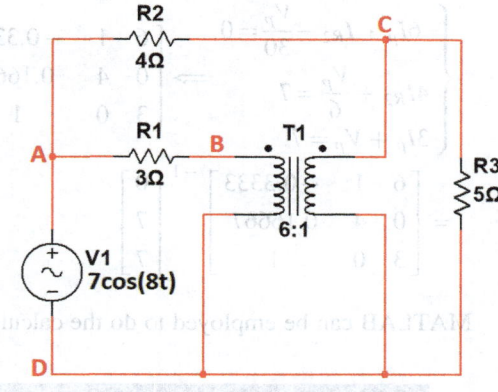

Solution:

The sinusoidal steady-state equivalent circuit for the circuit depicted in Fig. 2.110 is
illustrated in Fig. 2.111.

Fig. 2.111 Inusoidal
steady-state equivalent for
Fig. 2.110

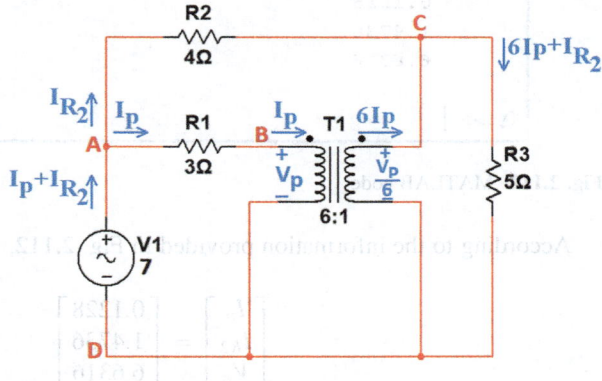

From Fig. 2.111, we have

$$V_{R3} = \frac{V_p}{6} \Rightarrow I_{R3} = \frac{V_{R3}}{R_3} \Rightarrow 6I_p + I_{R2} = \frac{\frac{V_p}{6}}{5} \Rightarrow 6I_p + I_{R2} = \frac{V_p}{30} \Rightarrow 6I_p + I_{R2} - \frac{V_p}{30} = 0$$

$$KVL@DACD : -7 + 4I_{R2} + \frac{V_p}{6} = 0 \Rightarrow 4I_{R2} + \frac{V_p}{6} = 7$$

$$KVL@DABD : -7 + 3I_p + V_p = 0 \Rightarrow 3I_p + V_p = 7$$

$$\begin{cases} 6I_p + I_{R2} - \dfrac{V_p}{30} = 0 \\ 4I_{R2} + \dfrac{V_p}{6} = 7 \\ 3I_p + V_p = 7 \end{cases} \Rightarrow \begin{bmatrix} 6 & 1 & -0.3333 \\ 0 & 4 & 0.1667 \\ 3 & 0 & 1 \end{bmatrix} \begin{bmatrix} I_p \\ I_{R2} \\ V_p \end{bmatrix} = \begin{bmatrix} 0 \\ 7 \\ 7 \end{bmatrix} \Rightarrow \begin{bmatrix} I_p \\ I_{R2} \\ V_p \end{bmatrix}$$

$$= \begin{bmatrix} 6 & 1 & -0.3333 \\ 0 & 4 & 0.16667 \\ 3 & 0 & 1 \end{bmatrix}^{-1} \begin{bmatrix} 0 \\ 7 \\ 7 \end{bmatrix}$$

MATLAB can be employed to do the calculations (Fig. 2.112).

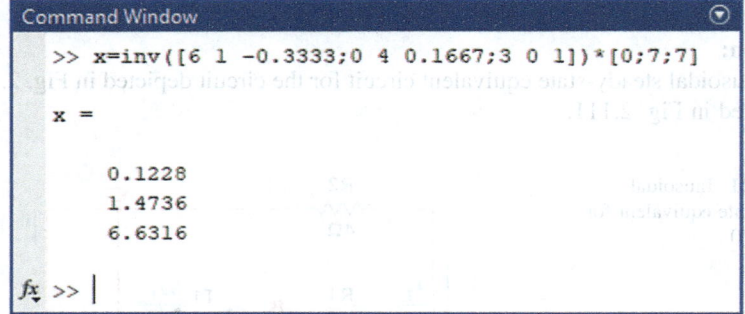

```
Command Window                                              ⊙
>> x=inv([6 1 -0.3333;0 4 0.1667;3 0 1])*[0;7;7]

x =

    0.1228
    1.4736
    6.6316

fx >> |
```

Fig. 2.112 MATLAB code

According to the information provided in Fig. 2.112,

$$\begin{bmatrix} I_p \\ I_{R2} \\ V_p \end{bmatrix} = \begin{bmatrix} 0.1228 \\ 1.4736 \\ 6.6316 \end{bmatrix}$$

Therefore,

$$I_{R3} = 6I_p + I_{R2} \Rightarrow I_{R3} = 6 \times 0.1228 + 1.4736 \Rightarrow I_{R3}$$
$$= 2.2104\ A \rightarrow I_{R3}(t) = 2.2104\cos(8t)$$

$$P_{R_3} = R_3 I_{R3\,RMS}^2 \Rightarrow P_{R_3} = 5\left(\frac{2.2104}{\sqrt{2}}\right)^2 \Rightarrow P_{R_3} = 12.2147\ W$$

2.2.39 Example 2.39

Calculate the voltage difference between terminals A and B (V_{AB}) for the circuit depicted in Fig. 2.113.

Fig. 2.113 Circuit for
Example 2.39

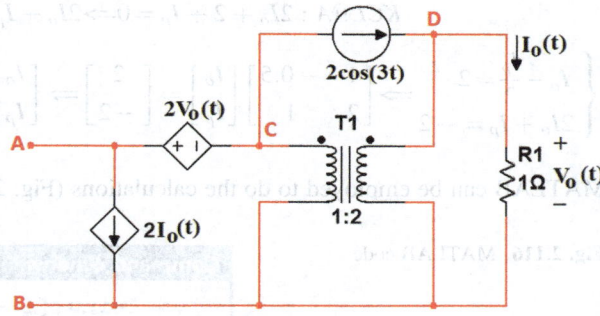

Solution:

The sinusoidal steady-state equivalent circuit for the circuit depicted in Fig. 2.113 is illustrated in Fig. 2.114.

Fig. 2.114 Sinusoidal
steady-state equivalent for
Fig. 2.113

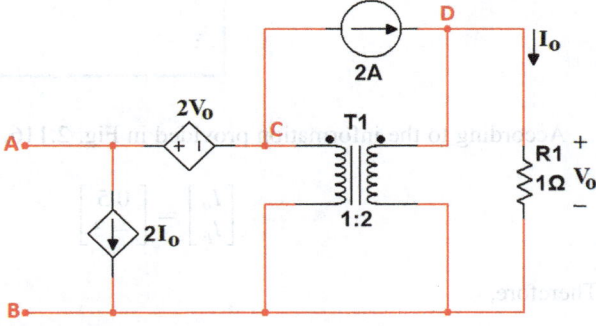

The current flow in the distinct branches is depicted in Fig. 2.115.

Fig. 2.115 The current flow
in the distinct branches

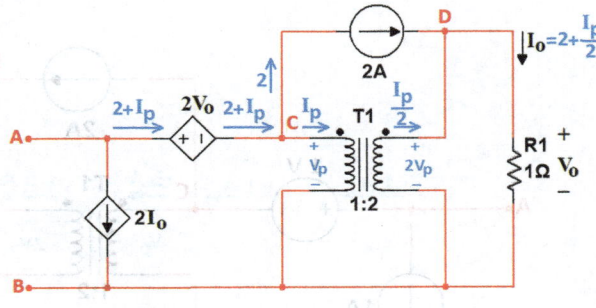

From Fig. 2.115, we have

$$I_o = 2 + \frac{I_p}{2} \Rightarrow I_o - \frac{I_p}{2} = 2$$

$$KCL@A : 2I_o + 2 + I_p = 0 \Longrightarrow 2I_o + I_p = -2$$

$$\begin{cases} I_o - \dfrac{I_p}{2} = 2 \\ 2I_o + I_p = -2 \end{cases} \Longrightarrow \begin{bmatrix} 1 & -0.5 \\ 2 & 1 \end{bmatrix} \begin{bmatrix} I_o \\ I_p \end{bmatrix} = \begin{bmatrix} 2 \\ -2 \end{bmatrix} \Longrightarrow \begin{bmatrix} I_o \\ I_p \end{bmatrix} = \begin{bmatrix} 1 & -0.5 \\ 2 & 1 \end{bmatrix}^{-1} \begin{bmatrix} 2 \\ -2 \end{bmatrix}$$

MATLAB can be employed to do the calculations (Fig. 2.116).

Fig. 2.116 MATLAB code

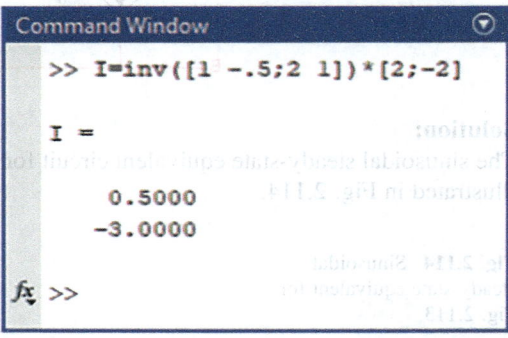

According to the information provided in Fig. 2.116,

$$\begin{bmatrix} I_o \\ I_p \end{bmatrix} = \begin{bmatrix} 0.5 \\ -3 \end{bmatrix}$$

Therefore,

$$V_o = R_1 I_o \Longrightarrow V_o = 1 \times 0.5 \Longrightarrow V_o = 0.5 \ V$$

The results obtained are illustrated in Fig. 2.117.

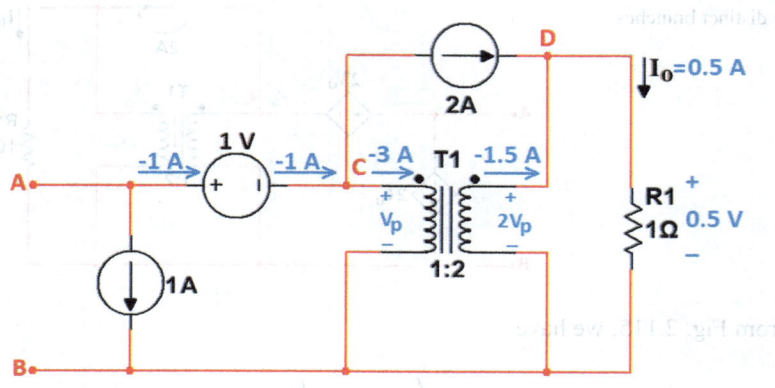

Fig. 2.117 Obtained results

An alternative representation of the circuit in Fig. 2.117 is provided in Fig. 2.118.

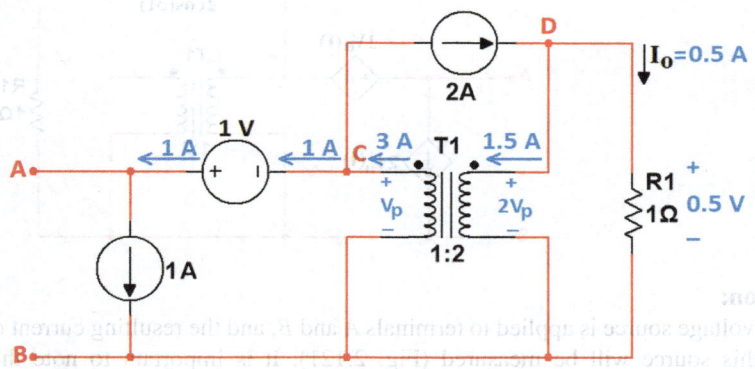

Fig. 2.118 Alternative representation for Fig. 2.117

From Fig. 2.118, we have $2V_p = 0.5 \Rightarrow V_p = 0.25\ V$ (Fig. 2.119).

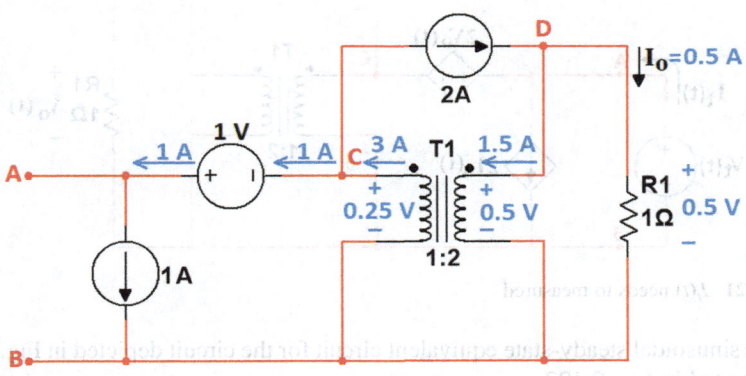

Fig. 2.119 $V_p = 0.25\ V$

From Fig. 2.119, we have $KVL\ @\ CABC: -1 + V_{AB} - 0.25 = 0 \Rightarrow V_{AB} = 1.25\ V$. Therefore, $V_{AB}(t) = 1.25 \cos(3t)$.

2.2.40 Example 2.40

Determine the impedance observed between terminals A and B for the circuit depicted in Fig. 2.120 at $\omega = 3\frac{Rad}{s}$.

Fig. 2.120 Circuit for
Example 2.40

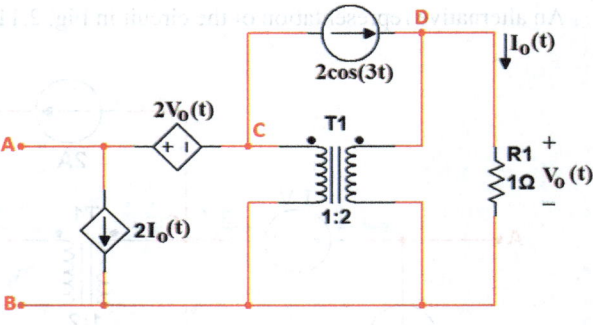

Solution:
A test voltage source is applied to terminals A and B, and the resulting current drawn
from this source will be measured (Fig. 2.121). It is important to note that the
independent current source is deactivated.

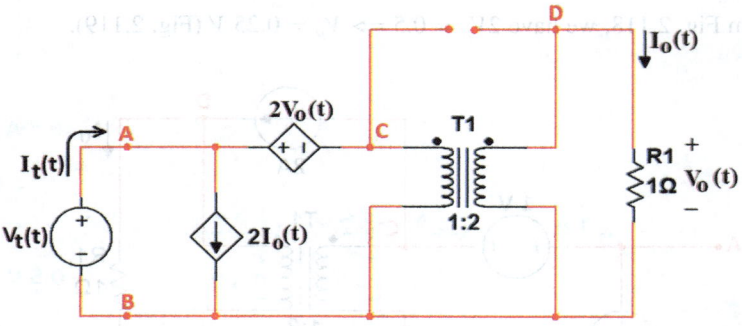

Fig. 2.121 $I_t(t)$ needs to measured

The sinusoidal steady-state equivalent circuit for the circuit depicted in Fig. 2.121
is illustrated in Fig. 2.122.

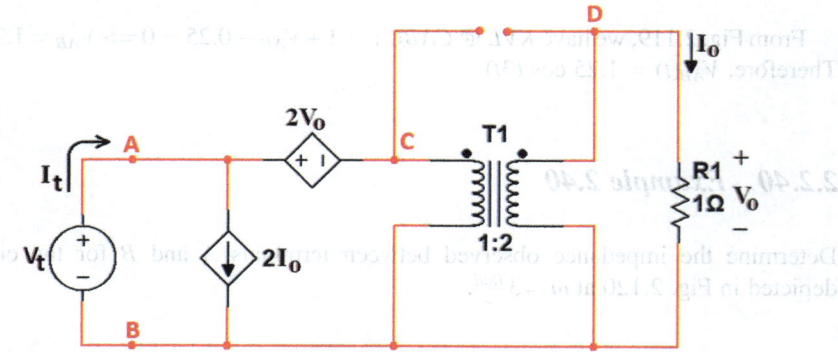

Fig. 2.122 The sinusoidal steady-state equivalent for Fig. 2.121

You can simplify the circuit in Fig. 2.122 to the circuit shown in Fig. 2.123.

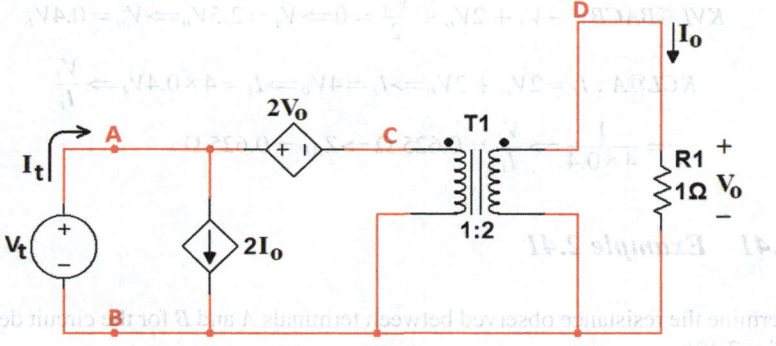

Fig. 2.123 Equivalent circuit for Fig. 2.122

From Fig. 2.123, we have $I_o = \frac{V_o}{R_1} \Rightarrow I_o = \frac{V_o}{1} \Rightarrow I_o = V_o$. Therefore, the circuit depicted in Fig. 2.123 can be simplified to the equivalent circuit illustrated in Fig. 2.124.

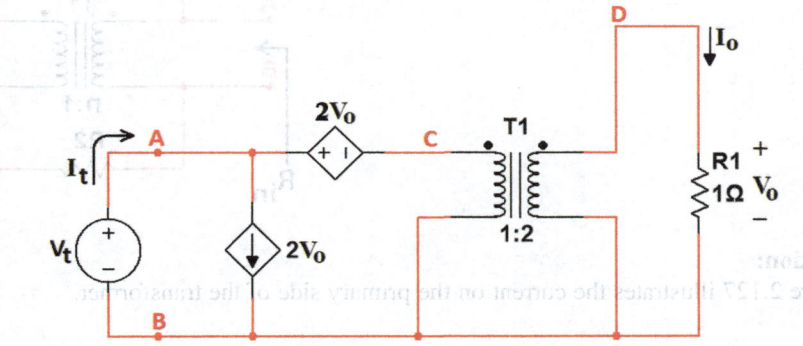

Fig. 2.124 Equivalent circuit for Fig. 2.123

Figure 2.125 illustrates the voltages and currents associated with the transformer.

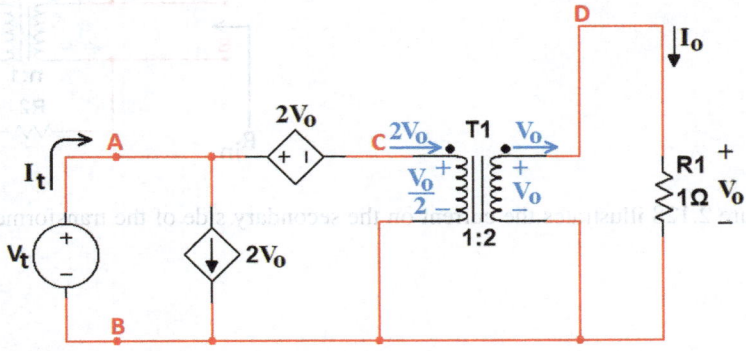

Fig. 2.125 The transformer voltages and currents

From Fig. 2.125, we have

$$KVL@BACB: \ -V_t + 2V_o + \frac{V_o}{2} = 0 \Rightarrow V_t = 2.5V_o \Rightarrow V_o = 0.4V_t$$

$$KCL@A: I_t = 2V_o + 2V_o \Rightarrow I_t = 4V_o \Rightarrow I_t = 4 \times 0.4V_t \Rightarrow \frac{V_t}{I_t}$$

$$= \frac{1}{4 \times 0.4} \Rightarrow \frac{V_t}{I_t} = 0.625 \ \Omega \Rightarrow Z_{AB} = 0.625 \ \Omega$$

2.2.41 Example 2.41

Determine the resistance observed between terminals A and B for the circuit depicted in Fig. 2.126.

Fig. 2.126 Circuit for Example 2.41

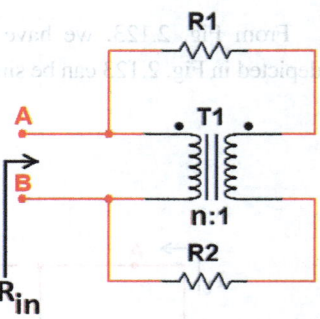

Solution:

Figure 2.127 illustrates the current on the primary side of the transformer.

Fig. 2.127 The current on the primary side

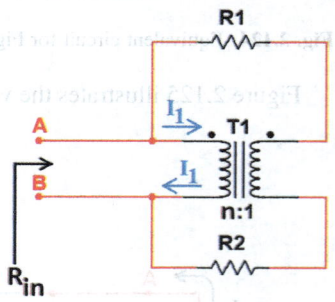

Figure 2.128 illustrates the current on the secondary side of the transformer.

Fig. 2.128 The current on the secondary side

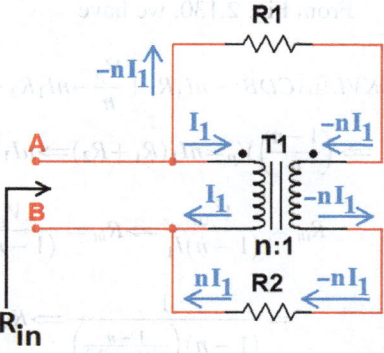

Kirchhoff's current law (KCL) can be employed to calculate the current leaving terminal A (Fig. 2.129).

Fig. 2.129 The current leaving terminal A

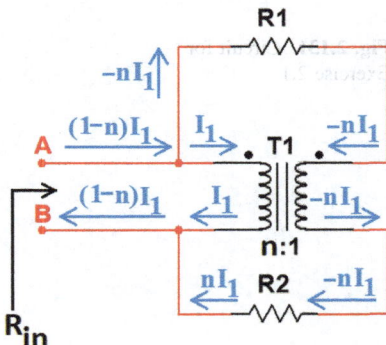

The primary and secondary voltages of the transformer are depicted in Fig. 2.130.

Fig. 2.130 The primary and secondary voltages of the transformer

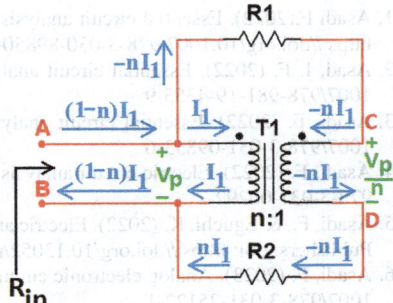

From Fig. 2.130, we have

$$KVL@ACDB: -nI_1R_1 + \frac{V_p}{n} - nI_1R_2 - V_p = 0 \Rightarrow \left(\frac{1}{n} - 1\right)V_p = nI_1(R_1 + R_2)$$

$$\Rightarrow \left(\frac{1-n}{n}\right)V_p = nI_1(R_1 + R_2) \Rightarrow nI_1(R_1 + R_2) = \left(\frac{1-n}{n}\right)V_p \Rightarrow I_1 = \left(\frac{1-n}{n^2(R_1+R_2)}\right)V_p$$

$$R_{in} = \frac{V_{AB}}{(1-n)I_1} \Rightarrow R_{in} = \frac{V_p}{(1-n)I_1} \Rightarrow R_{in} = \frac{V_p}{(1-n)\left(\frac{1-n}{n^2(R_1+R_2)}\right)V_p} \Rightarrow R_{in}$$

$$= \frac{1}{(1-n)\left(\frac{1-n}{n^2(R_1+R_2)}\right)} \Rightarrow R_{in} = \frac{n^2}{(1-n)^2}(R_1 + R_2)$$

Exercise 2.1: Prove that the equivalent resistance seen at the input terminals of the circuit illustrated in Fig. 2.131 is given by $R_{in} = \frac{n^2}{(1+n)^2}(R_1 + R_2)$.

Fig. 2.131 Circuit for
Exercise 2.1

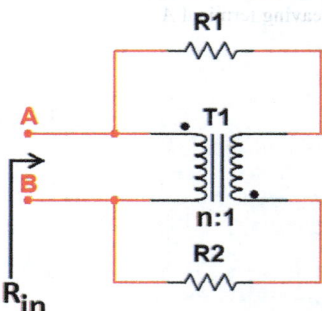

Further Reading

1. Asadi F.(2022). Essential circuit analysis using NI Multisim™ and MATLAB®. Springer. doi: https://doi.org/10.1007/978-3-030-89850-2
2. Asad, I. F. (2022). Essential circuit analysis using Proteus®. Springer. doi: https://doi.org/10.1007/978-981-19-4353-9
3. Asadi, F. (2022). Essential circuit analysis using LTspice®. Springer. doi: https://doi.org/10.1007/978-3-031-09853-6
4. Asadi, F. (2022). Electric circuit analysis with EasyEDA.Springer. doi: https://doi.org/10.1007/978-3-031-00292-2
5. Asadi, F., & Eguchi, K. (2022). Electric and electronic circuit simulation using TINA-TI®. River Publishers. doi: https://doi.org/10.13052/rp-9788770226851
6. Asadi, F. (2023). Analog electronic circuits laboratory manual. Springer. doi: https://doi.org/10.1007/978-3-031-25122-1

Chapter 3
Two-Port Networks

3.1 Introduction

A two-port network is a mathematical model in electrical circuit analysis that represents a system with two pairs of terminals, designated as input and output ports (Fig. 3.1).

Fig. 3.1 Two-port symbol

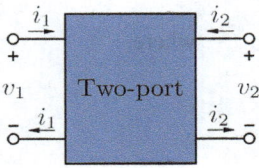

Two-port networks are characterized by four external variables: the voltage and current at the input port (V_1, I_1) and the voltage and current at the output port (V_2, I_2). Assuming the network is linear, these variables are related through a set of linear equations:

- Open circuit impedance (Z) parameters

$$\begin{bmatrix} V_1 \\ V_2 \end{bmatrix} = \begin{bmatrix} Z_{11} & Z_{12} \\ Z_{21} & Z_{22} \end{bmatrix} \begin{bmatrix} I_1 \\ I_2 \end{bmatrix}$$

where,

$$Z_{11} = \frac{V_1}{I_1}\bigg|_{I_2 = 0}, Z_{12} = \frac{V_1}{I_2}\bigg|_{I_1 = 0}$$

© The Author(s), under exclusive license to Springer Nature Switzerland AG 2025
F. Asadi, *A Problem-Solving Approach to Electric Circuits*,
https://doi.org/10.1007/978-3-031-95493-1_3

$$Z_{21} = \frac{V_2}{I_1}\bigg|_{I_2=0} , Z_{22} = \frac{V_2}{I_2}\bigg|_{I_1=0}$$

- Short circuit admittance (Y) parameters

$$\begin{bmatrix} I_1 \\ I_2 \end{bmatrix} = \begin{bmatrix} Y_{11} & Y_{12} \\ Y_{21} & Y_{22} \end{bmatrix} \begin{bmatrix} V_1 \\ V_2 \end{bmatrix}$$

where,

$$Y_{11} = \frac{I_1}{V_1}\bigg|_{V_2=0} , Y_{12} = \frac{I_1}{V_2}\bigg|_{V_1=0}$$

$$Y_{21} = \frac{I_2}{V_1}\bigg|_{V_2=0} , Y_{22} = \frac{I_2}{V_2}\bigg|_{V_1=0}$$

- Hybrid (h) parameters

$$\begin{bmatrix} V_1 \\ I_2 \end{bmatrix} = \begin{bmatrix} h_{11} & h_{12} \\ h_{21} & h_{22} \end{bmatrix} \begin{bmatrix} I_1 \\ V_2 \end{bmatrix}$$

where,

$$h_{11} = \frac{V_1}{I_1}\bigg|_{V_2=0} , h_{12} = \frac{V_1}{V_2}\bigg|_{I_1=0}$$

$$h_{21} = \frac{I_2}{I_1}\bigg|_{V_2=0} , h_{22} = \frac{I_2}{V_2}\bigg|_{I_1=0}$$

- Inverse hybrid (g) parameters

$$\begin{bmatrix} I_1 \\ V_2 \end{bmatrix} = \begin{bmatrix} g_{11} & g_{12} \\ g_{21} & g_{22} \end{bmatrix} \begin{bmatrix} V_1 \\ I_2 \end{bmatrix}$$

where,

$$g_{11} = \frac{I_1}{V_1}\bigg|_{I_2=0} , g_{12} = \frac{I_1}{I_2}\bigg|_{V_1=0}$$

$$g_{21} = \frac{V_2}{V_1}\bigg|_{I_2=0} , g_{22} = \frac{V_2}{I_2}\bigg|_{V_1=0}$$

- Transmission (t or ABCD) parameters

$$\begin{bmatrix} V_1 \\ I_1 \end{bmatrix} = \begin{bmatrix} t_{11} & t_{21} \\ t_{21} & t_{22} \end{bmatrix} \begin{bmatrix} V_2 \\ -I_2 \end{bmatrix}$$

where,

$$t_{11} = \left.\frac{V_1}{V_2}\right|_{I_2=0}, t_{12} = \left.\frac{V_1}{-I_2}\right|_{V_2=0}$$

$$t_{21} = \left.\frac{I_1}{V_2}\right|_{I_2=0}, t_{22} = \left.\frac{I_1}{-I_2}\right|_{V_2=0}$$

Calculations presented in this book are truncated to four decimal places. Consequently, minor discrepancies may arise between results obtained from different solution methodologies. These variations are solely attributable to accumulated rounding errors and do not reflect inaccuracies in the methodologies themselves.

3.2 Solved Examples

3.2.1 Example 3.1

Solve the system of equations given by $\begin{cases} 3x + 2y = 7 \\ x + 6y = 13 \end{cases}$.

Solution:

$$\begin{cases} 3x + 2y = 7 \\ x + 6y = 13 \end{cases} \Rightarrow \begin{bmatrix} 3 & 2 \\ 1 & 6 \end{bmatrix} \begin{bmatrix} x \\ y \end{bmatrix} = \begin{bmatrix} 7 \\ 13 \end{bmatrix} \Rightarrow \begin{bmatrix} x \\ y \end{bmatrix} = \begin{bmatrix} 3 & 2 \\ 1 & 6 \end{bmatrix}^{-1} \begin{bmatrix} 7 \\ 13 \end{bmatrix} \Rightarrow \begin{bmatrix} x \\ y \end{bmatrix}$$

$$= \frac{1}{3 \times 6 - 1 \times 2} \begin{bmatrix} 6 & -2 \\ -1 & 3 \end{bmatrix} \begin{bmatrix} 7 \\ 13 \end{bmatrix} \Rightarrow \begin{bmatrix} x \\ y \end{bmatrix} = \frac{1}{16} \begin{bmatrix} 6 \times 7 - 2 \times 13 \\ -1 \times 7 + 3 \times 13 \end{bmatrix} \Rightarrow \begin{bmatrix} x \\ y \end{bmatrix}$$

$$= \frac{1}{16} \begin{bmatrix} 6 \times 7 - 2 \times 13 \\ -1 \times 7 + 3 \times 13 \end{bmatrix} \Rightarrow \begin{bmatrix} x \\ y \end{bmatrix} = \frac{1}{16} \begin{bmatrix} 16 \\ 32 \end{bmatrix} \Rightarrow \begin{bmatrix} x \\ y \end{bmatrix} = \begin{bmatrix} 1 \\ 2 \end{bmatrix} \Rightarrow x = 1, y = 2$$

3.2.2 Example 3.2

Use Cramer's rule to solve the system of equations given by $\begin{cases} 3x + 2y = 7 \\ x + 6y = 13 \end{cases}$.

Solution:

$$\begin{cases} 3x + 2y = 7 \\ x + 6y = 13 \end{cases} \Rightarrow \begin{bmatrix} 3 & 2 \\ 1 & 6 \end{bmatrix} \begin{bmatrix} x \\ y \end{bmatrix} = \begin{bmatrix} 7 \\ 13 \end{bmatrix}$$

$$x = \frac{\begin{vmatrix} 7 & 2 \\ 13 & 6 \end{vmatrix}}{\begin{vmatrix} 3 & 2 \\ 1 & 6 \end{vmatrix}} \Rightarrow x = \frac{7 \times 6 - 2 \times 13}{3 \times 6 - 2 \times 1} \Rightarrow x = \frac{16}{16} \Rightarrow x = 1$$

$$y = \frac{\begin{vmatrix} 3 & 7 \\ 1 & 13 \end{vmatrix}}{\begin{vmatrix} 3 & 2 \\ 1 & 6 \end{vmatrix}} \Rightarrow y = \frac{3 \times 13 - 7 \times 1}{3 \times 6 - 2 \times 1} \Rightarrow y = \frac{32}{16} \Rightarrow y = 2$$

3.2.3 Example 3.3

Use Cramer's rule to solve the system of equations given by
$$\begin{cases} x + y - 2z = -12 \\ x + 3y = -2 \\ 2x - y + 16z = 122 \end{cases}.$$

Solution:

$$\begin{cases} x + y - 2z = -12 \\ x + 3y = -2 \\ 2x - y + 16z = 122 \end{cases} \Rightarrow \begin{bmatrix} 1 & 1 & -2 \\ 1 & 3 & 0 \\ 2 & -1 & 16 \end{bmatrix} \begin{bmatrix} x \\ y \\ z \end{bmatrix} = \begin{bmatrix} -12 \\ -2 \\ 122 \end{bmatrix}$$

$$x = \frac{\begin{vmatrix} -12 & 1 & -2 \\ -2 & 3 & 0 \\ 122 & -1 & 16 \end{vmatrix}}{\begin{vmatrix} 1 & 1 & -2 \\ 1 & 3 & 0 \\ 2 & -1 & 16 \end{vmatrix}} \Rightarrow x = \frac{-12\begin{vmatrix} 3 & 0 \\ -1 & 16 \end{vmatrix} - 1\begin{vmatrix} -2 & 0 \\ 122 & 16 \end{vmatrix} - 2\begin{vmatrix} -2 & 3 \\ 122 & -1 \end{vmatrix}}{1\begin{vmatrix} 3 & 0 \\ -1 & 16 \end{vmatrix} - 1\begin{vmatrix} 1 & 0 \\ 2 & 16 \end{vmatrix} - 2\begin{vmatrix} 1 & 3 \\ 2 & -1 \end{vmatrix}} \Rightarrow x$$

$$= \frac{-12(3 \times 16 - 0 \times -1) - 1(-2 \times 16 - 0 \times 122) - 2(-2 \times -1 - 3 \times 122)}{1(3 \times 16 - 0 \times -1) - 1(1 \times 16 - 0 \times 2) - 2(1 \times -1 - 3 \times 2)} \Rightarrow x$$

$$= \frac{-12 \times 48 - 1 \times -32 - 2 \times -364}{48 - 1 \times 16 - 2 \times -7} \Rightarrow x = \frac{-576 + 32 + 728}{48 - 16 + 14} \Rightarrow x = \frac{184}{46} \Rightarrow x = 4$$

$$y = \frac{\begin{vmatrix} 1 & -12 & -2 \\ 1 & -2 & 0 \\ 2 & 122 & 16 \end{vmatrix}}{\begin{vmatrix} 1 & 1 & -2 \\ 1 & 3 & 0 \\ 2 & -1 & 16 \end{vmatrix}} \Rightarrow y = \frac{-92}{46} \Rightarrow y = -2$$

$$z = \frac{\begin{vmatrix} 1 & 1 & -12 \\ 1 & 3 & -2 \\ 2 & -1 & 122 \end{vmatrix}}{\begin{vmatrix} 1 & 1 & -2 \\ 1 & 3 & 0 \\ 2 & -1 & 16 \end{vmatrix}} \Rightarrow z = \frac{322}{46} \Rightarrow z = 7$$

3.2.4 Example 3.4

Use MATLAB to solve the system of equations given by $\begin{cases} 45I_1 + 25I_2 = V_1 \\ I_1 + 3I_2 = 0 \end{cases}$.

Solution:

$$\begin{cases} 45I_1 + 25I_2 = V_1 \\ I_1 + 3I_2 = 0 \end{cases} \Rightarrow \begin{bmatrix} 45 & 25 \\ 1 & 3 \end{bmatrix} \begin{bmatrix} I_1 \\ I_2 \end{bmatrix} = \begin{bmatrix} V_1 \\ 0 \end{bmatrix} \Rightarrow \begin{bmatrix} I_1 \\ I_2 \end{bmatrix} = \begin{bmatrix} 45 & 25 \\ 1 & 3 \end{bmatrix}^{-1} \begin{bmatrix} V_1 \\ 0 \end{bmatrix}$$

The MATLAB code depicted in Fig. 3.2 computes the solution for this system.

Fig. 3.2 MATLAB code

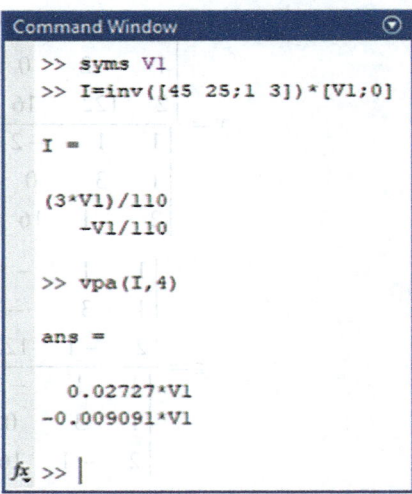

An alternative approach to solving this system is provided by the code illustrated in Fig. 3.3.

Fig. 3.3 MATLAB code

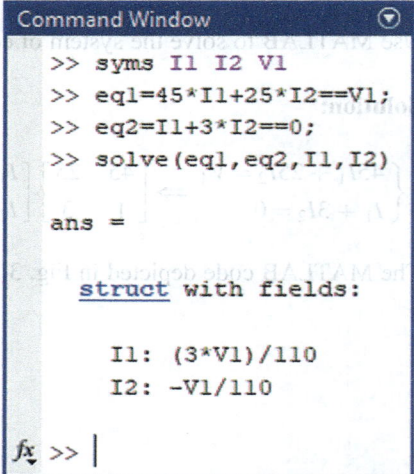

3.2.5   Example 3.5

Use MATLAB to solve the system of equations given by $\begin{cases} x + y - 2z = -12 \\ x + 3y = -2 \\ 2x - y + 16z = 122 \end{cases}$.

Solution:

$$\begin{cases} x+y-2z=-12 \\ x+3y=-2 \\ 2x-y+16z=122 \end{cases} \Rightarrow \begin{bmatrix} 1 & 1 & -2 \\ 1 & 3 & 0 \\ 2 & -1 & 16 \end{bmatrix} \begin{bmatrix} x \\ y \\ z \end{bmatrix} = \begin{bmatrix} -12 \\ -2 \\ 122 \end{bmatrix} \Rightarrow \begin{bmatrix} x \\ y \\ z \end{bmatrix}$$

$$= \begin{bmatrix} 1 & 1 & -2 \\ 1 & 3 & 0 \\ 2 & -1 & 16 \end{bmatrix}^{-1} \begin{bmatrix} -12 \\ -2 \\ 122 \end{bmatrix}$$

The MATLAB code depicted in Fig. 3.4 computes the solution for this system.

Fig. 3.4 MATLAB code

```
Command Window

>> sol=inv([1 1 -2;1 3 0;2 -1 16])*[-12;-2;122]

sol =

    4.0000
   -2.0000
    7.0000

fx >> |
```

An alternative approach to solving this system is provided by the code illustrated in Fig. 3.5.

Fig. 3.5 MATLAB code

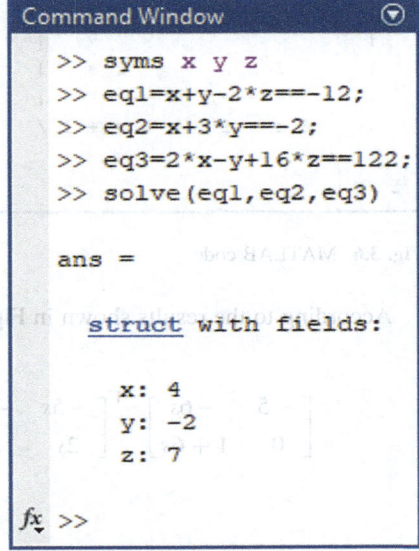

```
Command Window

>> syms x y z
>> eq1=x+y-2*z==-12;
>> eq2=x+3*y==-2;
>> eq3=2*x-y+16*z==122;
>> solve(eq1,eq2,eq3)

ans =

  struct with fields:

    x: 4
    y: -2
    z: 7

fx >>
```

3.2.6   *Example 3.6*

Use MATLAB to calculate $\begin{bmatrix} -5 & -6s \\ 0 & 1+6s \end{bmatrix}^{-1} \begin{bmatrix} -5s & -2s \\ 2s & 2s \end{bmatrix}$.

Solution:

The required MATLAB code is shown in Fig. 3.6.

```
Command Window                                                                    ⊙

   >> syms s
   >> inv([-5 -6*s;0 1+6*s])*[-5*s -2*s;2*s 2*s]

   ans =

   [s - (12*s^2)/(5*(6*s + 1)), (2*s)/5 - (12*s^2)/(5*(6*s + 1))]
   [            (2*s)/(6*s + 1),                   (2*s)/(6*s + 1)]

   >> simplify(ans)

   ans =

   [(s*(18*s + 5))/(5*(6*s + 1)), (2*s)/(30*s + 5)]
   [            (2*s)/(6*s + 1),    (2*s)/(6*s + 1)]

   >> pretty(ans)
   / s (18 s + 5)        2 s     \
   | -------------,    --------   |
   | (6 s + 1) 5      30 s + 5    |
   |                             |
   |                             |
   |     2 s             2 s     |
   |   --------,       -------    |
   \    6 s + 1        6 s + 1   /

fx >> |
```

Fig. 3.6 MATLAB code

According to the results shown in Fig. 3.6,

$$\begin{bmatrix} -5 & -6s \\ 0 & 1+6s \end{bmatrix}^{-1} \begin{bmatrix} -5s & -2s \\ 2s & 2s \end{bmatrix} = \begin{bmatrix} \dfrac{s(18s+5)}{(6s+1)5} & \dfrac{2s}{30s+5} \\ \dfrac{2s}{6s+1} & \dfrac{2s}{6s+1} \end{bmatrix}$$

3.2.7 Example 3.7

Determine the resistance observed between terminals *A* and *B* for the circuit depicted in Fig. 3.7.

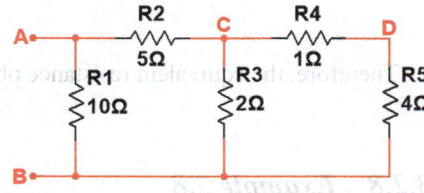

Fig. 3.7 Circuit for Example 3.7

Solution:
This example serves as a preparatory exercise for Example 3.8. An equivalent representation of the given circuit is shown in Fig. 3.8.

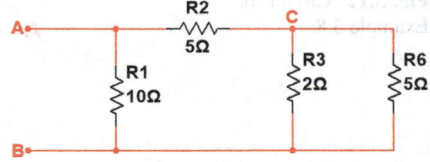

Fig. 3.8 Equivalent circuit for Fig. 3.7

Figure 3.9 presents an equivalent representation of the circuit depicted in Fig. 3.8.

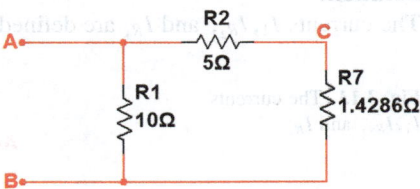

Fig. 3.9 Equivalent circuit for Fig. 3.8

Figure 3.10 presents an equivalent representation of the circuit depicted in Fig. 3.9.

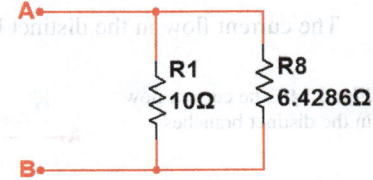

Fig. 3.10 Equivalent circuit for Fig. 3.9

Figure 3.11 presents an equivalent representation of the circuit depicted in Fig. 3.10.

Fig. 3.11 Equivalent circuit
for Fig. 3.10

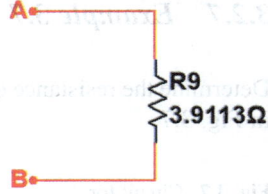

Therefore, the equivalent resistance observed between terminals A and B is 3.9 Ω.

3.2.8   Example 3.8

Determine the input impedance, that is, impedance observed between terminals A and B, for the circuit depicted in Fig. 3.12.

Fig. 3.12 Circuit for
Example 3.8

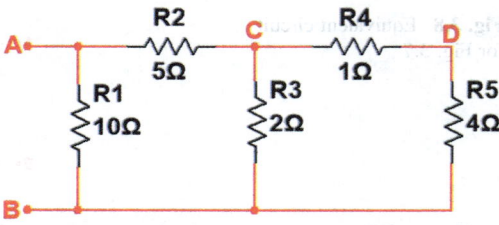

Solution:
The currents I_1, I_{R_2}, and I_{R_4} are defined as illustrated in Fig. 3.13.

Fig. 3.13 The currents
I_1, I_{R_2}, and I_{R_4}

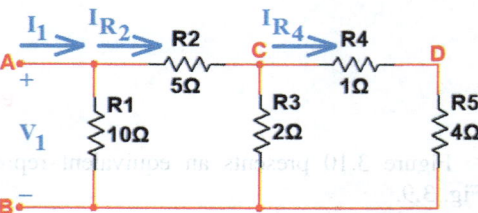

The current flow in the distinct branches is depicted in Fig. 3.14.

Fig. 3.14 The current flow
in the distinct branches

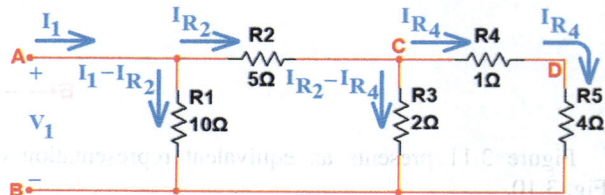

From Fig. 3.14, we have

$$KVL@ABA : R_1(I_1 - I_{R_2}) - V_1 = 0 \Rightarrow R_1(I_1 - I_{R_2}) = V_1 \Rightarrow 10I_1 - 10I_{R_2} = V_1$$

$$KVL@ACBA : R_2 I_{R_2} + R_3(I_{R_2} - I_{R_4}) - R_1(I_1 - I_{R_2}) = 0 \Rightarrow 5I_{R_2} + 2(I_{R_2} - I_{R_4})$$
$$- 10(I_1 - I_{R_2}) = 0 \Rightarrow -10I_1 + 17I_{R_2} - 2I_{R_4} = 0$$

$$KVL@CDBC : R_4 I_{R_4} + R_5 I_{R_4} - R_3(I_{R_2} - I_{R_4}) = 0 \Rightarrow I_{R_4} + 4I_{R_4} - 2(I_{R_2} - I_{R_4})$$
$$= 0 \Rightarrow -2I_{R_2} + 7I_{R_4} = 0$$

$$\begin{cases} 10I_1 - 10I_{R_2} = V_1 \\ -10I_1 + 17I_{R_2} - 2I_{R_4} = 0 \\ -2I_{R_2} + 7I_{R_4} = 0 \end{cases} \Rightarrow \begin{bmatrix} 10 & -10 & 0 \\ -10 & 17 & -2 \\ 0 & -2 & 7 \end{bmatrix} \begin{bmatrix} I_1 \\ I_{R_2} \\ I_{R_4} \end{bmatrix} = \begin{bmatrix} V_1 \\ 0 \\ 0 \end{bmatrix}$$

$$I_1 = \frac{\begin{vmatrix} V_1 & -10 & 0 \\ 0 & 17 & -2 \\ 0 & -2 & 7 \end{vmatrix}}{\begin{vmatrix} 10 & -10 & 0 \\ -10 & 17 & -2 \\ 0 & -2 & 7 \end{vmatrix}} \Rightarrow I_1$$

$$= \frac{V_1 \begin{vmatrix} 17 & -2 \\ -2 & 7 \end{vmatrix} - 0 \begin{vmatrix} -10 & 0 \\ -2 & 7 \end{vmatrix} + 0 \begin{vmatrix} -10 & 0 \\ 17 & -2 \end{vmatrix}}{10 \begin{vmatrix} 17 & -2 \\ -2 & 7 \end{vmatrix} - (-10) \begin{vmatrix} -10 & -2 \\ 0 & 7 \end{vmatrix} + 0 \begin{vmatrix} -10 & 17 \\ 0 & -2 \end{vmatrix}} \Rightarrow I_1$$

$$= \frac{V_1 \begin{vmatrix} 17 & -2 \\ -2 & 7 \end{vmatrix}}{10 \begin{vmatrix} 17 & -2 \\ -2 & 7 \end{vmatrix} - (-10) \begin{vmatrix} -10 & -2 \\ 0 & 7 \end{vmatrix}} \Rightarrow I_1$$

$$= \frac{V_1(17 \times 7 - (-2) \times -2)}{10(17 \times 7 - (-2) \times -2) + 10(-10 \times 7 - (-2) \times 0)} \Rightarrow I_1 = \frac{115 V_1}{450} \Rightarrow \frac{V_1}{I_1}$$

$$= \frac{450}{115} \Rightarrow \frac{V_1}{I_1} = 3.913 \Rightarrow Z_{in} = 3.913 \ \Omega$$

The result obtained is identical to that presented in the previous example.

3.2.9 Example 3.9

Determine the input impedance, that is, impedance observed between terminals A and B, for the circuit depicted in Fig. 3.15.

Fig. 3.15 Circuit for
Example 3.9

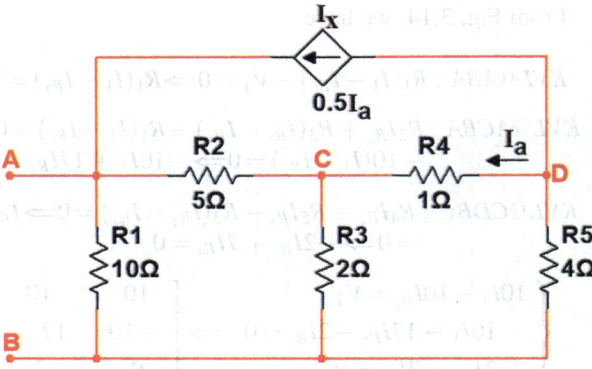

Solution:
The currents I_1 and I_{R_2} are defined as illustrated in Fig. 3.16.

Fig. 3.16 The currents I_1
and I_{R_2}

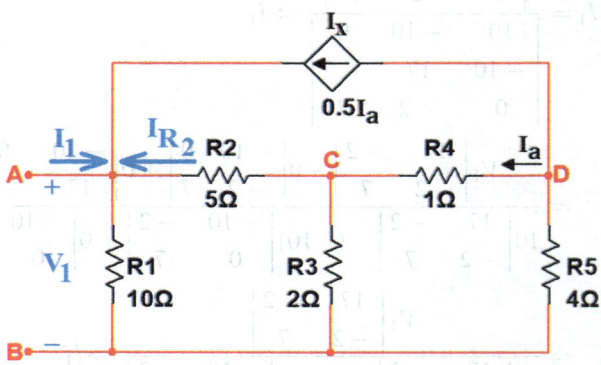

The current flow in the distinct branches is depicted in Fig. 3.17.

Fig. 3.17 The current flow
in the distinct branches

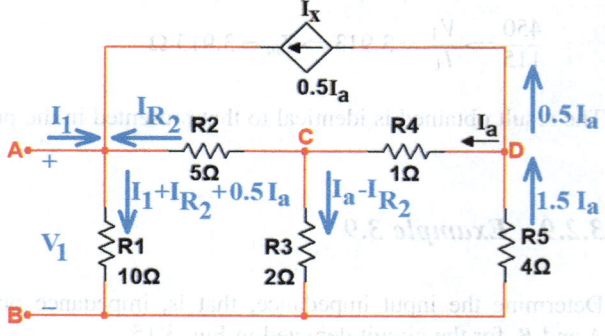

From Fig. 3.17, we have

$$KVL@ABA : R_1(I_1 + I_{R_2} + 0.5I_a) - V_1 = 0 \Rightarrow 10I_1 + 5I_a + 10I_{R_2} = V_1$$

$$KVL@CABC : R_2I_{R_2} + R_1(I_1 + I_{R_2} + 0.5I_a) - R_3(I_a - I_{R_2}) = 0 \Rightarrow 5I_{R_2} + 10I_1$$
$$+ 10I_{R_2} + 5I_a - 2I_a + 2I_{R_2} = 0 \Rightarrow 10I_1 + 3I_a + 17I_{R_2} = 0$$

$$KVL@DCBD : R_4I_a + R_3(I_a - I_{R_2}) + R_5 \times 1.5I_a = 0 \Rightarrow I_a + 2I_a - 2I_{R_2} + 6I_a$$
$$= 0 \Rightarrow 9I_a - 2I_{R_2} = 0$$

$$\begin{cases} 10I_1 + 5I_a + 10I_{R_2} = V_1 \\ 10I_1 + 3I_a + 17I_{R_2} = 0 \\ 9I_a - 2I_{R_2} = 0 \end{cases} \Rightarrow \begin{bmatrix} 10 & 5 & 10 \\ 10 & 3 & 17 \\ 0 & 9 & -2 \end{bmatrix} \begin{bmatrix} I_1 \\ I_a \\ I_{R_2} \end{bmatrix} = \begin{bmatrix} V_1 \\ 0 \\ 0 \end{bmatrix}$$

$$I_1 = \frac{\begin{vmatrix} V_1 & 5 & 10 \\ 0 & 3 & 17 \\ 0 & 9 & -2 \end{vmatrix}}{\begin{vmatrix} 10 & 5 & 10 \\ 10 & 3 & 17 \\ 0 & 9 & -2 \end{vmatrix}} \Rightarrow I_1 = \frac{V_1 \begin{vmatrix} 3 & 17 \\ 9 & -2 \end{vmatrix}}{-590} \Rightarrow I_1 = \frac{-159V_1}{-590} \Rightarrow \frac{V_1}{I_1}$$

$$= \frac{-590}{-159} \Rightarrow Z_{in} = 3.7107 \ \Omega$$

3.2.10 Example 3.10

Determine the impedance parameters (Z parameters) of the circuit depicted in Fig. 3.18.

Fig. 3.18 Circuit for Example 3.10

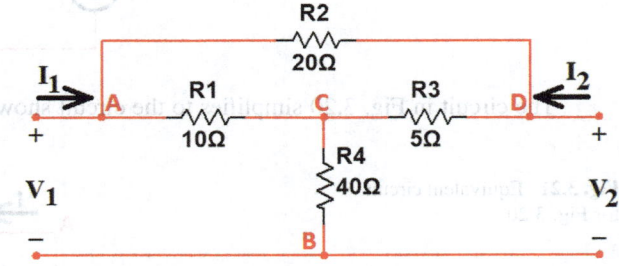

Solution:
In this example, we will rely on the direct definition of impedance parameters. Remember that

$$\begin{bmatrix} V_1 \\ V_2 \end{bmatrix} = \begin{bmatrix} Z_{11} & Z_{12} \\ Z_{21} & Z_{22} \end{bmatrix} \begin{bmatrix} I_1 \\ I_2 \end{bmatrix}$$

where,

$$Z_{11} = \frac{V_1}{I_1}\bigg|_{I_2=0}, Z_{12} = \frac{V_1}{I_2}\bigg|_{I_1=0}$$

$$Z_{21} = \frac{V_2}{I_1}\bigg|_{I_2=0}, Z_{22} = \frac{V_2}{I_2}\bigg|_{I_1=0}$$

- Calculation of Z_{11}.

   $Z_{11} = \frac{V_1}{I_1}\bigg|_{I_2=0}$ is calculated using the circuit shown in Fig. 3.19.

Fig. 3.19 Circuit for calculating Z_{11}

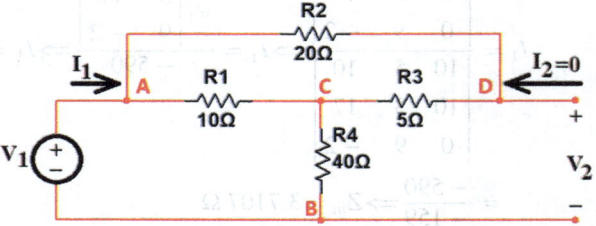

The circuit in Fig. 3.19 simplifies to the circuit shown in Fig. 3.20.

Fig. 3.20 Equivalent circuit for Fig. 3.19

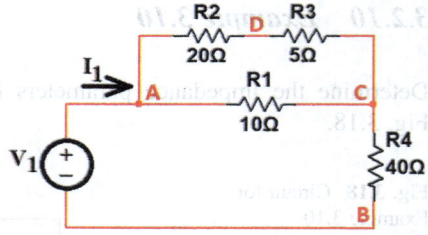

The circuit in Fig. 3.20 simplifies to the circuit shown in Fig. 3.21.

Fig. 3.21 Equivalent circuit for Fig. 3.20

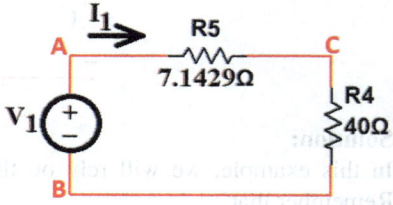

The circuit in Fig. 3.21 simplifies to the circuit shown in Fig. 3.22.

Fig. 3.22 Equivalent circuit for Fig. 3.21

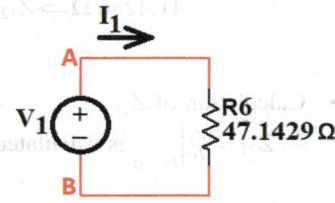

From Fig. 3.22, we have $V_1 = R_6 I_1 \Rightarrow \frac{V_1}{I_1} = R_6 \Rightarrow Z_{11} = R_6 \Rightarrow Z_{11} = 47.1429\,\Omega$.

- Calculation of Z_{12}

$Z_{12} = \frac{V_1}{I_2}\Big|_{I_1=0}$ is calculated using the circuit shown in Fig. 3.23.

Fig. 3.23 Circuit for calculating Z_{12}

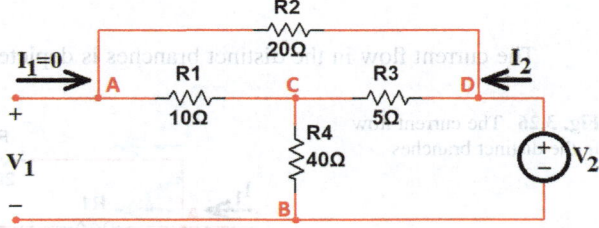

The current flow in the distinct branches is depicted in Fig. 3.24.

Fig. 3.24 The current flow in the distinct branches

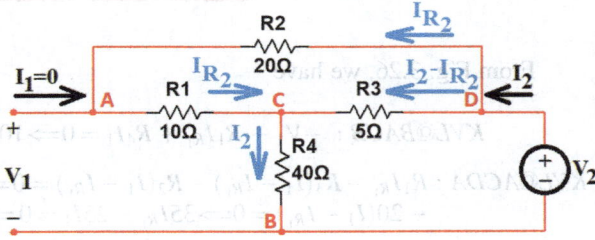

From Fig. 3.24, we have

$$V_1 = V_A - V_B = R_1 I_{R_2} + R_4 I_2 \Rightarrow V_1 = 10 I_{R_2} + 40 I_2$$

$$KVL@DACD: R_2 I_{R_2} + R_1 I_{R_2} - R_3 (I_2 - I_{R_2}) = 0 \Rightarrow 20 I_{R_2} + 10 I_{R_2} - 5(I_2 - I_{R_2})$$

$$= 0 \Rightarrow 35 I_{R_2} - 5 I_2 = 0 \Rightarrow 7 I_{R_2} - I_2 = 0 \Rightarrow I_{R_2} = \frac{I_2}{7}$$

$$V_1 = 10I_{R_2} + 40I_2 \Rightarrow V_1 = 10\frac{I_2}{7} + 40I_2 \Rightarrow V_1 = 41.4286I_2 \Rightarrow \frac{V_1}{I_2}$$

$$= 41.4286\ \Omega \Rightarrow Z_{12} = 41.4286\ \Omega$$

- Calculation of Z_{21}

  $Z_{21} = \frac{V_2}{I_1}\Big|_{I_2=0}$ is calculated using the circuit shown in Fig. 3.25.

Fig. 3.25 Circuit for calculating Z_{21}

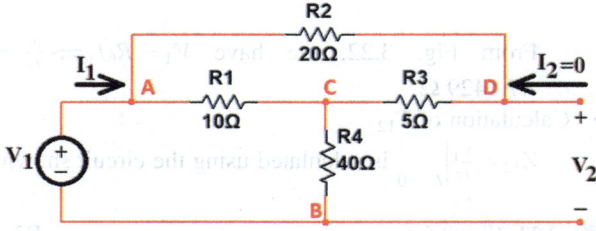

The current flow in the distinct branches is depicted in Fig. 3.26.

Fig. 3.26 The current flow in the distinct branches

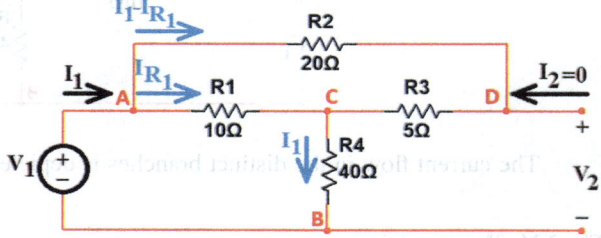

From Fig. 3.26, we have

$$KVL@BACB: \ -V_1 + R_1I_{R_1} + R_4I_1 = 0 \Rightarrow 10I_{R_1} + 40I_1 = V_1$$

$$KVL@ACDA: \ R_1I_{R_1} - R_3(I_1 - I_{R_1}) - R_2(I_1 - I_{R_1}) = 0 \Rightarrow 10I_{R_1} - 5(I_1 - I_{R_1})$$
$$- 20(I_1 - I_{R_1}) = 0 \Rightarrow 35I_{R_1} - 25I_1 = 0 \Rightarrow 7I_{R_1} - 5I_1 = 0 \Rightarrow 7I_{R_1}$$

$$= 5I_1 \Rightarrow I_{R_1} = \frac{5}{7}I_1$$

$$V_2 = V_{DB} = R_3(I_1 - I_{R_1}) + R_4I_1 = 5(I_1 - I_{R_1}) + 40I_1 = 45I_1 - 5I_{R_1}$$

$$= 45I_1 - 5 \times \frac{5}{7}I_1 = 41.4286I_1 \Rightarrow \frac{V_2}{I_1} = 41.4286\ \Omega \Rightarrow Z_{21} = 41.4286\ \Omega$$

- Calculation of Z_{22}

  $Z_{22} = \frac{V_2}{I_2}\Big|_{I_1=0}$ is calculated using the circuit shown in Fig. 3.27.

Fig. 3.27 Circuit for calculating Z_{22}

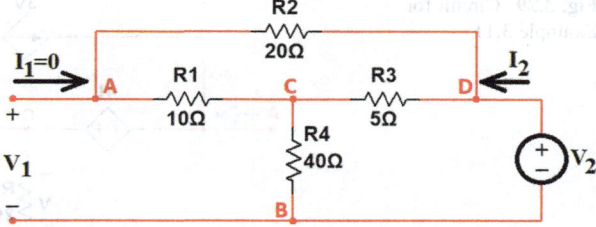

The current flow in the distinct branches is depicted in Fig. 3.28.

Fig. 3.28 The current flow in the distinct branches

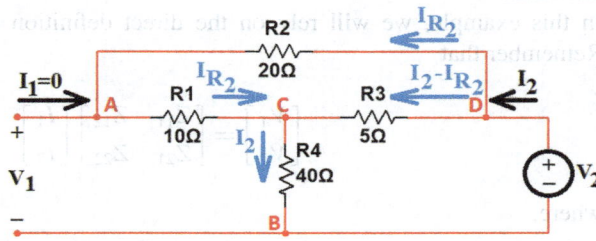

From Fig. 3.28, we have

$$KVL@DCBD : R_3(I_2 - I_{R_2}) + R_4 I_2 - V_2 = 0 \Rightarrow 5(I_2 - I_{R_2}) + 40I_2 - V_2$$
$$= 0 \Rightarrow 5I_2 - 5I_{R_2} + 40I_2 = V_2$$

$$KVL@DACD : R_2 I_{R_2} + R_1 I_{R_2} - R_3(I_2 - I_{R_2}) = 0 \Rightarrow 20I_{R_2} + 10I_{R_2} - 5(I_2 - I_{R_2})$$
$$= 0 \Rightarrow 35I_{R_2} - 5I_2 = 0 \Rightarrow 7I_{R_2} - I_2 = 0 \Rightarrow I_{R_2} = \frac{I_2}{7}$$

$$5I_2 - 5I_{R_2} + 40I_2 = V_2 \Rightarrow 5I_2 - 5\frac{I_2}{7} + 40I_2 = V_2 \Rightarrow 44.2857I_2 = V_2 \Rightarrow \frac{V_2}{I_2}$$

$$= 44.2857 \Rightarrow Z_{22} = 44.2857\ \Omega$$

Therefore,

$$\begin{bmatrix} V_1 \\ V_2 \end{bmatrix} = \begin{bmatrix} 47.1429 & 41.4286 \\ 41.4286 & 44.2857 \end{bmatrix} \begin{bmatrix} I_1 \\ I_2 \end{bmatrix} \Rightarrow Z = \begin{bmatrix} 47.1429 & 41.4286 \\ 41.4286 & 44.2857 \end{bmatrix}$$

3.2.11 Example 3.11

Determine the impedance parameters (Z parameters) of the circuit depicted in Fig. 3.29.

Fig. 3.29 Circuit for
Example 3.11

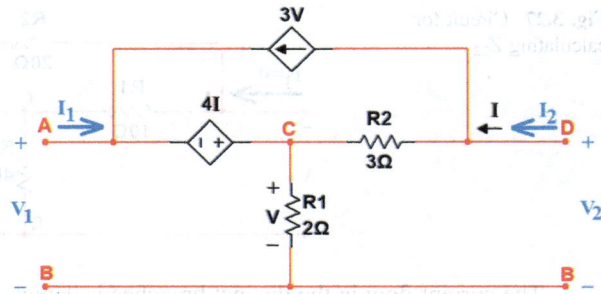

Solution:
In this example, we will rely on the direct definition of impedance parameters.
Remember that

$$\begin{bmatrix} V_1 \\ V_2 \end{bmatrix} = \begin{bmatrix} Z_{11} & Z_{12} \\ Z_{21} & Z_{22} \end{bmatrix} \begin{bmatrix} I_1 \\ I_2 \end{bmatrix}$$

where,

$$Z_{11} = \frac{V_1}{I_1}\bigg|_{I_2=0}, Z_{12} = \frac{V_1}{I_2}\bigg|_{I_1=0}$$

$$Z_{21} = \frac{V_2}{I_1}\bigg|_{I_2=0}, Z_{22} = \frac{V_2}{I_2}\bigg|_{I_1=0}$$

The circuit depicted in Fig. 3.29 can be redrawn as shown in Fig. 3.30.

Fig. 3.30 Equivalent circuit
for Fig. 3.29

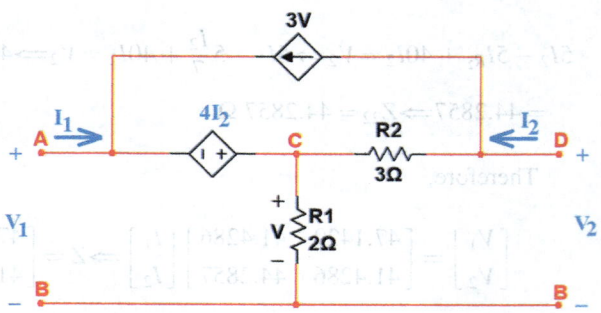

- Calculation of Z_{11}

    $Z_{11} = \frac{V_1}{I_1}\big|_{I_2=0}$ is calculated using the circuit shown in Fig. 3.31.

Fig. 3.31 Circuit for calculating Z_{11}

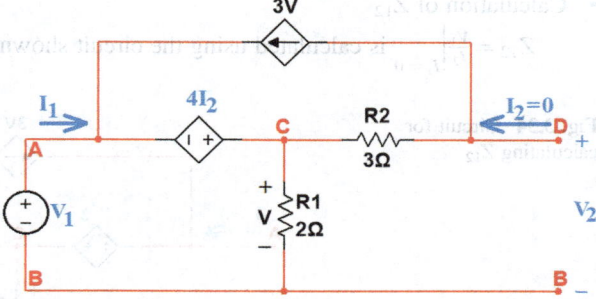

The current flow in the distinct branches is depicted in Fig. 3.32.

Fig. 3.32 The current flow in the distinct branches

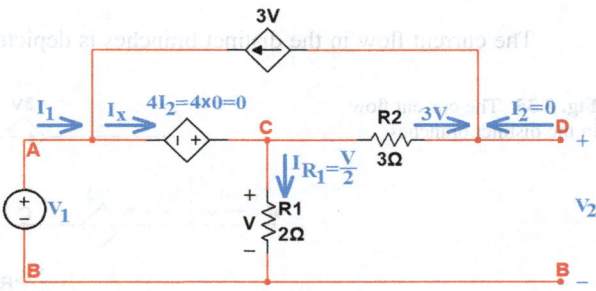

The circuit in Fig. 3.32 simplifies to the circuit shown in Fig. 3.33.

Fig. 3.33 Equivalent circuit for Fig. 3.32

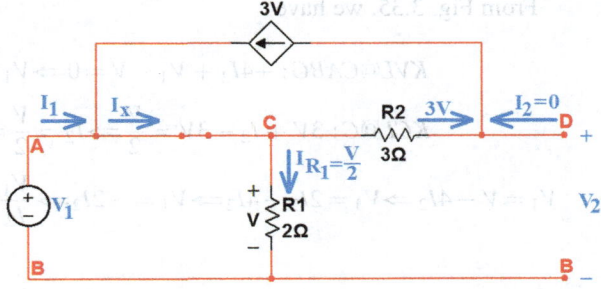

From Fig. 3.33, we have

$$KCL@A: I_1 + 3V = I_x$$

$$KCL@C: I_x = \frac{V}{2} + 3V \Rightarrow I_1 + 3V = \frac{V}{2} + 3V \Rightarrow I_1 = \frac{V}{2} \Rightarrow I_1 = \frac{V_1}{2} \Rightarrow \frac{V_1}{I_1} = 2\ \Omega$$

- Calculation of Z_{12}

$Z_{12} = \left. \dfrac{V_1}{I_2} \right|_{I_1=0}$ is calculated using the circuit shown in Fig. 3.34.

Fig. 3.34 Circuit for calculating Z_{12}

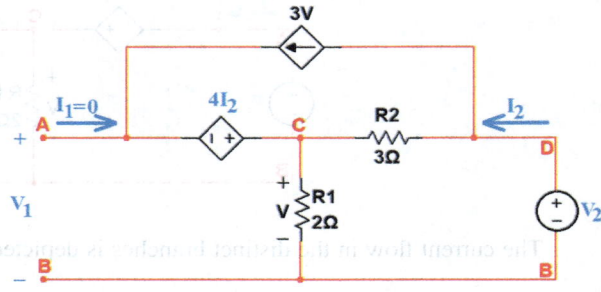

The current flow in the distinct branches is depicted in Fig. 3.35.

Fig. 3.35 The current flow in the distinct branches

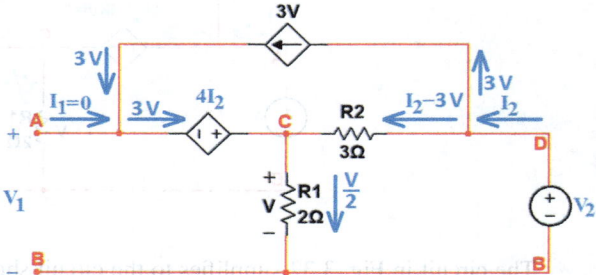

From Fig. 3.35, we have

$$KVL@CABC: +4I_2 + V_1 - V = 0 \Rightarrow V_1 = V - 4I_2$$

$$KCL@C: 3V + I_2 - 3V = \frac{V}{2} \Rightarrow I_2 = \frac{V}{2} \Rightarrow V = 2I_2$$

$$V_1 = V - 4I_2 \Rightarrow V_1 = 2I_2 - 4I_2 \Rightarrow V_1 = -2I_2 \Rightarrow \frac{V_1}{I_2} = -2 \Rightarrow Z_{12} = -2\ \Omega$$

- Calculation of Z_{21}

$Z_{21} = \left. \dfrac{V_2}{I_1} \right|_{I_2=0}$ is calculated using the circuit shown in Fig. 3.36.

Fig. 3.36 Circuit for calculating Z_{21}

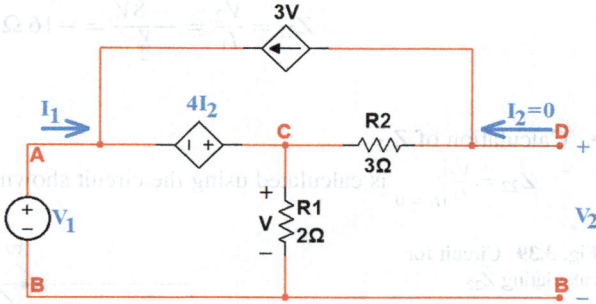

The current flow in the distinct branches is depicted in Fig. 3.37.

Fig. 3.37 The current flow in the distinct branches

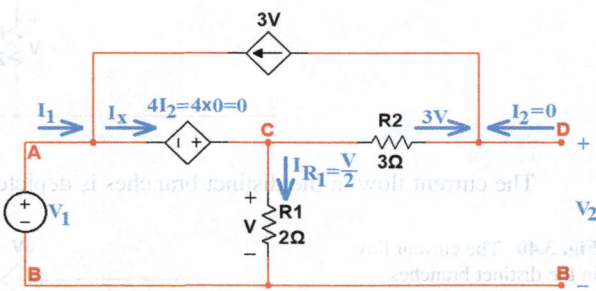

The circuit in Fig. 3.37 simplifies to the circuit shown in Fig. 3.38.

Fig. 3.38 Equivalent circuit for Fig. 3.37

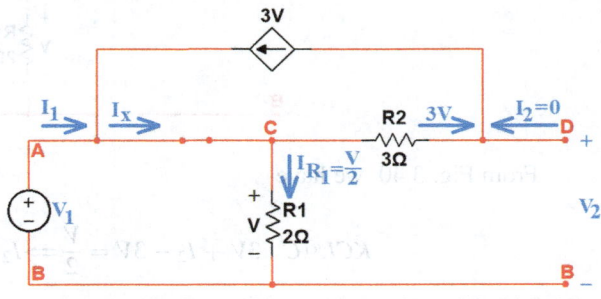

From Fig. 3.38, we have

$$KVL@CDBC : R_2 \times 3V + V_2 - V = 0 \Longrightarrow 3 \times 3V + V_2 - V = 0 \Longrightarrow 8V + V_2$$
$$= 0 \Longrightarrow V_2 = -8V$$

$$KCL@A : I_1 + 3V = I_x$$

$$KCL@C : I_x = \frac{V}{2} + 3V \Longrightarrow I_1 + 3V = \frac{V}{2} + 3V \Longrightarrow I_1 = \frac{V}{2}$$

$$Z_{21} = \frac{V_2}{I_1} = \frac{-8V}{\frac{V}{2}} = -16\,\Omega$$

- Calculation of Z_{22}

  $Z_{22} = \frac{V_2}{I_2}\Big|_{I_1=0}$ is calculated using the circuit shown in Fig. 3.39.

Fig. 3.39 Circuit for calculating Z_{22}

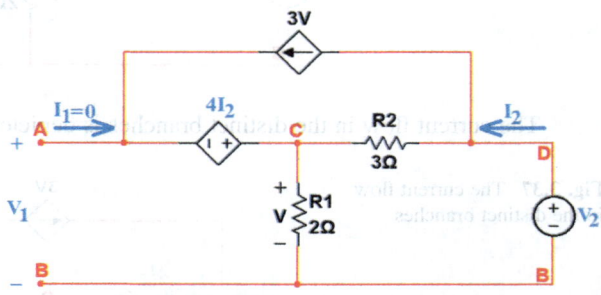

The current flow in the distinct branches is depicted in Fig. 3.40.

Fig. 3.40 The current flow in the distinct branches

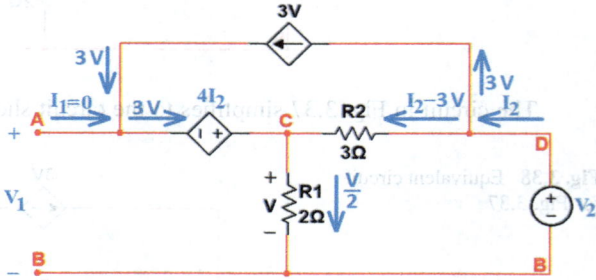

From Fig. 3.40, we have

$$KCL@C : 3V + I_2 - 3V = \frac{V}{2} \Rightarrow I_2 = \frac{V}{2}$$

$$KVL@BDCB : \; -V_2 + 3(I_2 - 3V) + V = 0 \Rightarrow -V_2 + 3I_2 - 9V + V = 0 \Rightarrow -V_2$$

$$+ 3I_2 - 8V = 0 \Rightarrow -V_2 + 3 \times \frac{V}{2} - 8V = 0 \Rightarrow V_2 = -6.5V \Rightarrow V$$

$$= -\frac{V_2}{6.5}$$

$$I_2 = \frac{V}{2} \Rightarrow I_2 = \frac{-\frac{V_2}{6.5}}{2} \Rightarrow I_2 = -\frac{V_2}{13} \Rightarrow \frac{V_2}{I_2} = -13 \Rightarrow Z_{22} = -13\,\Omega$$

Therefore,

$$\begin{cases} V_1 = 2I_1 - 2I_2 \\ V_2 = -16I_1 - 13I_2 \end{cases} \Rightarrow \begin{bmatrix} V_1 \\ V_2 \end{bmatrix} = \begin{bmatrix} 2 & -2 \\ -16 & -13 \end{bmatrix} \begin{bmatrix} I_1 \\ I_2 \end{bmatrix} \Rightarrow Z = \begin{bmatrix} 2 & -2 \\ -16 & -13 \end{bmatrix}$$

3.2.12   Example 3.12

Determine the admittance parameters (Y parameters) of the circuit depicted in Fig. 3.41.

Fig. 3.41 Circuit for Example 3.12

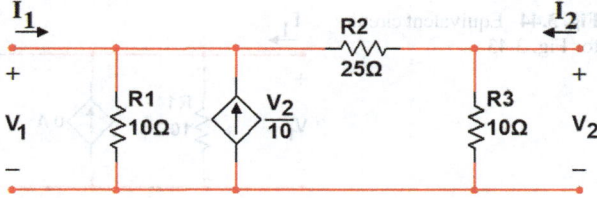

Solution:
In this example, we will rely on the direct definition of admittance parameters. Remember that

$$\begin{bmatrix} I_1 \\ I_2 \end{bmatrix} = \begin{bmatrix} Y_{11} & Y_{12} \\ Y_{21} & Y_{22} \end{bmatrix} \begin{bmatrix} V_1 \\ V_2 \end{bmatrix}$$

where,

$$Y_{11} = \frac{I_1}{V_1}\bigg|_{V_2 = 0}, \quad Y_{12} = \frac{I_1}{V_2}\bigg|_{V_1 = 0}$$

$$Y_{21} = \frac{I_2}{V_1}\bigg|_{V_2 = 0}, \quad Y_{22} = \frac{I_2}{V_2}\bigg|_{V_1 = 0}$$

- Calculation of Y_{11} and Y_{21}

  $Y_{11} = \frac{I_1}{V_1}\big|_{V_2 = 0}$ and $Y_{21} = \frac{I_2}{V_1}\big|_{V_2 = 0}$ is calculated using the circuit shown in Fig. 3.42.

Fig. 3.42 Circuit for calculating Y_{11} and Y_{21}

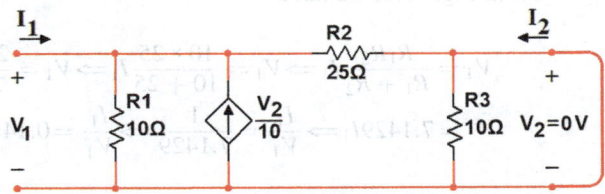

The circuit in Fig. 3.42 simplifies to the circuit shown in Fig. 3.43.

Fig. 3.43 Equivalent circuit for Fig. 3.42

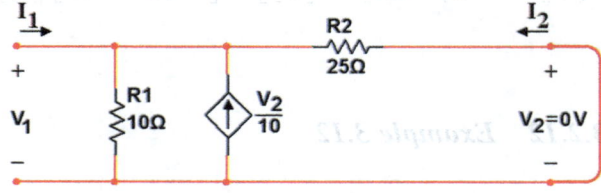

The circuit in Fig. 3.43 simplifies to the circuit shown in Fig. 3.44.

Fig. 3.44 Equivalent circuit for Fig. 3.43

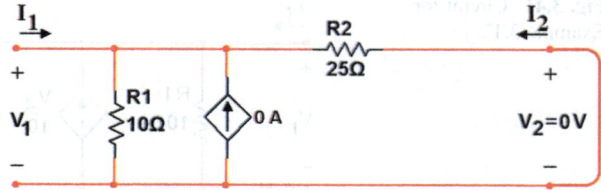

The circuit in Fig. 3.44 simplifies to the circuit shown in Fig. 3.45.

Fig. 3.45 Equivalent circuit for Fig. 3.44

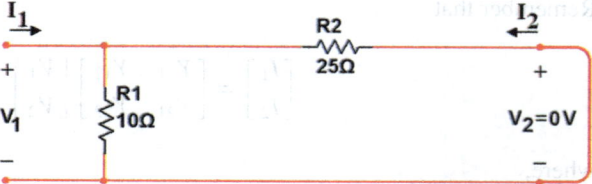

The circuit in Fig. 3.45 simplifies to the circuit shown in Fig. 3.46.

Fig. 3.46 Equivalent circuit for Fig. 3.45

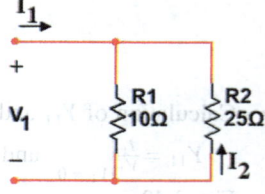

From Fig. 3.46, we have

$$V_1 = \frac{R_1 R_2}{R_1 + R_2} I_1 \Rightarrow V_1 = \frac{10 \times 25}{10 + 25} I_1 \Rightarrow V_1 = \frac{250}{35} I_1 \Rightarrow V_1$$

$$= 7.1429 I_1 \Rightarrow \frac{I_1}{V_1} = \frac{1}{7.1429} \Rightarrow \frac{I_1}{V_1} = 0.14 \Rightarrow Y_{11} = 0.14 \ S$$

$$I_2 = \frac{-V_1}{R_2} \Rightarrow I_2 = \frac{-V_1}{25} \Rightarrow \frac{I_2}{V_1} = -0.04 \Rightarrow Y_{21} = -0.04 \ S$$

- Calculation of Y_{12} and Y_{22}

  $Y_{12} = \frac{I_1}{V_2}\Big|_{V_1=0}$ and $Y_{22} = \frac{I_2}{V_2}\Big|_{V_1=0}$ is calculated using the circuit shown in Fig. 3.47.

Fig. 3.47 Circuit for calculating Y_{12} and Y_{22}

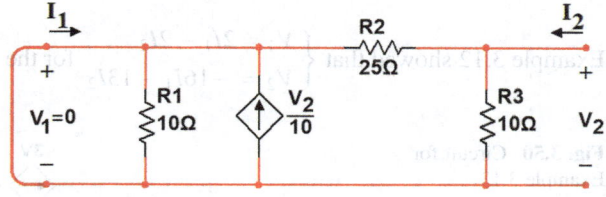

The circuit in Fig. 3.47 simplifies to the circuit shown in Fig. 3.48.

Fig. 3.48 Equivalent circuit for Fig. 3.47

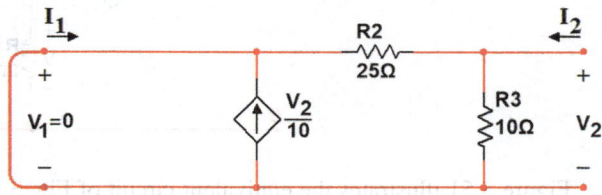

The circuit in Fig. 3.48 simplifies to the circuit shown in Fig. 3.49.

Fig. 3.49 Equivalent circuit for Fig. 3.48

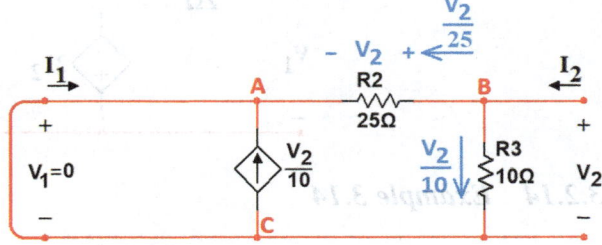

From Fig. 3.49, we have

$$KCL@A: I_1 + \frac{V_2}{10} + \frac{V_2}{25} = 0 \Rightarrow I_1 + 0.14V_2 = 0 \Rightarrow \frac{I_1}{V_2} = -0.14 \Rightarrow Y_{12} = -0.14 \ S$$

$$KCL@B: I_2 = \frac{V_2}{10} + \frac{V_2}{25} \Rightarrow I_2 = 0.14V_2 \Rightarrow \frac{I_2}{V_2} = 0.14 \Rightarrow Y_{22} = 0.14 \ S$$

Therefore,

$$\begin{bmatrix} I_1 \\ I_2 \end{bmatrix} = \begin{bmatrix} 0.14 & -0.14 \\ -0.04 & 0.14 \end{bmatrix} \begin{bmatrix} V_1 \\ V_2 \end{bmatrix} \Rightarrow Y = \begin{bmatrix} 0.14 & -0.14 \\ -0.04 & 0.14 \end{bmatrix}$$

3.2.13   Example 3.13

Example 3.12 showed that $\begin{cases} V_1 = 2I_1 - 2I_2 \\ V_2 = -16I_1 - 13I_2 \end{cases}$ for the circuit shown in Fig. 3.50.

Fig. 3.50 Circuit for
Example 3.13

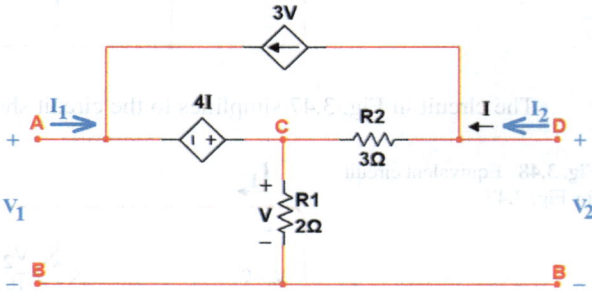

Figure 3.51 illustrates the equivalent circuit of Fig. 3.50.

Fig. 3.51 Equivalent circuit
for Fig. 3.50

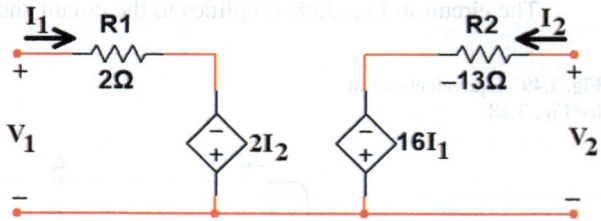

3.2.14   Example 3.14

The admittance matrix of a two-port network is given by $\begin{bmatrix} 0.1192 & -0.1115 \\ -0.1115 & 0.1269 \end{bmatrix}$.

This implies $\begin{cases} I_1 = 0.1192I_1 - 0.1115I_2 \\ I_2 = -0.1115I_1 + 0.1269I_2 \end{cases}$. Such a two-port network can be
represented by the circuit shown in Fig. 3.52.

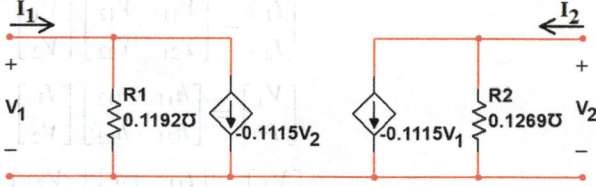

Fig. 3.52 Circuit corresponding to the admittance matrix in Example 3.14

The circuit shown in Fig. 3.52 is equivalent to the one in Fig. 3.53.

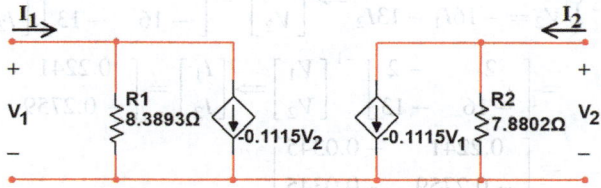

Fig. 3.53 Equivalent circuit for Fig. 3.52

The circuit shown in Fig. 3.53 is equivalent to the one in Fig. 3.54.

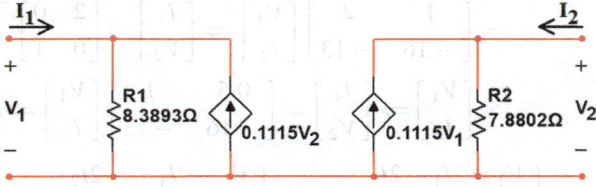

Fig. 3.54 Equivalent circuit for Fig. 3.53

3.2.15 Example 3.15

The impedance parameters of a two-port network are given by $Z = \begin{bmatrix} 2 & -2 \\ -16 & -13 \end{bmatrix}$.
Determine the (**A**) admittance (y) parameters, (**B**) hybrid (h) parameters, and (**C**) transmission (T) parameters.

Solution:
Because the impedance parameters are given, it follows that

$$\begin{bmatrix} V_1 \\ V_2 \end{bmatrix} = Z \begin{bmatrix} I_1 \\ I_2 \end{bmatrix} \Rightarrow \begin{bmatrix} V_1 \\ V_2 \end{bmatrix} = \begin{bmatrix} 2 & -2 \\ -16 & -13 \end{bmatrix} \begin{bmatrix} I_1 \\ I_2 \end{bmatrix} \Rightarrow \begin{cases} V_1 = 2I_1 - 2I_2 \\ V_2 = -16I_1 - 13I_2 \end{cases}$$

We will now review the equations for two-port networks represented by admittance, hybrid, and T-models.

$$\begin{bmatrix} I_1 \\ I_2 \end{bmatrix} = \begin{bmatrix} Y_{11} & Y_{12} \\ Y_{21} & Y_{22} \end{bmatrix} \begin{bmatrix} V_1 \\ V_2 \end{bmatrix}$$

$$\begin{bmatrix} V_1 \\ I_2 \end{bmatrix} = \begin{bmatrix} h_{11} & h_{12} \\ h_{21} & h_{22} \end{bmatrix} \begin{bmatrix} I_1 \\ V_2 \end{bmatrix}$$

$$\begin{bmatrix} V_1 \\ I_1 \end{bmatrix} = \begin{bmatrix} t_{11} & t_{21} \\ t_{21} & t_{22} \end{bmatrix} \begin{bmatrix} V_2 \\ -I_2 \end{bmatrix}$$

$$\begin{cases} V_1 = 2I_1 - 2I_2 \\ V_2 = -16I_1 - 13I_2 \end{cases} \Rightarrow \begin{bmatrix} V_1 \\ V_2 \end{bmatrix} = \begin{bmatrix} 2 & -2 \\ -16 & -13 \end{bmatrix} \begin{bmatrix} I_1 \\ I_2 \end{bmatrix} \Rightarrow \begin{bmatrix} I_1 \\ I_2 \end{bmatrix}$$

$$= \begin{bmatrix} 2 & -2 \\ -16 & -13 \end{bmatrix}^{-1} \begin{bmatrix} V_1 \\ V_2 \end{bmatrix} \Rightarrow \begin{bmatrix} I_1 \\ I_2 \end{bmatrix} = \begin{bmatrix} 0.2241 & -0.0345 \\ -0.2759 & -0.0345 \end{bmatrix} \begin{bmatrix} V_1 \\ V_2 \end{bmatrix} \Rightarrow Y$$

$$= \begin{bmatrix} 0.2241 & -0.0345 \\ -0.2759 & -0.0345 \end{bmatrix}$$

$$\begin{cases} V_1 = 2I_1 - 2I_2 \\ V_2 = -16I_1 - 13I_2 \end{cases} \Rightarrow \begin{cases} 2I_1 = V_1 + 2I_2 \\ V_2 = -16I_1 - 13I_2 \end{cases} \Rightarrow \begin{bmatrix} 2 & 0 \\ 0 & 1 \end{bmatrix} \begin{bmatrix} I_1 \\ V_2 \end{bmatrix}$$

$$= \begin{bmatrix} 1 & 2 \\ -16 & -13 \end{bmatrix} \begin{bmatrix} V_1 \\ I_2 \end{bmatrix} \Rightarrow \begin{bmatrix} I_1 \\ V_2 \end{bmatrix} = \begin{bmatrix} 2 & 0 \\ 0 & 1 \end{bmatrix}^{-1} \begin{bmatrix} 1 & 2 \\ -16 & -13 \end{bmatrix}$$

$$\times \begin{bmatrix} V_1 \\ I_2 \end{bmatrix} \Rightarrow \begin{bmatrix} I_1 \\ V_2 \end{bmatrix} = \begin{bmatrix} 0.5 & 1 \\ -16 & -13 \end{bmatrix} \begin{bmatrix} V_1 \\ I_2 \end{bmatrix} \Rightarrow h = \begin{bmatrix} 0.5 & 1 \\ -16 & -13 \end{bmatrix}$$

$$\begin{cases} V_1 = 2I_1 - 2I_2 \\ V_2 = -16I_1 - 13I_2 \end{cases} \Rightarrow \begin{cases} V_1 - 2I_1 = -2I_2 \\ 16I_1 = -V_2 - 13I_2 \end{cases} \Rightarrow \begin{bmatrix} 1 & -2 \\ 0 & 16 \end{bmatrix} \begin{bmatrix} V_1 \\ I_1 \end{bmatrix}$$

$$= \begin{bmatrix} 0 & 2 \\ -1 & 13 \end{bmatrix} \begin{bmatrix} V_2 \\ -I_2 \end{bmatrix} \Rightarrow \begin{bmatrix} V_1 \\ I_1 \end{bmatrix} = \begin{bmatrix} 1 & -2 \\ 0 & 16 \end{bmatrix}^{-1} \begin{bmatrix} 0 & 2 \\ -1 & 13 \end{bmatrix}$$

$$\times \begin{bmatrix} V_2 \\ -I_2 \end{bmatrix} \Rightarrow \begin{bmatrix} V_1 \\ I_1 \end{bmatrix} = \begin{bmatrix} -0.1250 & 3.6250 \\ -0.0625 & 0.8125 \end{bmatrix} \begin{bmatrix} V_2 \\ -I_2 \end{bmatrix} \Rightarrow T$$

$$= \begin{bmatrix} -0.1250 & 3.6250 \\ -0.0625 & 0.8125 \end{bmatrix}$$

3.2.16   Example 3.16

The hybrid parameters of a two-port network are given by
$h = \begin{bmatrix} 36.6667 & 0.6667 \\ -0.3333 & 0.0067 \end{bmatrix}$. Determine the transmission (T) parameters.

Solution:

$$\begin{bmatrix} V_1 \\ I_2 \end{bmatrix} = \begin{bmatrix} h_{11} & h_{12} \\ h_{21} & h_{22} \end{bmatrix} \begin{bmatrix} I_1 \\ V_2 \end{bmatrix} \Rightarrow \begin{bmatrix} V_1 \\ I_2 \end{bmatrix} = \begin{bmatrix} 36.6667 & 0.6667 \\ -0.3333 & 0.0067 \end{bmatrix}$$

$$\times \begin{bmatrix} I_1 \\ V_2 \end{bmatrix} \Rightarrow \begin{cases} V_1 = 36.6667 I_1 + 0.6667 V_2 \\ I_2 = -0.3333 I_1 + 0.0067 V_2 \end{cases}$$

$$\begin{bmatrix} V_1 \\ I_1 \end{bmatrix} = \begin{bmatrix} t_{11} & t_{21} \\ t_{21} & t_{22} \end{bmatrix} \begin{bmatrix} V_2 \\ -I_2 \end{bmatrix}$$

$$\begin{cases} V_1 = 36.6667 I_1 + 0.6667 V_2 \\ I_2 = -0.3333 I_1 + 0.0067 V_2 \end{cases} \Rightarrow \begin{cases} V_1 - 36.6667 I_1 = 0.6667 V_2 \\ 0.3333 I_1 = 0.0067 V_2 - I_2 \end{cases} \Rightarrow \begin{bmatrix} 1 & -36.6667 \\ 0 & 0.3333 \end{bmatrix}$$

$$\times \begin{bmatrix} V_1 \\ I_1 \end{bmatrix} = \begin{bmatrix} 0.6667 & 0 \\ 0.0067 & 1 \end{bmatrix} \begin{bmatrix} V_2 \\ -I_2 \end{bmatrix} \Rightarrow \begin{bmatrix} V_1 \\ I_1 \end{bmatrix} = \begin{bmatrix} 1 & -36.6667 \\ 0 & 0.3333 \end{bmatrix}^{-1} \begin{bmatrix} 0.6667 & 0 \\ 0.0067 & 1 \end{bmatrix}$$

$$\times \begin{bmatrix} V_2 \\ -I_2 \end{bmatrix} \Rightarrow \begin{bmatrix} V_1 \\ I_1 \end{bmatrix} = \begin{bmatrix} 1.4038 & 110.0111 \\ 0.0201 & 3.0003 \end{bmatrix} \begin{bmatrix} V_2 \\ -I_2 \end{bmatrix} \Rightarrow T = \begin{bmatrix} 1.4038 & 110.0111 \\ 0.0201 & 3.0003 \end{bmatrix}$$

3.2.17 Example 3.17

Determine the impedance parameters (Z parameters) of the circuit depicted in Fig. 3.55.

Fig. 3.55 Circuit for Example 3.17

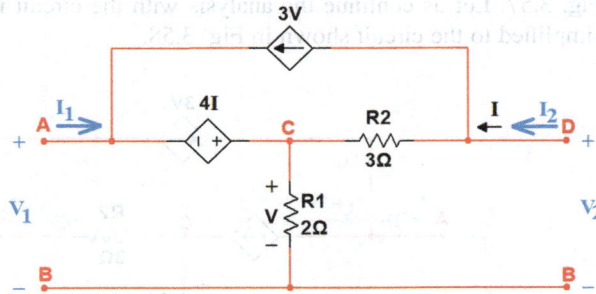

Solution:
For the purpose of analysis, let us connect two sources (voltage or current) to the input and output terminals of the circuit (Figs. 3.56 and 3.57).

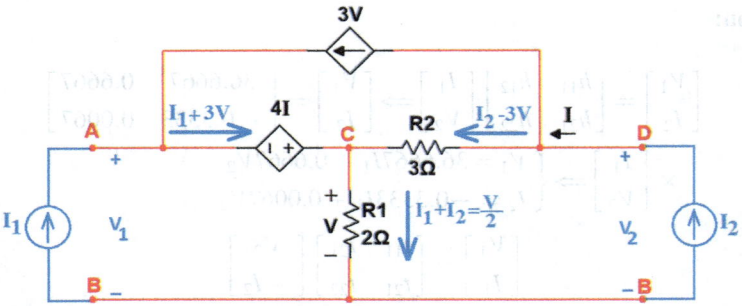

Fig. 3.56 Input-output ports are driven by two current sources

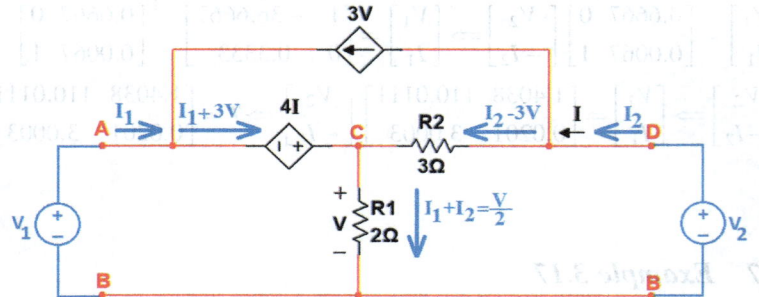

Fig. 3.57 Input-output ports are driven by two voltage sources

Continue the analysis using your choice of circuit from either Fig. 3.56 or Fig. 3.57. Let us continue the analysis with the circuit in Fig. 3.57, which can be simplified to the circuit shown in Fig. 3.58.

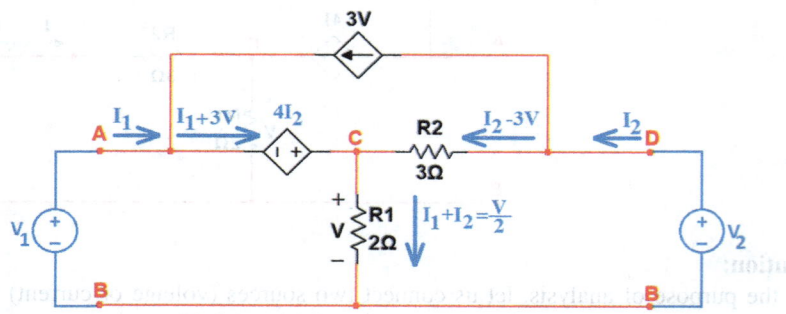

Fig. 3.58 Voltages and currents from Fig. 3.57

From Fig. 3.58, we have

$$\frac{V}{2}=I_1+I_2\Rightarrow V=2(I_1+I_2)$$

KVL@BACB: $-V_1-4I_2+V=0\Rightarrow -V_1-4I_2+2(I_1+I_2)=0\Rightarrow V_1=2I_1-2I_2$

KVL@BDCB: $-V_2+3(I_2-3V)+V=0\Rightarrow -V_2+3I_2-8V=0\Rightarrow -V_2+3I_2-8(2I_1+2I_2)=0\Rightarrow V_2=3I_2-8(2I_1+2I_2)\Rightarrow V_2=-16I_1-13I_2$

$$\begin{cases} V_1=2I_1-2I_2 \\ V_2=-16I_1-13I_2 \end{cases}\Rightarrow \begin{bmatrix} V_1 \\ V_2 \end{bmatrix}=\begin{bmatrix} 2 & -2 \\ -16 & -13 \end{bmatrix}\begin{bmatrix} I_1 \\ I_2 \end{bmatrix}\Rightarrow \begin{bmatrix} V_1 \\ V_2 \end{bmatrix}$$

$$=\begin{bmatrix} 2 & -2 \\ -16 & -13 \end{bmatrix}\begin{bmatrix} I_1 \\ I_2 \end{bmatrix}\Rightarrow Z=\begin{bmatrix} 2 & -2 \\ -16 & -13 \end{bmatrix}$$

For practice and to reinforce understanding, solve this problem following the method presented in Examples 3.10, 3.11, and 3.12. The expected outcome should match.

3.2.18 Example 3.18

Determine the impedance parameters (Z parameters) of the circuit depicted in Fig. 3.59.

Fig. 3.59 Circuit for Example 3.18

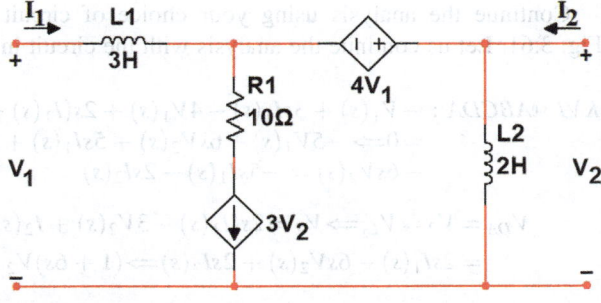

Solution:

For the purpose of analysis, let us connect two sources (voltage or current) to the input and output terminals of the circuit (Figs. 3.60 and 3.61). Since the circuit includes an inductor, we have transformed it into the frequency domain for analysis.

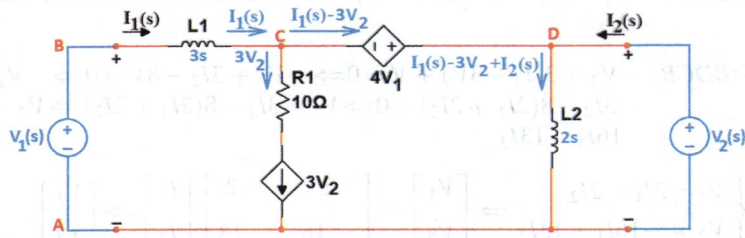

Fig. 3.60 Input-output ports are driven by two voltage sources

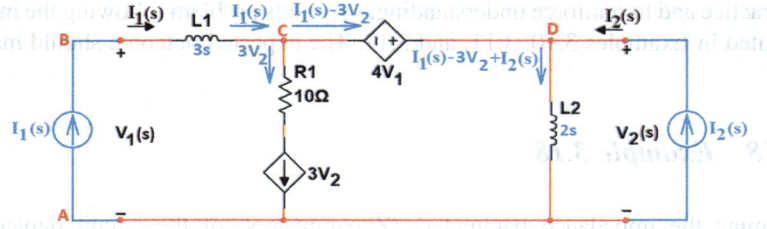

Fig. 3.61 Input-output ports are driven by two current sources

Continue the analysis using your choice of circuit from either Fig. 3.60 or Fig. 3.61. Let us continue the analysis with the circuit in Fig. 3.61.

$$KVL@ABCDA: \ -V_1(s) + 3sI_1(s) - 4V_1(s) + 2s(I_1(s) - 3V_2(s) + I_2(s))$$
$$= 0 \Longrightarrow -5V_1(s) - 6sV_2(s) + 5sI_1(s) + 2sI_2(s) = 0 \Longrightarrow -5V_1(s)$$
$$- 6sV_2(s) = -5sI_1(s) - 2sI_2(s)$$

$$V_{DA} = V_2 = V_{L_2} \Longrightarrow V_2 = 2s(I_1(s) - 3V_2(s) + I_2(s)) \Longrightarrow V_2$$
$$= 2sI_1(s) - 6sV_2(s) + 2sI_2(s) \Longrightarrow (1 + 6s)V_2 = 2sI_1(s) + 2sI_2(s)$$

$$\begin{cases} -5V_1(s) - 6sV_2(s) = -5sI_1(s) - 2sI_2(s) \\ (1+6s)V_2 = 2sI_1(s) + 2sI_2(s) \end{cases} \Rightarrow \begin{bmatrix} -5 & -6s \\ 0 & 1+6s \end{bmatrix} \begin{bmatrix} V_1(s) \\ V_2(s) \end{bmatrix}$$

$$= \begin{bmatrix} -5s & -2s \\ 2s & 2s \end{bmatrix} \begin{bmatrix} I_1(s) \\ I_2(s) \end{bmatrix} \Rightarrow \begin{bmatrix} V_1(s) \\ V_2(s) \end{bmatrix} = \begin{bmatrix} -5 & -6s \\ 0 & 1+6s \end{bmatrix}^{-1} \begin{bmatrix} -5s & -2s \\ 2s & 2s \end{bmatrix}$$

$$\times \begin{bmatrix} I_1(s) \\ I_2(s) \end{bmatrix} \Rightarrow \begin{bmatrix} V_1(s) \\ V_2(s) \end{bmatrix} = \frac{1}{-5(1+6s)-0} \begin{bmatrix} 1+6s & 6s \\ 0 & -5 \end{bmatrix} \begin{bmatrix} -5s & -2s \\ 2s & 2s \end{bmatrix}$$

$$\times \begin{bmatrix} I_1(s) \\ I_2(s) \end{bmatrix} \Rightarrow \begin{bmatrix} V_1(s) \\ V_2(s) \end{bmatrix} = \frac{1}{-5(1+6s)}$$

$$\times \begin{bmatrix} (1+6s) \times -5s + 6s \times 2s & (1+6s) \times -2s + 6s \times 2s \\ 0 - 5 \times 2s & 0 - 5 \times 2s \end{bmatrix}$$

$$\times \begin{bmatrix} I_1(s) \\ I_2(s) \end{bmatrix} \Rightarrow \begin{bmatrix} V_1(s) \\ V_2(s) \end{bmatrix} = \frac{1}{-5(1+6s)} \begin{bmatrix} -5s - 18s^2 & -2s \\ -10s & -10s \end{bmatrix}$$

$$\times \begin{bmatrix} I_1(s) \\ I_2(s) \end{bmatrix} \Rightarrow \begin{bmatrix} V_1(s) \\ V_2(s) \end{bmatrix} = \begin{bmatrix} \dfrac{5s+18s^2}{5(1+6s)} & \dfrac{2s}{5(1+6s)} \\ \dfrac{10s}{5(1+6s)} & \dfrac{10s}{5(1+6s)} \end{bmatrix} \begin{bmatrix} I_1(s) \\ I_2(s) \end{bmatrix} \Rightarrow \begin{bmatrix} V_1(s) \\ V_2(s) \end{bmatrix}$$

$$= \begin{bmatrix} \dfrac{5s+18s^2}{5(1+6s)} & \dfrac{2s}{5(1+6s)} \\ \dfrac{2s}{(1+6s)} & \dfrac{2s}{(1+6s)} \end{bmatrix} \begin{bmatrix} I_1(s) \\ I_2(s) \end{bmatrix} \Rightarrow Z = \begin{bmatrix} \dfrac{5s+18s^2}{5(1+6s)} & \dfrac{2s}{5(1+6s)} \\ \dfrac{2s}{(1+6s)} & \dfrac{2s}{(1+6s)} \end{bmatrix}$$

3.2.19 Example 3.19

Obtain $V_O(t)$ for the configuration shown in Fig. 3.62.

Fig. 3.62 Circuit for Example 3.19

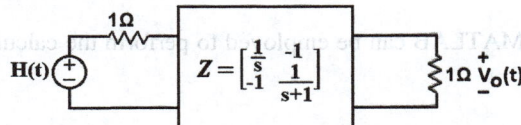

Solution:
Figure 3.63 shows the frequency domain representation of the circuit.

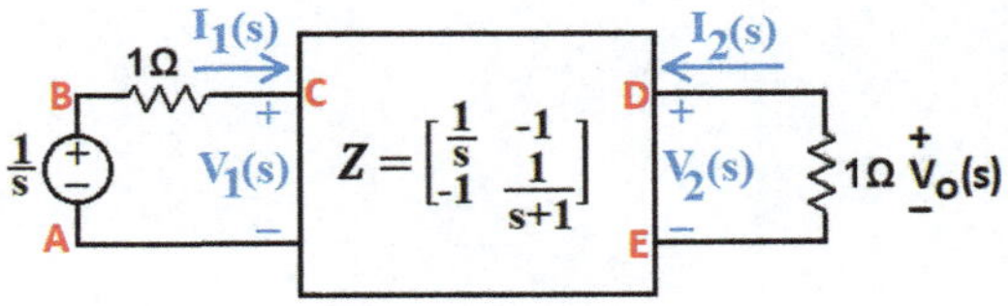

Fig. 3.63 The frequency domain representation of Fig. 3.62

From Fig. 3.63, we have

$$V_o(s) = V_2(s)$$

$$\begin{bmatrix} V_1(s) \\ V_2(s) \end{bmatrix} = \begin{bmatrix} \dfrac{1}{s} & -1 \\ -1 & \dfrac{1}{s+1} \end{bmatrix} \begin{bmatrix} I_1(s) \\ I_2(s) \end{bmatrix} \Rightarrow \begin{bmatrix} V_1(s) \\ V_2(s) \end{bmatrix}$$

$$= \begin{bmatrix} \dfrac{I_1(s)}{s} - I_2(s) \\ -I_1(s) + \dfrac{I_2(s)}{s+1} \end{bmatrix} \Rightarrow \begin{cases} V_1(s) = \dfrac{I_1(s)}{s} - I_2(s) \\ V_2(s) = -I_1(s) + \dfrac{I_2(s)}{s+1} \end{cases}$$

$$KVL@ABCA : -H(t) + I_1(s) + V_1(s) = 0 \Rightarrow -H(t) + I_1(s) + \dfrac{I_1(s)}{s} - I_2(s)$$

$$= 0 \Rightarrow \dfrac{s+1}{s} I_1(s) - I_2(s) = \dfrac{1}{s}$$

$$KVL@EDE : I_2(s) + V_2(s) = 0$$

$$\begin{cases} V_1(s) = \dfrac{I_1(s)}{s} - I_2(s) \\ V_2(s) = -I_1(s) + \dfrac{I_2(s)}{s+1} \\ \dfrac{s+1}{s} I_1(s) - I_2(s) = \dfrac{1}{s} \\ I_2(s) + V_2(s) = 0 \end{cases} \Rightarrow \begin{bmatrix} 1 & 0 & -\dfrac{1}{s} & 1 \\ 0 & 1 & 1 & -\dfrac{1}{s+1} \\ 0 & 0 & \dfrac{s+1}{s} & -1 \\ 0 & 1 & 0 & 1 \end{bmatrix} \begin{bmatrix} V_1(s) \\ V_2(s) \\ I_1(s) \\ I_2(s) \end{bmatrix}$$

$$= \begin{bmatrix} 0 \\ 0 \\ \dfrac{1}{s} \\ 0 \end{bmatrix} \Rightarrow \begin{bmatrix} V_1(s) \\ V_2(s) \\ I_1(s) \\ I_2(s) \end{bmatrix} = \begin{bmatrix} 1 & 0 & -\dfrac{1}{s} & 1 \\ 0 & 1 & 1 & -\dfrac{1}{s+1} \\ 0 & 0 & \dfrac{s+1}{s} & -1 \\ 0 & 1 & 0 & 1 \end{bmatrix}^{-1} \begin{bmatrix} 0 \\ 0 \\ \dfrac{1}{s} \\ 0 \end{bmatrix}$$

MATLAB can be employed to perform the calculations (Fig. 3.64).

```
Command Window                                                    ⊙
  >> syms s
  >> inv([1 0 -1/s 1;0 1 1 -1/(s+1);0 0 (s+1)/s -1;0 1 0 1])*[0;0;1/s;0]


  ans =

   -(s^2 - 2)/(2*s*(s + 1))
                      -1/2
   (s^2 + 2*s)/(2*s*(s + 1))
                       1/2

fx >>
```

Fig. 3.64 MATLAB code

According to the results shown in Fig. 3.64,

$$
\begin{bmatrix} V_1(s) \\ V_2(s) \\ I_1(s) \\ I_2(s) \end{bmatrix} = \begin{bmatrix} -\dfrac{s^2 - 2}{2s(s+1)} \\ -\dfrac{1}{2} \\ \dfrac{s^2 + 2s}{2s(s+1)} \\ \dfrac{1}{2} \end{bmatrix}
$$

Therefore,

$$
V_o(s) = V_2(s) = -\frac{1}{2} \Rightarrow v_0(t) = \mathcal{L}^{-1}\left\{-\frac{1}{2}\right\} = -\frac{1}{2}\delta(t)
$$

3.2.20 Example 3.20

Determine the hybrid parameters (h parameters) of the circuit depicted in Fig. 3.65.

Fig. 3.65 Circuit for
Example 3.20

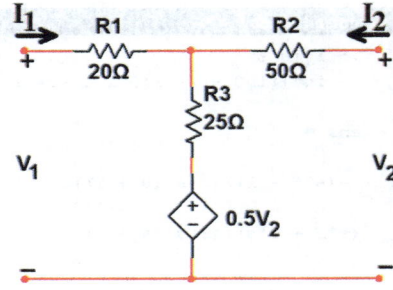

Solution:

For analysis, we will connect two sources to the input and output terminals of the circuit. The type of source (voltage or current) is not restricted. Figures 3.66, 3.67, 3.68, and 3.69 all show acceptable circuits.

Fig. 3.66 Input-output
ports are driven by two
voltage sources

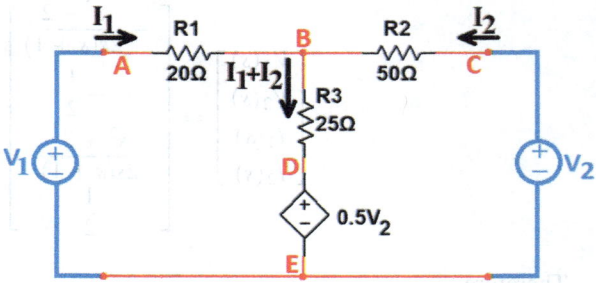

Fig. 3.67 Input-output
ports are driven by a current
source and a voltage source

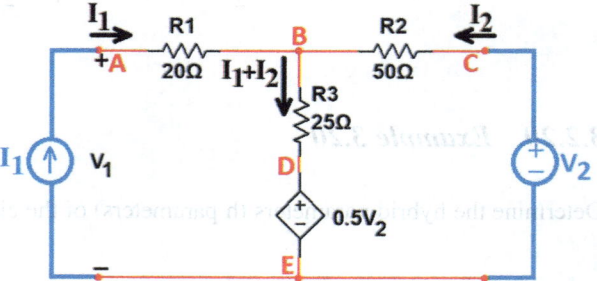

Fig. 3.68 Input-output ports are driven by two current sources

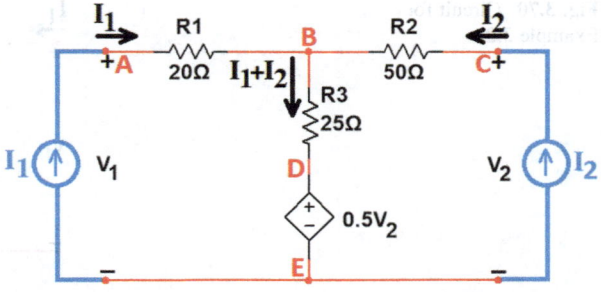

Fig. 3.69 Input-output ports are driven by a voltage source and a current source

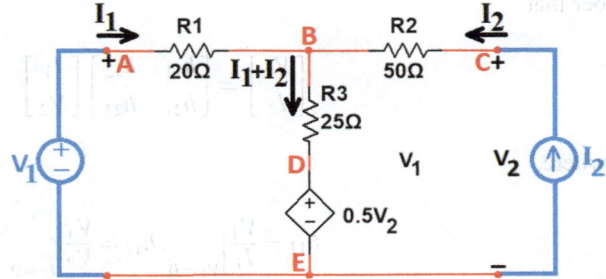

From Figs. 3.66, 3.67, 3.68, or 3.69, we have

$$KVL@EABDE: \ -V_1 + 20I_1 + 25(I_1 + I_2) + 0.5V_2 = 0 \Longrightarrow -V_1 + 25I_2 = -45I_1 - 0.5V_2$$

$$KVL@CBAEC: 50I_2 - 20I_1 + V_1 - V_2 = 0 \Longrightarrow 50I_2 + V_1 = 20I_1 + V_2$$

$$\begin{cases} -V_1 + 25I_2 = -45I_1 - 0.5V_2 \\ 50I_2 + V_1 = 20I_1 + V_2 \end{cases} \Longrightarrow \begin{bmatrix} -1 & 25 \\ 1 & 50 \end{bmatrix} \begin{bmatrix} V_1 \\ I_2 \end{bmatrix} = \begin{bmatrix} -45 & -0.5 \\ 20 & 1 \end{bmatrix}$$

$$\times \begin{bmatrix} I_1 \\ V_2 \end{bmatrix} \Longrightarrow \begin{bmatrix} V_1 \\ I_2 \end{bmatrix} = \begin{bmatrix} -1 & 25 \\ 1 & 50 \end{bmatrix}^{-1} \begin{bmatrix} -45 & -0.5 \\ 20 & 1 \end{bmatrix} \begin{bmatrix} I_1 \\ V_2 \end{bmatrix} \Longrightarrow \begin{bmatrix} V_1 \\ I_2 \end{bmatrix}$$

$$= \begin{bmatrix} 36.6667 & 0.6667 \\ -0.3333 & 0.0067 \end{bmatrix} \begin{bmatrix} I_1 \\ V_2 \end{bmatrix} \Longrightarrow h = \begin{bmatrix} 36.6667 & 0.6667 \\ -0.3333 & 0.0067 \end{bmatrix}$$

3.2.21 Example 3.21

Determine the hybrid parameters (h parameters) of the circuit depicted in Fig. 3.70.

Fig. 3.70 Circuit for
Example 3.21

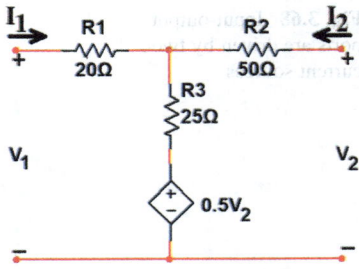

Solution:
In this example, we will rely on the direct definition of hybrid parameters. Remember that

$$\begin{bmatrix} V_1 \\ I_2 \end{bmatrix} = \begin{bmatrix} h_{11} & h_{12} \\ h_{21} & h_{22} \end{bmatrix} \begin{bmatrix} I_1 \\ V_2 \end{bmatrix}$$

where,

$$h_{11} = \frac{V_1}{I_1}\bigg|_{V_2=0}, h_{12} = \frac{V_1}{V_2}\bigg|_{I_1=0}$$

$$h_{21} = \frac{I_2}{I_1}\bigg|_{V_2=0}, h_{22} = \frac{I_2}{V_2}\bigg|_{I_1=0}$$

- Calculation of h_{11} and h_{21}

  $h_{11} = \frac{V_1}{I_1}\big|_{V_2=0}$ and $h_{21} = \frac{I_2}{I_1}\big|_{V_2=0}$ is calculated using the circuit shown in
  Fig. 3.71.

Fig. 3.71 Circuit for
calculating h_{11} and h_{21}

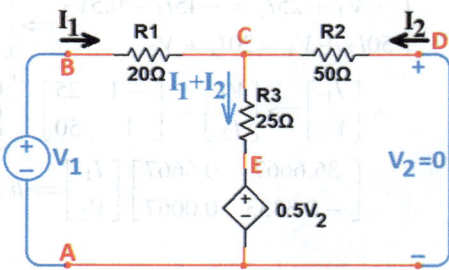

The circuit in Fig. 3.71 simplifies to the circuit shown in Fig. 3.72.

Fig. 3.72 Equivalent circuit for Fig. 3.71

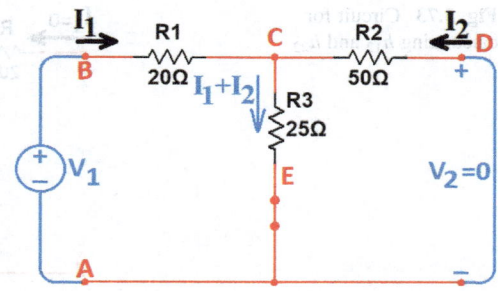

From Fig. 3.72, we have

$$KVL@ABCEA: -V_1 + 20I_1 + 25(I_1 + I_2) = 0 \Longrightarrow 45I_1 + 25I_2 = V_1$$

$$KVL@DCED: 50I_2 + 25(I_1 + I_2) = 0 \Longrightarrow 2I_2 + (I_1 + I_2) = 0 \Longrightarrow I_1 + 3I_2 = 0$$

$$\begin{cases} 45I_1 + 25I_2 = V_1 \\ I_1 + 3I_2 = 0 \end{cases} \Rightarrow \begin{bmatrix} 45 & 25 \\ 1 & 3 \end{bmatrix}\begin{bmatrix} I_1 \\ I_2 \end{bmatrix} = \begin{bmatrix} V_1 \\ 0 \end{bmatrix} \Rightarrow I_1 = \frac{\begin{vmatrix} V_1 & 25 \\ 0 & 3 \end{vmatrix}}{\begin{vmatrix} 45 & 25 \\ 1 & 3 \end{vmatrix}} \Rightarrow I_1$$

$$= \frac{V_1 \times 3 - 25 \times 0}{45 \times 3 - 25 \times 1} \Rightarrow I_1 = \frac{3V_1}{110} \Rightarrow \frac{V_1}{I_1} = \frac{110}{3} \Rightarrow \frac{V_1}{I_1} = 36.6667 \Rightarrow h_{11}$$

$$= 36.6667\ \Omega$$

$$\begin{cases} 45I_1 + 25I_2 = V_1 \\ I_1 + 3I_2 = 0 \end{cases} \Rightarrow \frac{I_2}{I_1} = \frac{\dfrac{\begin{vmatrix} 45 & V_1 \\ 1 & 0 \end{vmatrix}}{\begin{vmatrix} 45 & 25 \\ 1 & 3 \end{vmatrix}}}{\dfrac{\begin{vmatrix} V_1 & 25 \\ 0 & 3 \end{vmatrix}}{\begin{vmatrix} 45 & 25 \\ 1 & 3 \end{vmatrix}}} \Rightarrow \frac{I_2}{I_1} = \frac{\begin{vmatrix} 45 & V_1 \\ 1 & 0 \end{vmatrix}}{\begin{vmatrix} V_1 & 25 \\ 0 & 3 \end{vmatrix}} \Rightarrow \frac{I_2}{I_1}$$

$$= \frac{-V_1}{3V_1} \Rightarrow \frac{I_2}{I_1} = \frac{-1}{3} \Rightarrow \frac{I_2}{I_1} = -0.3333 \Rightarrow h_{21} = -0.3333$$

- Calculation of h_{12} and h_{22}

  $h_{12} = \left.\frac{V_1}{V_2}\right|_{I_1=0}$ and $h_{22} = \left.\frac{I_2}{V_2}\right|_{I_1=0}$ is calculated using the circuit shown in Fig. 3.73.

Fig. 3.73 Circuit for calculating h_{12} and h_{22}

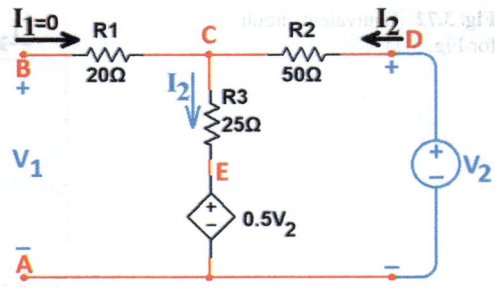

From Fig. 3.73, we have

$$I_1 = 0 \Rightarrow I_1 = \frac{V_B - V_C}{R_1} = 0 \Rightarrow V_B - V_C = 0 \Rightarrow V_B = V_C$$

$$V_1 = V_{BA} = V_{CA} = 25I_2 + 0.5V_2$$

$$KVL@ADCEA: \; -V_2 + 50I_2 + V_1 = 0 \Rightarrow -V_2 + 50I_2 + 25I_2 + 0.5V_2$$

$$= 0 \Rightarrow 75I_2 - 0.5V_2 = 0 \Rightarrow 75I_2 = 0.5V_2 \Rightarrow 150I_2 = V_2 \Rightarrow \frac{I_2}{V_2}$$

$$= \frac{1}{150} \Rightarrow h_{22} = 0.0067 \text{ S}$$

$$V_1 = V_{BA} = V_{CA} = 25I_2 + 0.5V_2 = 25\frac{V_2}{150} + 0.5V_2 \Rightarrow V_1 = 0.1667V_2$$

$$+ 0.5V_2 \Rightarrow V_1 = 0.6667V_2 \Rightarrow \frac{V_1}{V_2} = 0.6667 \Rightarrow h_{12} = 0.6667$$

Therefore,

$$h = \begin{bmatrix} h_{11} & h_{12} \\ h_{21} & h_{22} \end{bmatrix} = \begin{bmatrix} 36.6667 & 0.6667 \\ -0.3333 & 0.0067 \end{bmatrix}$$

3.2.22 Example 3.22

Determine the transmission parameters (T parameters) of the circuit depicted in Fig. 3.74.

Fig. 3.74 Circuit for
Example 3.22

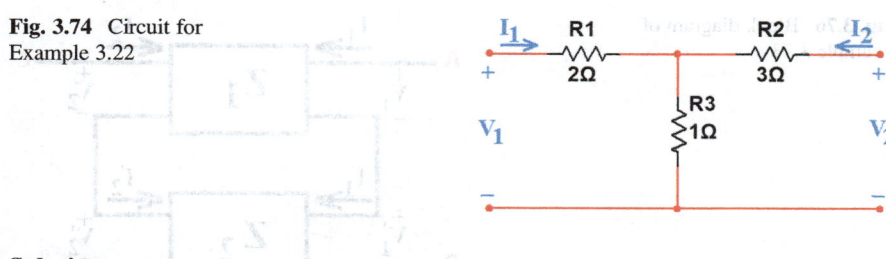

Solution:

Remember that T parameters have the following form: $\begin{bmatrix} V_1 \\ I_1 \end{bmatrix} = \begin{bmatrix} t_{11} & t_{21} \\ t_{21} & t_{22} \end{bmatrix} \begin{bmatrix} V_2 \\ -I_2 \end{bmatrix}$.

For analysis, we will connect two sources to the input and output terminals of the circuit (Fig. 3.75). The type of source (voltage or current) is not restricted.

Fig. 3.75 Input-output
ports are driven by two
voltage sources

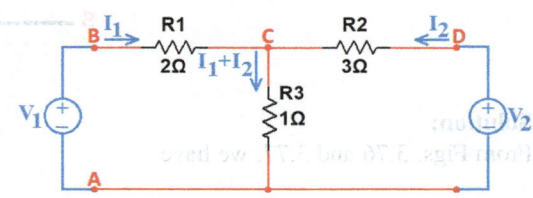

From Fig. 3.75, we have

$$KVL@ABCA: \ -V_1 + 2I_1 + 1(I_1 + I_2) = 0 \Longrightarrow V_1 - 3I_1 = I_2$$

$$KVL@DCAD: \ 3I_2 + 1(I_1 + I_2) - V_2 = 0 \Longrightarrow I_1 = V_2 - 4I_2$$

$$\begin{cases} V_1 - 3I_1 = I_2 \\ I_1 = V_2 - 4I_2 \end{cases} \Longrightarrow \begin{bmatrix} 1 & -3 \\ 0 & 1 \end{bmatrix} \begin{bmatrix} V_1 \\ I_1 \end{bmatrix} = \begin{bmatrix} 0 & -1 \\ 1 & 4 \end{bmatrix} \begin{bmatrix} V_2 \\ -I_2 \end{bmatrix} \Longrightarrow \begin{bmatrix} V_1 \\ I_1 \end{bmatrix}$$

$$= \begin{bmatrix} 1 & -3 \\ 0 & 1 \end{bmatrix}^{-1} \begin{bmatrix} 0 & -1 \\ 1 & 4 \end{bmatrix} \begin{bmatrix} V_2 \\ -I_2 \end{bmatrix} \Longrightarrow \begin{bmatrix} V_1 \\ I_1 \end{bmatrix} = \begin{bmatrix} 3 & 11 \\ 1 & 4 \end{bmatrix} \begin{bmatrix} V_2 \\ -I_2 \end{bmatrix} \Longrightarrow T$$

$$= \begin{bmatrix} 3 & 11 \\ 1 & 4 \end{bmatrix}$$

3.2.23 Example 3.23

Given two two-port networks with Z parameters Z_1 and Z_2 connected as in Fig. 3.76, find the equivalent Z parameters (Z_T) of the combined network (Fig. 3.77).

Fig. 3.76 Block diagram of Example 3.23

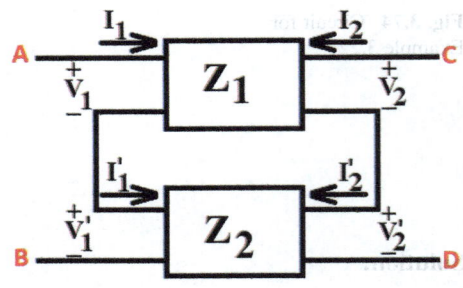

Fig. 3.77 Equivalent of the block diagram in Fig. 3.76

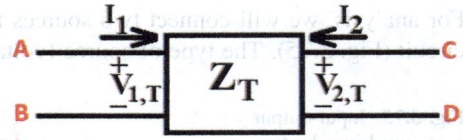

Solution:

From Figs. 3.76 and 3.77, we have

$$I_1 = I'_1$$

$$I_2 = I'_2$$

$$\begin{bmatrix} V_{1,T} \\ V_{2,T} \end{bmatrix} = Z_T \begin{bmatrix} I_1 \\ I_2 \end{bmatrix}$$

$$\begin{bmatrix} V_1 \\ V_2 \end{bmatrix} = Z_1 \begin{bmatrix} I_1 \\ I_2 \end{bmatrix}$$

$$\begin{bmatrix} V'_1 \\ V'_2 \end{bmatrix} = Z_2 \begin{bmatrix} I'_1 \\ I'_2 \end{bmatrix} \Rightarrow \begin{bmatrix} V'_1 \\ V'_2 \end{bmatrix} = Z_2 \begin{bmatrix} I_1 \\ I_2 \end{bmatrix}$$

$$\begin{cases} V_{1,T} = V_1 + V'_1 \\ V_{2,T} = V_2 + V'_2 \end{cases} \Rightarrow \begin{bmatrix} V_{1,T} \\ V_{2,T} \end{bmatrix} = \begin{bmatrix} V_1 + V'_1 \\ V_2 + V'_2 \end{bmatrix} \Rightarrow \begin{bmatrix} V_{1,T} \\ V_{2,T} \end{bmatrix} = \begin{bmatrix} V_1 \\ V_2 \end{bmatrix} + \begin{bmatrix} V'_1 \\ V'_2 \end{bmatrix}$$

$$\Rightarrow \begin{bmatrix} V_{1,T} \\ V_{2,T} \end{bmatrix} = Z_1 \begin{bmatrix} I_1 \\ I_2 \end{bmatrix} + Z_2 \begin{bmatrix} I_1 \\ I_2 \end{bmatrix} \Rightarrow \begin{bmatrix} V_{1,T} \\ V_{2,T} \end{bmatrix} = (Z_1 + Z_2) \begin{bmatrix} I_1 \\ I_2 \end{bmatrix} \Rightarrow Z_T = Z_1 + Z_2$$

3.2.24   Example 3.24

Given two two-port networks with Y parameters Y_1 and Y_2 connected as in Fig. 3.78, find the equivalent Y parameters (Y_T) of the combined network (Fig. 3.79).

Fig. 3.78 Block diagram of Example 3.24

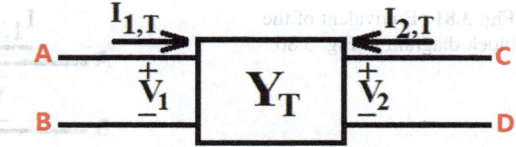

Fig. 3.79 Equivalent of the block diagram in Fig. 3.78

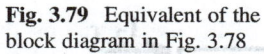

Solution:

From Figs. 3.78 and 3.79, we have

$$V_1 = V'_1$$

$$V_2 = V'_2$$

$$\begin{bmatrix} I_{1,T} \\ I_{2,T} \end{bmatrix} = Y_T \begin{bmatrix} V_1 \\ V_2 \end{bmatrix}$$

$$\begin{bmatrix} I_1 \\ I_2 \end{bmatrix} = Y_1 \begin{bmatrix} V_1 \\ V_2 \end{bmatrix}$$

$$\begin{bmatrix} I'_1 \\ I'_2 \end{bmatrix} = Y_2 \begin{bmatrix} V'_1 \\ V'_2 \end{bmatrix} = Y_2 \begin{bmatrix} V_1 \\ V_2 \end{bmatrix}$$

$$\begin{cases} I_{1,T} = I_1 + I'_1 \\ I_{2,T} = I_2 + I'_2 \end{cases} \Rightarrow \begin{bmatrix} I_{1,T} \\ I_{2,T} \end{bmatrix} = \begin{bmatrix} I_1 + I'_1 \\ I_2 + I'_2 \end{bmatrix} \Rightarrow \begin{bmatrix} I_{1,T} \\ I_{2,T} \end{bmatrix} = \begin{bmatrix} I_1 \\ I_2 \end{bmatrix} + \begin{bmatrix} I'_1 \\ I'_2 \end{bmatrix} \Rightarrow \begin{bmatrix} I_{1,T} \\ I_{2,T} \end{bmatrix}$$

$$= Y_1 \begin{bmatrix} V_1 \\ V_2 \end{bmatrix} + Y_2 \begin{bmatrix} V_1 \\ V_2 \end{bmatrix} \Rightarrow \begin{bmatrix} I_{1,T} \\ I_{2,T} \end{bmatrix} = (Y_1 + Y_2) \begin{bmatrix} V_1 \\ V_2 \end{bmatrix} \Rightarrow Y_T = Y_1 + Y_2$$

3.2.25 *Example 3.25*

Given two two-port networks with Y parameters Y_1 and Y_2 connected as in Fig. 3.80, find the equivalent Y parameters (Y_T) of the combined network (Fig. 3.81).

Fig. 3.80 Block diagram of Example 3.25

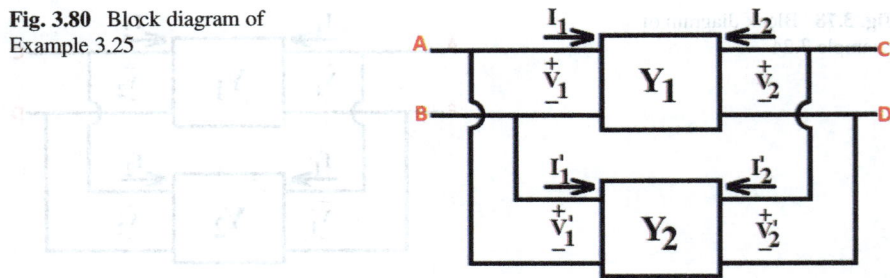

Fig. 3.81 Equivalent of the block diagram in Fig. 3.80

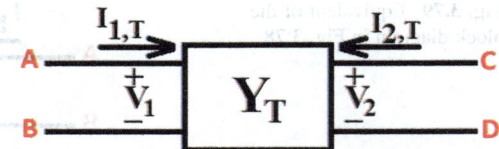

Solution:

From Figs. 3.80 and 3.81, we have

$$V_1 = -V'_1$$

$$V_2 = V'_2$$

$$\begin{bmatrix} I_{1,T} \\ I_{2,T} \end{bmatrix} = Y_T \begin{bmatrix} V_1 \\ V_2 \end{bmatrix}$$

$$\begin{bmatrix} I_1 \\ I_2 \end{bmatrix} = Y_1 \begin{bmatrix} V_1 \\ V_2 \end{bmatrix}$$

$$\begin{bmatrix} I'_1 \\ I'_2 \end{bmatrix} = Y_2 \begin{bmatrix} V'_1 \\ V'_2 \end{bmatrix} = Y_2 \begin{bmatrix} -V_1 \\ V_2 \end{bmatrix} = \begin{bmatrix} y_{2,11} & y_{2,12} \\ y_{2,21} & y_{2,22} \end{bmatrix} \begin{bmatrix} -V_1 \\ V_2 \end{bmatrix}$$

$$= \begin{bmatrix} -y_{2,11}V_1 + y_{2,12}V_2 \\ -y_{2,21}V_1 + y_{2,22}V_2 \end{bmatrix} = \begin{bmatrix} -y_{2,11} & y_{2,12} \\ -y_{2,21} & y_{2,22} \end{bmatrix} \begin{bmatrix} V_1 \\ V_2 \end{bmatrix} = Y_3 \begin{bmatrix} V_1 \\ V_2 \end{bmatrix}$$

$$\begin{cases} I_{1,T} = I_1 + I'_1 \\ I_{2,T} = I_2 + I'_2 \end{cases} \Rightarrow \begin{bmatrix} I_{1,T} \\ I_{2,T} \end{bmatrix} = \begin{bmatrix} I_1 + I'_1 \\ I_2 + I'_2 \end{bmatrix} \Rightarrow \begin{bmatrix} I_{1,T} \\ I_{2,T} \end{bmatrix} = \begin{bmatrix} I_1 \\ I_2 \end{bmatrix} + \begin{bmatrix} I'_1 \\ I'_2 \end{bmatrix} \Rightarrow \begin{bmatrix} I_{1,T} \\ I_{2,T} \end{bmatrix}$$

$$= Y_1 \begin{bmatrix} V_1 \\ V_2 \end{bmatrix} + Y_3 \begin{bmatrix} V_1 \\ V_2 \end{bmatrix} \Rightarrow \begin{bmatrix} I_{1,T} \\ I_{2,T} \end{bmatrix} = (Y_1 + Y_3) \begin{bmatrix} V_1 \\ V_2 \end{bmatrix} \Rightarrow Y_T = Y_1 + Y_3$$

where $Y_1 = \begin{bmatrix} y_{1,11} & y_{1,12} \\ y_{1,21} & y_{1,22} \end{bmatrix}$, $Y_2 = \begin{bmatrix} y_{2,11} & y_{2,12} \\ y_{2,21} & y_{2,22} \end{bmatrix}$ and $Y_3 = \begin{bmatrix} -y_{2,11} & y_{2,12} \\ -y_{2,21} & y_{2,22} \end{bmatrix}$.

3.2.26 Example 3.26

Given two two-port networks with h parameters h_1 and h_2 connected as in Fig. 3.82, find the equivalent h parameters (h_T) of the combined network (Fig. 3.83).

Fig. 3.82 Block diagram of Example 3.26

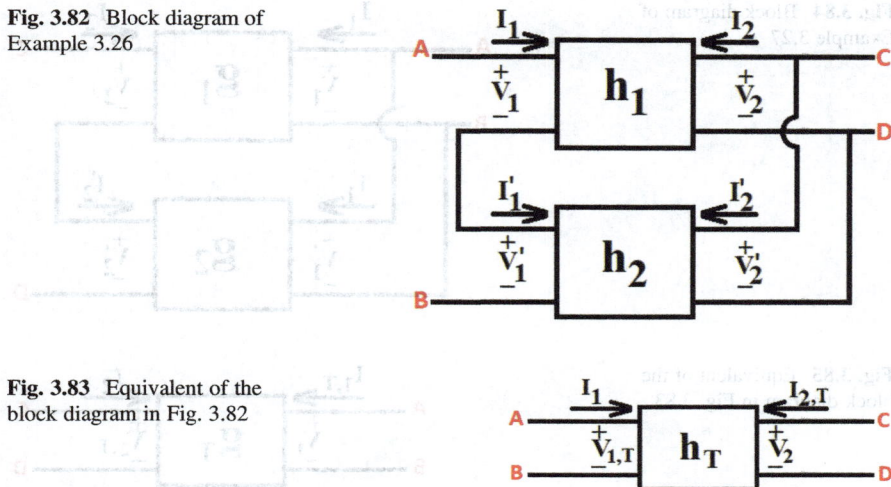

Fig. 3.83 Equivalent of the block diagram in Fig. 3.82

Solution:
From Figs. 3.82 and 3.83, we have

$$I_1 = -I'_1$$

$$V_2 = V'_2$$

$$\begin{bmatrix} V_{1,T} \\ I_{2,T} \end{bmatrix} = h_T \begin{bmatrix} I_1 \\ V_2 \end{bmatrix}$$

$$\begin{bmatrix} V_1 \\ I_2 \end{bmatrix} = h_1 \begin{bmatrix} I_1 \\ V_2 \end{bmatrix}$$

$$\begin{bmatrix} V'_1 \\ I'_2 \end{bmatrix} = h_2 \begin{bmatrix} I'_1 \\ V'_2 \end{bmatrix} = h_2 \begin{bmatrix} I_1 \\ V_2 \end{bmatrix}$$

$$\begin{cases} V_{1,T} = V_1 + V'_1 \\ I_{2,T} = I_2 + I'_2 \end{cases} \Rightarrow \begin{bmatrix} V_{1,T} \\ I_{2,T} \end{bmatrix} = \begin{bmatrix} V_1 + V'_1 \\ I_2 + I'_2 \end{bmatrix} \Rightarrow \begin{bmatrix} V_{1,T} \\ I_{2,T} \end{bmatrix} = \begin{bmatrix} V_1 \\ I_2 \end{bmatrix}$$

$$+ \begin{bmatrix} V'_1 \\ I'_2 \end{bmatrix} \Rightarrow \begin{bmatrix} V_{1,T} \\ I_{2,T} \end{bmatrix} = h_1 \begin{bmatrix} I_1 \\ V_2 \end{bmatrix} + h_2 \begin{bmatrix} I_1 \\ V_2 \end{bmatrix} \Rightarrow \begin{bmatrix} V_{1,T} \\ I_{2,T} \end{bmatrix} = (h_1 + h_2) \begin{bmatrix} I_1 \\ V_2 \end{bmatrix} \Rightarrow h_T$$

$$= h_1 + h_2$$

3.2.27 Example 3.27

Given two two-port networks with g parameters g_1 and g_2 connected as in Fig. 3.84, find the equivalent g parameters (g_T) of the combined network (Fig. 3.85).

Fig. 3.84 Block diagram of Example 3.27

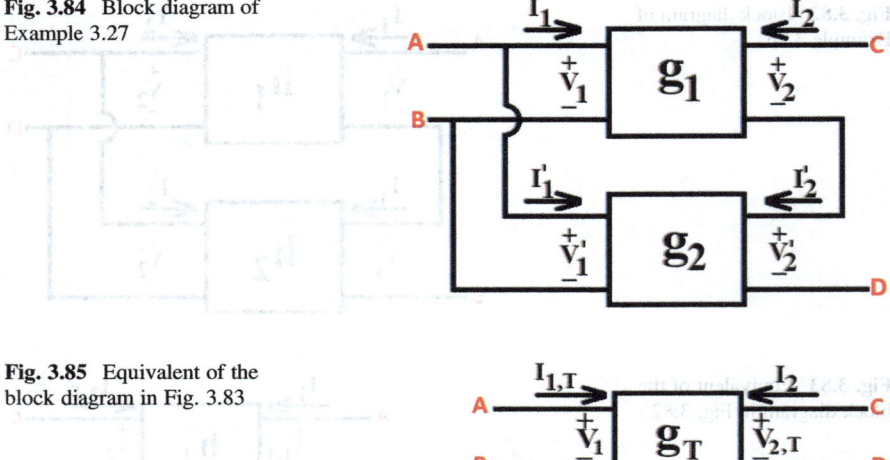

Fig. 3.85 Equivalent of the block diagram in Fig. 3.83

Solution:

From Figs. 3.84 and 3.85, we have

$$\begin{bmatrix} I_1 \\ V_2 \end{bmatrix} = \begin{bmatrix} g_{11} & g_{12} \\ g_{21} & g_{22} \end{bmatrix} \begin{bmatrix} V_1 \\ I_2 \end{bmatrix}$$

$$I_2 = I'_2$$

$$V_1 = V'_1$$

$$\begin{bmatrix} I_{1,T} \\ V_{2,T} \end{bmatrix} = g_T \begin{bmatrix} V_1 \\ I_2 \end{bmatrix}$$

$$\begin{bmatrix} I_1 \\ V_2 \end{bmatrix} = g_1 \begin{bmatrix} V_1 \\ I_2 \end{bmatrix}$$

$$\begin{bmatrix} I'_1 \\ V'_2 \end{bmatrix} = g_2 \begin{bmatrix} V'_1 \\ I'_2 \end{bmatrix} = g_2 \begin{bmatrix} V_1 \\ I_2 \end{bmatrix}$$

$$\begin{cases} I_{1,T} = I_1 + I'_1 \\ V_{2,T} = V_2 + V'_2 \end{cases} \Rightarrow \begin{bmatrix} I_{1,T} \\ V_{2,T} \end{bmatrix} = \begin{bmatrix} I_1 + I'_1 \\ V_2 + V'_2 \end{bmatrix} \Rightarrow \begin{bmatrix} I_{1,T} \\ V_{2,T} \end{bmatrix} = \begin{bmatrix} I_1 \\ V_2 \end{bmatrix} + \begin{bmatrix} I'_1 \\ V'_2 \end{bmatrix}$$

$$\Rightarrow \begin{bmatrix} I_{1,T} \\ V_{2,T} \end{bmatrix} = g_1 \begin{bmatrix} V_1 \\ I_2 \end{bmatrix} + g_2 \begin{bmatrix} V_1 \\ I_2 \end{bmatrix} \Rightarrow \begin{bmatrix} I_{1,T} \\ V_{2,T} \end{bmatrix} = (g_1 + g_2) \begin{bmatrix} V_1 \\ I_2 \end{bmatrix} \Rightarrow g_T = g_1 + g_2$$

3.2.28 Exercise 3.1

Given two two-port networks with t parameters t_1 and t_2 connected as in Fig. 3.86, find the equivalent t parameters (t_T) of the combined network (Fig. 3.87).

Fig. 3.86 Block diagram of Exercise 3.1

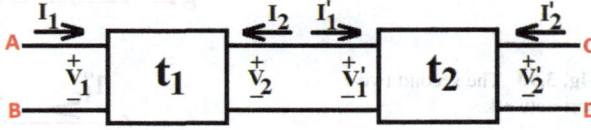

Fig. 3.87 Equivalent of the block diagram in Fig. 3.86

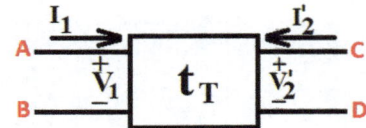

3.2.29 Example 3.28

Determine the admittance parameters (Y parameters) of the circuit depicted in Fig. 3.88.

Fig. 3.88 Circuit for Example 3.28

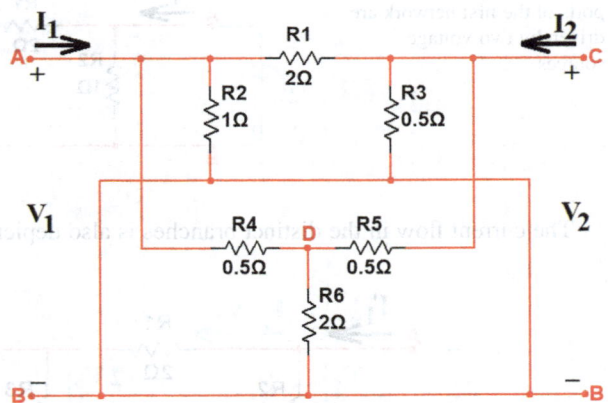

Solution:

Figure 3.88 can be viewed as a parallel-parallel combination of the two networks depicted in Figs. 3.89 and 3.90.

Fig. 3.89 The first two-port network

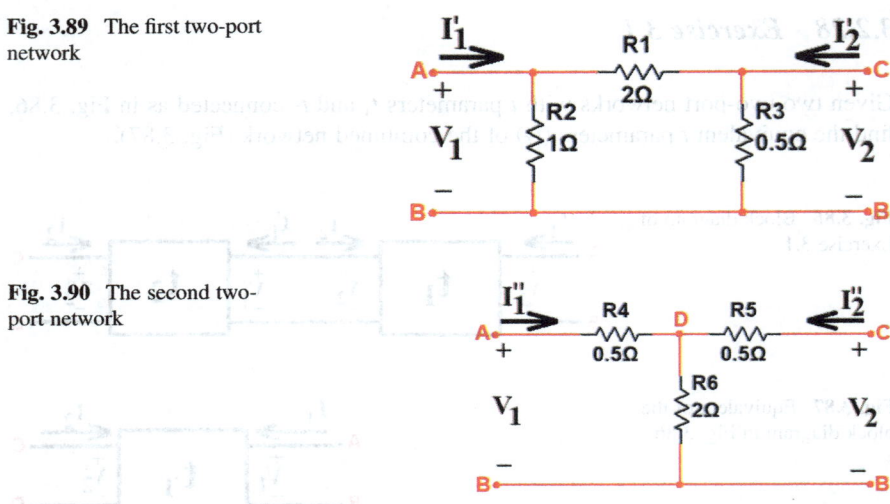

Fig. 3.90 The second two-port network

We will now determine the Y parameters of the two-port network in Fig. 3.90. For the purpose of analysis, let us connect two sources to the input and output terminals of the circuit (Fig. 3.91). The input and output terminals may be connected to current sources, if desired.

Fig. 3.91 Input-output ports of the first network are driven by two voltage sources

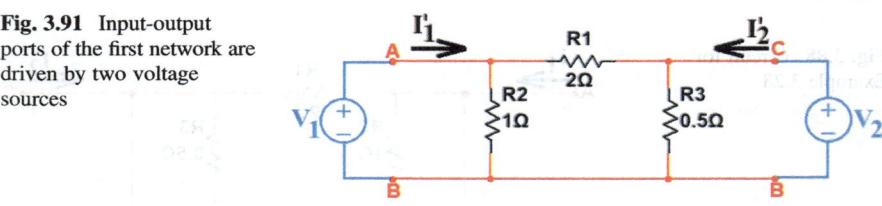

The current flow in the distinct branches is also depicted in Fig. 3.92.

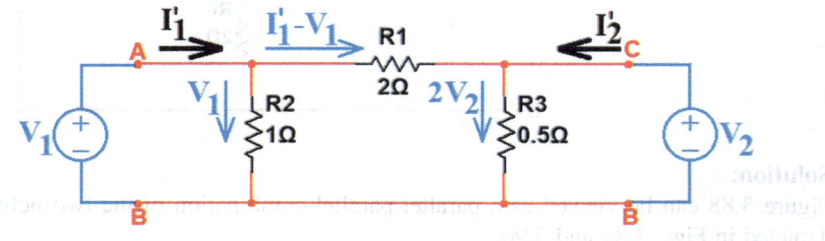

Fig. 3.92 The current flow in the distinct branches

From Fig. 3.92, we have

$$KVL@BACB: \ -V_1 + 2(I_1' - V_1) + V_2 = 0 \Rightarrow 2I_1' = 3V_1 - V_2$$

$$KCL@C: \ I'_2 + I'_1 - V_1 = 2V_2 \Rightarrow I_2' + I_1' = V_1 + 2V_2$$

$$\begin{cases} 2I_1' = 3V_1 - V_2 \\ I_1' + I_2' = V_1 + 2V_2 \end{cases} \Rightarrow \begin{bmatrix} 2 & 0 \\ 1 & 1 \end{bmatrix} \begin{bmatrix} I'_1 \\ I'_2 \end{bmatrix} = \begin{bmatrix} 3 & -1 \\ 1 & 2 \end{bmatrix} \begin{bmatrix} V_1 \\ V_2 \end{bmatrix} \Rightarrow \begin{bmatrix} I'_1 \\ I'_2 \end{bmatrix}$$

$$= \begin{bmatrix} 2 & 0 \\ 1 & 1 \end{bmatrix}^{-1} \begin{bmatrix} 3 & -1 \\ 1 & 2 \end{bmatrix} \begin{bmatrix} V_1 \\ V_2 \end{bmatrix} \Rightarrow \begin{bmatrix} I'_1 \\ I'_2 \end{bmatrix} = \begin{bmatrix} 0.5 & 0 \\ -0.5 & 1 \end{bmatrix} \begin{bmatrix} 3 & -1 \\ 1 & 2 \end{bmatrix}$$

$$\times \begin{bmatrix} V_1 \\ V_2 \end{bmatrix} \Rightarrow \begin{bmatrix} I'_1 \\ I'_2 \end{bmatrix} = \begin{bmatrix} 1.5 & -0.5 \\ -0.5 & 2.5 \end{bmatrix} \begin{bmatrix} V_1 \\ V_2 \end{bmatrix}$$

The analysis of the first network is complete. We will now analyze the second network (Fig. 3.90). For the purpose of analysis, let us connect two sources to the input and output terminals of the circuit (Fig. 3.93). The input and output terminals may be connected to current sources, if desired.

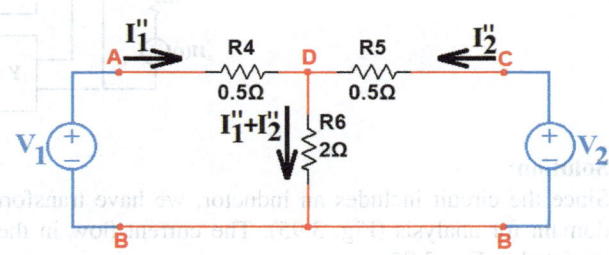

Fig. 3.93 Input-output ports of the second network are driven by two voltage sources

From Fig. 3.93, we have

$$KVL@BADB: \ -V_1 + 0.5I''_1 + 2(I''_1 + I''_2) = 0 \Rightarrow 2.5I''_1 + 2I''_2 = V_1$$

$$KVL@BCDB: \ -V_2 + 0.5I''_2 + 2(I''_1 + I''_2) = 0 \Rightarrow 2I''_1 + 2.5I''_2 = V_2$$

$$\begin{cases} 2.5I''_1 + 2I''_2 = V_1 \\ 2I''_1 + 2.5I''_2 = V_2 \end{cases} \Rightarrow \begin{bmatrix} 2.5 & 2 \\ 2 & 2.5 \end{bmatrix} \begin{bmatrix} I''_1 \\ I''_2 \end{bmatrix} = \begin{bmatrix} V_1 \\ V_2 \end{bmatrix} \Rightarrow \begin{bmatrix} I''_1 \\ I''_2 \end{bmatrix}$$

$$= \begin{bmatrix} 2.5 & 2 \\ 2 & 2.5 \end{bmatrix}^{-1} \begin{bmatrix} V_1 \\ V_2 \end{bmatrix} \Rightarrow \begin{bmatrix} I''_1 \\ I''_2 \end{bmatrix} = \begin{bmatrix} 1.1111 & -0.8889 \\ -0.8889 & 1.1111 \end{bmatrix} \begin{bmatrix} V_1 \\ V_2 \end{bmatrix}$$

Therefore,

$$\begin{bmatrix} I_1 \\ I_2 \end{bmatrix} = \begin{bmatrix} I'_1 \\ I'_2 \end{bmatrix} + \begin{bmatrix} I''_1 \\ I''_2 \end{bmatrix} \Rightarrow \begin{bmatrix} I_1 \\ I_2 \end{bmatrix} = \begin{bmatrix} 1.5 & -0.5 \\ -0.5 & 2.5 \end{bmatrix} \begin{bmatrix} V_1 \\ V_2 \end{bmatrix}$$

$$+ \begin{bmatrix} 1.1111 & -0.8889 \\ -0.8889 & 1.1111 \end{bmatrix} \begin{bmatrix} V_1 \\ V_2 \end{bmatrix} \Rightarrow \begin{bmatrix} I_1 \\ I_2 \end{bmatrix} = \begin{bmatrix} 2.6111 & -1.3889 \\ -1.3889 & 3.6111 \end{bmatrix}$$

$$\times \begin{bmatrix} V_1 \\ V_2 \end{bmatrix} \Rightarrow Y = \begin{bmatrix} 2.6111 & -1.3889 \\ -1.3889 & 3.6111 \end{bmatrix}$$

3.2.30 Example 3.29

Find the capacitor voltage in the circuit of Fig. 3.94, given that its initial voltage is 0 V.

Fig. 3.94 Circuit for
Example 3.29

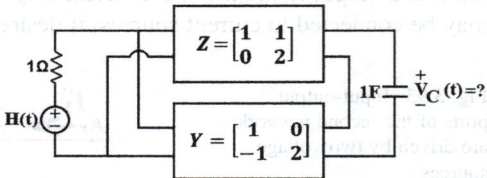

Solution:
Since the circuit includes an inductor, we have transformed it into the frequency domain for analysis (Fig. 3.95). The current flow in the distinct branches is also depicted in Fig. 3.95.

Fig. 3.95 The current flow
in the distinct branches

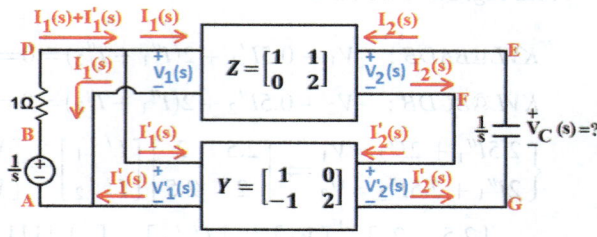

From Fig. 3.95, we have

$$KCL@F : I_2(s) = I'_2(s)$$

$$V_C(s) = V_2(s) + V'_2(s) = -I_2(s) \times \frac{1}{s} = -I'_2(s) \times \frac{1}{s}$$

$$\begin{bmatrix} V_1(s) \\ V_2(s) \end{bmatrix} = \begin{bmatrix} 1 & 1 \\ 0 & 2 \end{bmatrix}\begin{bmatrix} I_1(s) \\ I_2(s) \end{bmatrix} \Rightarrow \begin{cases} V_1(s) = I_1(s) + I_2(s) \\ V_2(s) = 2I_2(s) \end{cases} \Rightarrow \begin{cases} V_1(s) - I_1(s) - I_2(s) = 0 \\ V_2(s) - 2I_2(s) = 0 \end{cases}$$

$$\begin{bmatrix} I_1'(s) \\ I_2'(s) \end{bmatrix} = \begin{bmatrix} 1 & 0 \\ -1 & 2 \end{bmatrix}\begin{bmatrix} V_1'(s) \\ V_2'(s) \end{bmatrix} \Rightarrow \begin{cases} I_1'(s) = V_1'(s) \\ I_2'(s) = -V_1'(s) + 2V_2'(s) \end{cases}$$

$$\Rightarrow \begin{cases} I_1'(s) - V_1'(s) = 0 \\ I_2'(s) + V_1'(s) - 2V_2'(s) = 0 \end{cases} \Rightarrow \begin{cases} I_1'(s) - V_1'(s) = 0 \\ I_2(s) + V_1'(s) - 2V_2'(s) = 0 \end{cases}$$

$$KVL@ABDA: \quad -\frac{1}{s} + I_1(s) + I_1'(s) + V_1'(s) = 0$$

$$KVL@GEFG: \quad -V_C(s) + V_2(s) + V_2'(s) = 0 \Rightarrow \frac{I_2(s)}{s} + V_2(s) + V_2'(s) = 0$$

$$\begin{cases} V_1(s) - I_1(s) - I_2(s) = 0 \\ V_2(s) - 2I_2(s) = 0 \\ I_1'(s) - V_1'(s) = 0 \\ I_2'(s) + V_1'(s) - 2V_2'(s) = 0 \\ -\frac{1}{s} + I_1(s) + I_1'(s) + V_1'(s) = 0 \\ \frac{I_2(s)}{s} + V_2(s) + V_2'(s) = 0 \end{cases} \Rightarrow \begin{bmatrix} 1 & 0 & 0 & -1 & -1 & 0 \\ 0 & 1 & 0 & 0 & -2 & 0 \\ -1 & 0 & 0 & 0 & 0 & 1 \\ 1 & 0 & -2 & 0 & 1 & 0 \\ 1 & 0 & 0 & 1 & 0 & 1 \\ 0 & 1 & 1 & 0 & \frac{1}{s} & 0 \end{bmatrix}\begin{bmatrix} V_1(s) \\ V_2(s) \\ V_2'(s) \\ I_1(s) \\ I_2(s) \\ I_1'(s) \end{bmatrix}$$

$$\begin{bmatrix} 0 \\ 0 \\ 0 \\ 0 \\ \frac{1}{s} \\ 0 \end{bmatrix} = \begin{bmatrix} V_1(s) \\ V_2(s) \\ V_2'(s) \\ I_1(s) \\ I_2(s) \\ I_1'(s) \end{bmatrix} \Rightarrow = \begin{bmatrix} 1 & 0 & 0 & -1 & -1 & 0 \\ 0 & 1 & 0 & 0 & -2 & 0 \\ -1 & 0 & 0 & 0 & 0 & 1 \\ 1 & 0 & -2 & 0 & 1 & 0 \\ 1 & 0 & 0 & 1 & 0 & 1 \\ 0 & 1 & 1 & 0 & \frac{1}{s} & 0 \end{bmatrix}^{-1}\begin{bmatrix} 0 \\ 0 \\ 0 \\ 0 \\ \frac{1}{s} \\ 0 \end{bmatrix}$$

MATLAB can be employed to perform the calculations (Fig. 3.96).

```
Command Window
>> syms s
>> inv([1 0 0 -1 -1 0;0 1 0 0 -2 0;-1 0 0 0 0 1;1 0 -2 0 1 0;1 0 0 1 0 1;0 1 1 0 1/s 0])*[0;0;0;0;1/s;0]

ans =

(5*s + 2)/(2*s*(8*s + 3))
           -1/(8*s + 3)
(2*s + 1)/(2*s*(8*s + 3))
  (3*s + 1)/(s*(8*s + 3))
         -1/(2*(8*s + 3))
(5*s + 2)/(2*s*(8*s + 3))

fx >> |
```

Fig. 3.96 MATLAB code

According to the results shown in Fig. 3.96,

$$
\begin{bmatrix} V_1(s) \\ V_2(s) \\ V_2'(s) \\ I_1(s) \\ I_2(s) \\ I_1'(s) \end{bmatrix} = \begin{bmatrix} \dfrac{5s+2}{2s(8s+3)} \\ -\dfrac{1}{8s+3} \\ \dfrac{2s+1}{2s(8s+3)} \\ \dfrac{3s+1}{s(8s+3)} \\ \dfrac{-1}{2(8s+3)} \\ \dfrac{5s+2}{2s(8s+3)} \end{bmatrix}
$$

Therefore,

$$
V_2(s) = -\frac{1}{8s+3}
$$

$$
V'_2(s) = \frac{2s+1}{2s(8s+3)}
$$

$$
V_C(s) = V_2(s) + V_2'(s) = -\frac{1}{8s+3} + \frac{2s+1}{2s(8s+3)} = \frac{-2s+2s+1}{2s(8s+3)} = \frac{1}{2s(8s+3)}
$$

$$
= \frac{1}{16s\left(s+\frac{3}{8}\right)} = \frac{\frac{1}{6}}{s} + \frac{-\frac{1}{6}}{s+\frac{3}{8}}
$$

$$
v_C(t) = \mathcal{L}^{-1}\left\{ \frac{\frac{1}{6}}{s} + \frac{-\frac{1}{6}}{s+\frac{3}{8}} \right\} = \frac{1}{6}\left(1 - e^{-\frac{3}{8}t}\right) H(t)
$$

3.2.31   *Example 3.30*

Given a two-port network with Z parameters $\begin{bmatrix} 2s & 4 \\ 8 & 3s \end{bmatrix}$ and a coupled inductor connected as in Fig. 3.97, calculate the Z matrix of the combined network (Fig. 3.98).

Fig. 3.97 Circuit for
Example 3.30

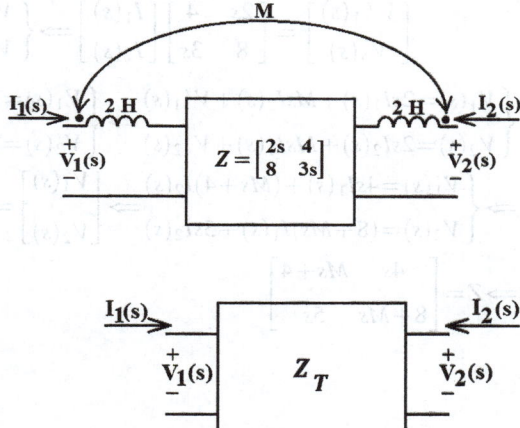

Fig. 3.98 Equivalent of the
block diagram in Fig. 3.97

Solution:

Let us connect two sources (voltage or current) to the input and output terminals of
the circuit (Figs. 3.99 and 3.100). Since the circuit includes an inductor, we have
transformed it into the frequency domain for analysis.

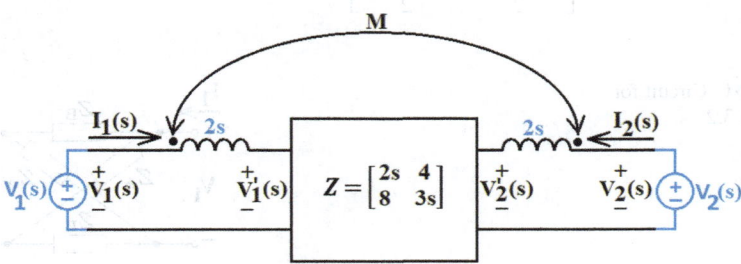

Fig. 3.99 Input-output ports are driven by two voltage sources

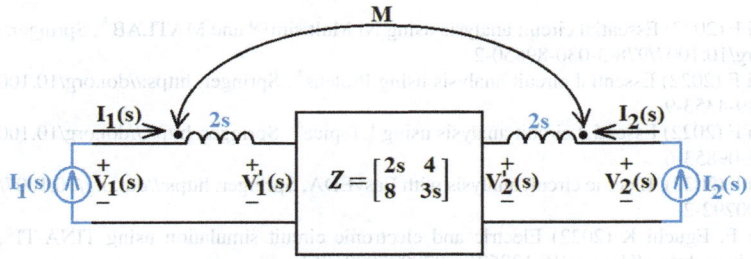

Fig. 3.100 Input-output ports are driven by two current sources

Continue the analysis using your choice of circuit from either Fig. 3.99 or
Fig. 3.100. Let us continue the analysis with the circuit in Fig. 3.100.

$$\begin{bmatrix} V'_1(s) \\ V'_2(s) \end{bmatrix} = \begin{bmatrix} 2s & 4 \\ 8 & 3s \end{bmatrix} \begin{bmatrix} I_1(s) \\ I_2(s) \end{bmatrix} \Rightarrow \begin{cases} V'_1(s) = 2sI_1(s) + 4I_2(s) \\ V'_2(s) = 8I_1(s) + 3sI_2(s) \end{cases}$$

$$\begin{cases} V_1(s) = 2sI_1(s) + MsI_2(s) + V'_1(s) \\ V_2(s) = 2sI_2(s) + MsI_1(s) + V'_2(s) \end{cases} \Rightarrow \begin{cases} V_1(s) = 2sI_1(s) + MsI_2(s) + 2sI_1(s) + 4I_2(s) \\ V_2(s) = 2sI_2(s) + MsI_1(s) + 8I_1(s) + 3sI_2(s) \end{cases}$$

$$\Rightarrow \begin{cases} V_1(s) = 4sI_1(s) + (Ms+4)I_2(s) \\ V_2(s) = (8+Ms)I_1(s) + 5sI_2(s) \end{cases} \Rightarrow \begin{bmatrix} V_1(s) \\ V_2(s) \end{bmatrix} = \begin{bmatrix} 4s & Ms+4 \\ 8+Ms & 5s \end{bmatrix} \begin{bmatrix} I_1(s) \\ I_2(s) \end{bmatrix}$$

$$\Rightarrow Z = \begin{bmatrix} 4s & Ms+4 \\ 8+Ms & 5s \end{bmatrix}$$

3.2.32   Exercise 3.2

Determine the impedance parameters (Z parameters) of the circuit depicted in Fig. 3.101. (Ans. $Z = \begin{bmatrix} \dfrac{Z_a+Z_b}{2} & \dfrac{Z_a-Z_b}{2} \\ \dfrac{Z_a-Z_b}{2} & \dfrac{Z_a+Z_b}{2} \end{bmatrix}$)

Fig. 3.101 Circuit for
Exercise 3.2

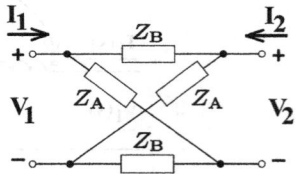

Further Reading

1. Asadi F (2022) Essential circuit analysis using NI Multisim™ and MATLAB®, Springer. https://doi.org/10.1007/978-3-030-89850-2
2. Asadi F (2022) Essential circuit analysis using Proteus®, Springer. https://doi.org/10.1007/978-981-19-4353-9
3. Asadi F (2022) Essential circuit analysis using LTspice®, Springer. https://doi.org/10.1007/978-3-031-09853-6
4. Asadi F (2022), Electric circuit analysis with EasyEDA, Springer. https://doi.org/10.1007/978-3-031-00292-2
5. Asadi F, Eguchi K (2022) Electric and electronic circuit simulation using TINA-TI®, River Publishers. https://doi.org/10.13052/rp-9788770226851
6. Asadi F (2023) Analog electronic circuits laboratory manual, Springer. https://doi.org/10.1007/978-3-031-25122-1

Chapter 4
State-Space Model of Electrical Circuits

4.1 Introduction

A state-space model is a mathematical representation of a physical system as a set of input, output, and state variables related by first-order differential or difference equations. It provides a powerful and flexible framework for analyzing and designing complex dynamic systems.

The linear state-space equation, representing the dynamic behavior of a system, is expressed as follows:

$$\begin{cases} \dot{x}(t) = Ax(t) + Bu(t) \\ y(t) = Cx(t) + Du(t) \end{cases}$$

where,

$x(t)$ is the state vector ($n \times 1$),

$\dot{x}(t)$ is the time derivative of the state vector ($n \times 1$),

$u(t)$ is the input vector ($r \times 1$),

$y(t)$ is the output vector ($m \times 1$),

A is the state matrix ($n \times n$),

B is the input matrix ($n \times r$),

C is the output matrix ($m \times n$),

D is the direct transmission (or feedthrough) matrix ($m \times r$).

For single-input single-output (SISO) systems, $r = m = 1$.

The transfer function associated with the state-space representation: $\begin{cases} \dot{x}(t) = Ax(t) + Bu(t) \\ y(t) = Cx(t) + Du(t) \end{cases}$ is given by $T(s) = \frac{Y(s)}{U(s)} = C(sI_{n \times n} - A)^{-1}B + D$.

The Laplace transform can be applied to solve the state-space equations: $\begin{cases} \dot{x}(t) = Ax(t) + Bu(t) \\ y(t) = Cx(t) + Du(t) \end{cases}$ as follows:

F. Asadi, *A Problem-Solving Approach to Electric Circuits*,
https://doi.org/10.1007/978-3-031-95493-1_4

$$\dot{x}(t) = Ax(t) + Bu(t) \Longrightarrow L\{\dot{x}(t)\} = L\{Ax(t) + Bu(t)\} \Longrightarrow sI_{n \times n}.X(s) - x(0)$$
$$= AX(s) + BU(s) \Longrightarrow sI_{n \times n}.X(s) - AX(s) = x(0) + BU(s) \Longrightarrow (sI_{n \times n} - A)X(s)$$
$$= x(0) + BU(s) \Longrightarrow X(s) = (sI_{n \times n} - A)^{-1}(x(0) + BU(s)) \Longrightarrow x(t)$$
$$= L^{-1}\left\{(sI_{n \times n} - A)^{-1}(x(0) + BU(s))\right\}$$

$$y(t) = Cx(t) + Du(t) \Longrightarrow y(t) = CL^{-1}\left\{(sI_{n \times n} - A)^{-1}(x(0) + BU(s))\right\} + Du(t)$$

Therefore,

$$\begin{cases} \dot{x}(t) = Ax(t) + Bu(t) \\ y(t) = Cx(t) + Du(t) \end{cases} \Longrightarrow \begin{cases} x(t) = L^{-1}\left\{(sI_{n \times n} - A)^{-1}(x(0) + BU(s))\right\} \\ y(t) = CL^{-1}\left\{(sI_{n \times n} - A)^{-1}(x(0) + BU(s))\right\} + Du(t) \end{cases}$$

With B and D equal to zero, the state-space equations become

$$\begin{cases} \dot{x}(t) = Ax(t) \\ y(t) = Cx(t) \end{cases} \Longrightarrow \begin{cases} x(t) = L^{-1}\left\{(sI_{n \times n} - A)^{-1}x(0)\right\} \\ y(t) = CL^{-1}\left\{(sI_{n \times n} - A)^{-1}x(0)\right\} \end{cases} \Longrightarrow \begin{cases} x(t) = L^{-1}\left\{(sI_{n \times n} - A)^{-1}\right\}x(0) \\ y(t) = CL^{-1}\left\{(sI_{n \times n} - A)^{-1}\right\}x(0) \end{cases}$$

This can be written as

$$\begin{cases} \dot{x}(t) = Ax(t) \\ y(t) = Cx(t) \end{cases} \Longrightarrow \begin{cases} x(t) = e^{At}x(0) \\ y(t) = Ce^{At}x(0) \end{cases}$$

Therefore,

$$e^{At} = L^{-1}\left\{(sI_{n \times n} - A)^{-1}\right\}$$

The poles of a state-space model can be found by calculating the eigenvalues of the matrix A. These eigenvalues determine the stability and dynamic behavior of the system. The stability of a linear time-invariant (LTI) system is fundamentally determined by the location of its poles in the complex s-plane. Stability is achieved when all poles lie in the left half of the s-plane; a system is considered stable if all its poles have negative real parts. Conversely, if any pole resides in the right half of the s-plane, or if there are poles on the imaginary axis with multiplicity greater than one, the system becomes unstable.

Inductors store energy in their magnetic fields, with the inductor current representing the magnitude of this stored energy, while capacitors store energy in their electric fields, with the capacitor voltage quantifying this stored energy. The inductor currents, denoted as and capacitor voltages are conventionally chosen as state variables due to their intrinsic relationship with the energy storage mechanisms within these components.

Each independent inductor's current and each independent capacitor's voltage constitute a state variable. Therefore, the total number of state variables in the state-space model is equal to the sum of the independent inductors and capacitors present in the circuit.

A loop composed solely of capacitors and voltage sources imposes a constraint on the capacitor voltages, as their sum must equal the sum of the voltage sources, effectively rendering one capacitor voltage dependent on the others. Similarly, a node composed solely of inductors and current sources results in a constraint on the inductor currents, where their sum must equal the sum of the current sources, making one inductor current dependent. In both scenarios, this dependency reduces the number of independent energy storage elements by one, thereby decreasing the required number of state variables in the state-space model by one. This reduction reflects the elimination of a redundant degree of freedom in the circuit's dynamic behavior, ensuring a minimal and sufficient representation.

For instance, in Fig. 4.1, a circuit containing three inductors and three capacitors might initially suggest $3 + 3 = 6$ state variables. However, node E, being composed solely of inductors and a current source, introduces a dependency among the inductor currents, reducing the number of independent inductor currents to two (Fig. 4.2). Consequently, the system requires only $3 + 2 = 5$ state variables.

Fig. 4.1 Node E connects three inductors and a current source

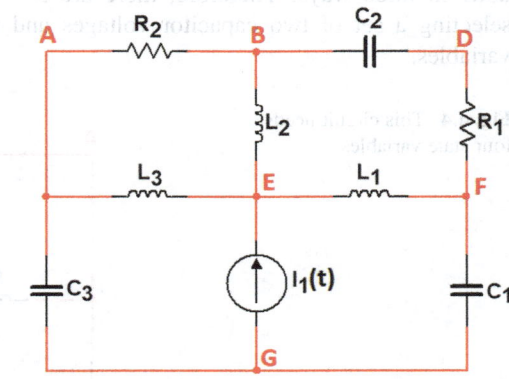

Fig. 4.2 L_3 current is a function of L_1 current, L_2 current, and $I_1(t)$

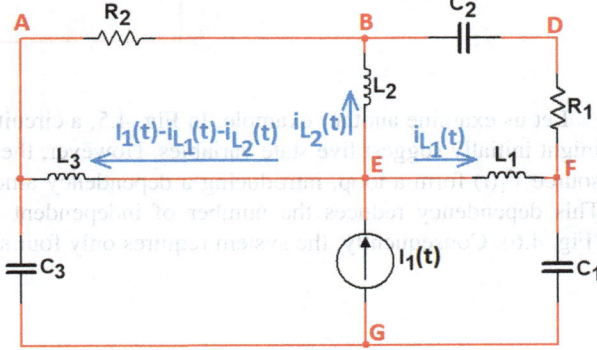

Since all inductor currents in the circuit of Fig. 4.3 are independent, a total of $3 + 3 = 6$ state variables are necessary.

Fig. 4.3 In this circuit, L_1, L_2, and L_3 currents are independent

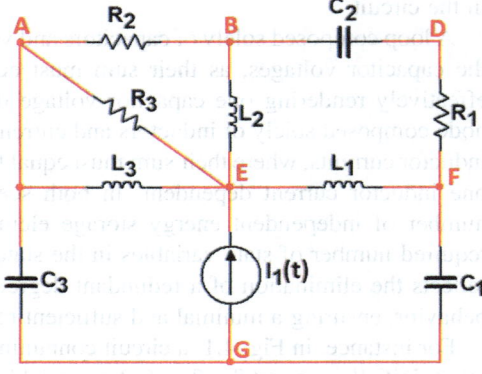

With two independent inductors and two independent capacitors, the circuit in Fig. 4.4 necessitates four state variables. Choosing two capacitors out of three can be done in three ways, and similarly, selecting two inductors out of three can also be done in three ways. Therefore, there are $3 \times 3 = 9$ different combinations for selecting a set of two capacitor voltages and two inductor currents as the state variables.

Fig. 4.4 This circuit needs four state variables

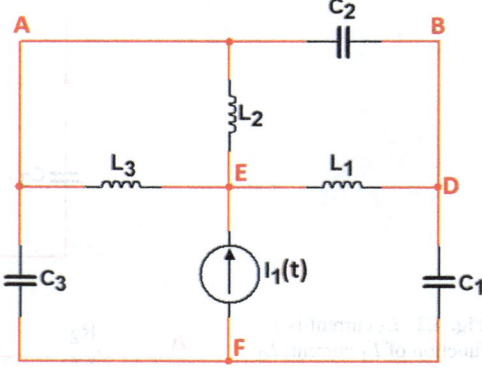

Let us examine another example. In Fig. 4.5, a circuit containing five capacitors might initially suggest five state variables. However, the capacitors and the voltage source $V_1(t)$ form a loop, introducing a dependency among the capacitor voltages. This dependency reduces the number of independent capacitor voltages to four (Fig. 4.6). Consequently, the system requires only four state variables.

Fig. 4.5 This circuit needs four state variables

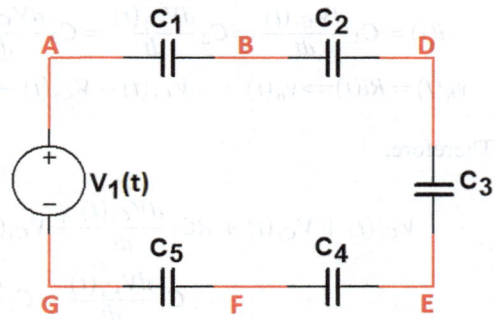

Fig. 4.6 The voltage across C_5 depends on the voltages of C_1, C_2, C_3, C_4, and $V_1(t)$

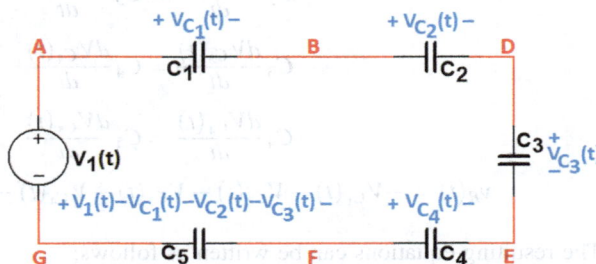

Since all capacitor voltages in the circuit of Fig. 4.7 are independent, a total of five state variables are necessary.

Fig. 4.7 This circuit needs five state variables

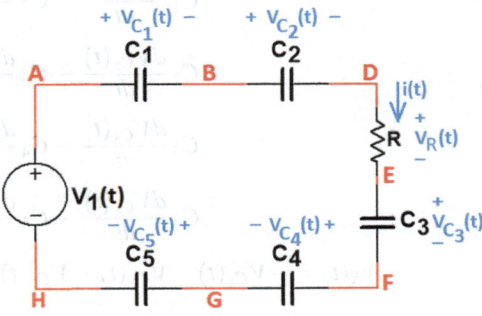

The state-space equations for the circuit in Fig. 4.7 are given below, with resistor voltage as the output.

$$V_{C_1}(t) + V_{C_2}(t) + Ri(t) + V_{C_3}(t) + V_{C_4}(t) + V_{C_5}(t) = V_1(t)$$

$$i(t) = C_1 \frac{dV_{C_1}(t)}{dt} = C_2 \frac{dV_{C_2}(t)}{dt} = C_3 \frac{dV_{C_3}(t)}{dt} = C_4 \frac{dV_{C_4}(t)}{dt} = C_5 \frac{dV_{C_5}(t)}{dt}$$

$$v_R(t) = Ri(t) \Rightarrow v_R(t) = -V_{C_1}(t) - V_{C_2}(t) - V_{C_3}(t) - V_{C_4}(t) - V_{C_5}(t) + V_1(t)$$

Therefore,

$$V_{C_1}(t) + V_{C_2}(t) + RC_1 \frac{dV_{C_1}(t)}{dt} + V_{C_3}(t) + V_{C_4}(t) + V_{C_5}(t) = V_1(t)$$

$$C_1 \frac{dV_{C_1}(t)}{dt} = C_2 \frac{dV_{C_2}(t)}{dt}$$

$$C_2 \frac{dV_{C_2}(t)}{dt} = C_3 \frac{dV_{C_3}(t)}{dt}$$

$$C_3 \frac{dV_{C_3}(t)}{dt} = C_4 \frac{dV_{C_4}(t)}{dt}$$

$$C_4 \frac{dV_{C_4}(t)}{dt} = C_5 \frac{dV_{C_5}(t)}{dt}$$

$$v_R(t) = -V_{C_1}(t) - V_{C_2}(t) - V_{C_3}(t) - V_{C_4}(t) - V_{C_5}(t) + V_1(t)$$

The resulting equations can be written as follows:

$$RC_1 \frac{dV_{C_1}(t)}{dt} = -V_{C_1}(t) - V_{C_2}(t) - V_{C_3}(t) - V_{C_4}(t) - V_{C_5}(t) + V_1(t)$$

$$C_1 \frac{dV_{C_1}(t)}{dt} - C_2 \frac{dV_{C_2}(t)}{dt} = 0$$

$$C_2 \frac{dV_{C_2}(t)}{dt} - C_3 \frac{dV_{C_3}(t)}{dt} = 0$$

$$C_3 \frac{dV_{C_3}(t)}{dt} - C_4 \frac{dV_{C_4}(t)}{dt} = 0$$

$$C_4 \frac{dV_{C_4}(t)}{dt} - C_5 \frac{dV_{C_5}(t)}{dt} = 0$$

$$v_R(t) = -V_{C_1}(t) - V_{C_2}(t) - V_{C_3}(t) - V_{C_4}(t) - V_{C_5}(t) + V_1(t)$$

We shall now proceed to convert the derived equations into matrix form.

$$\begin{cases} RC_1\dfrac{dV_{C_1}(t)}{dt} = -V_{C_1}(t) - V_{C_2}(t) - V_{C_3}(t) - V_{C_4}(t) - V_{C_5}(t) + V_1(t) \\[2mm] C_1\dfrac{dV_{C_1}(t)}{dt} - C_2\dfrac{dV_{C_2}(t)}{dt} = 0 \\[2mm] C_2\dfrac{dV_{C_2}(t)}{dt} - C_3\dfrac{dV_{C_3}(t)}{dt} = 0 \\[2mm] C_3\dfrac{dV_{C_3}(t)}{dt} - C_4\dfrac{dV_{C_4}(t)}{dt} = 0 \\[2mm] C_4\dfrac{dV_{C_4}(t)}{dt} - C_5\dfrac{dV_{C_5}(t)}{dt} = 0 \end{cases} \Rightarrow$$

$$\begin{bmatrix} RC_1 & 0 & 0 & 0 & 0 \\ +C_1 & -C_2 & 0 & 0 & 0 \\ 0 & +C_2 & -C_3 & 0 & 0 \\ 0 & 0 & +C_3 & -C_4 & 0 \\ 0 & 0 & 0 & +C_4 & -C_5 \end{bmatrix} \begin{bmatrix} \dfrac{dV_{C_1}(t)}{dt} \\[2mm] \dfrac{dV_{C_2}(t)}{dt} \\[2mm] \dfrac{dV_{C_3}(t)}{dt} \\[2mm] \dfrac{dV_{C_4}(t)}{dt} \\[2mm] \dfrac{dV_{C_5}(t)}{dt} \end{bmatrix}$$

$$= \begin{bmatrix} -1 & -1 & -1 & -1 & -1 \\ 0 & 0 & 0 & 0 & 0 \\ 0 & 0 & 0 & 0 & 0 \\ 0 & 0 & 0 & 0 & 0 \\ 0 & 0 & 0 & 0 & 0 \end{bmatrix} \begin{bmatrix} V_{C_1}(t) \\ V_{C_2}(t) \\ V_{C_3}(t) \\ V_{C_4}(t) \\ V_{C_5}(t) \end{bmatrix} + \begin{bmatrix} 1 \\ 0 \\ 0 \\ 0 \\ 0 \end{bmatrix} V_1(t) \Rightarrow \begin{bmatrix} \dfrac{dV_{C_1}(t)}{dt} \\[2mm] \dfrac{dV_{C_2}(t)}{dt} \\[2mm] \dfrac{dV_{C_3}(t)}{dt} \\[2mm] \dfrac{dV_{C_4}(t)}{dt} \\[2mm] \dfrac{dV_{C_5}(t)}{dt} \end{bmatrix}$$

$$= \begin{bmatrix} RC_1 & 0 & 0 & 0 & 0 \\ +C_1 & -C_2 & 0 & 0 & 0 \\ 0 & +C_2 & -C_3 & 0 & 0 \\ 0 & 0 & +C_3 & -C_4 & 0 \\ 0 & 0 & 0 & +C_4 & -C_5 \end{bmatrix}^{-1} \begin{bmatrix} -1 & -1 & -1 & -1 & -1 \\ 0 & 0 & 0 & 0 & 0 \\ 0 & 0 & 0 & 0 & 0 \\ 0 & 0 & 0 & 0 & 0 \\ 0 & 0 & 0 & 0 & 0 \end{bmatrix} \begin{bmatrix} V_{C_1}(t) \\ V_{C_2}(t) \\ V_{C_3}(t) \\ V_{C_4}(t) \\ V_{C_5}(t) \end{bmatrix}$$

$$+ \begin{bmatrix} RC_1 & 0 & 0 & 0 & 0 \\ +C_1 & -C_2 & 0 & 0 & 0 \\ 0 & +C_2 & -C_3 & 0 & 0 \\ 0 & 0 & +C_3 & -C_4 & 0 \\ 0 & 0 & 0 & +C_4 & -C_5 \end{bmatrix}^{-1} \begin{bmatrix} 1 \\ 0 \\ 0 \\ 0 \\ 0 \end{bmatrix} V_1(t)$$

$$v_R(t) = -V_{C_1}(t) - V_{C_2}(t) - V_{C_3}(t) - V_{C_4}(t) - V_{C_5}(t) + V_1(t) \Rightarrow v_R(t)$$

$$= \begin{bmatrix} -1 & -1 & -1 & -1 & -1 \end{bmatrix} \begin{bmatrix} V_{C_1}(t) \\ V_{C_2}(t) \\ V_{C_3}(t) \\ V_{C_4}(t) \\ V_{C_5}(t) \end{bmatrix} + [1]V_1(t)$$

For example, if $C_1 = C_2 = C_3 = C_4 = C_5 = 1\ F$ and $R = 1\ \Omega$, we have

$$\begin{bmatrix} \dfrac{dV_{C_1}(t)}{dt} \\ \dfrac{dV_{C_2}(t)}{dt} \\ \dfrac{dV_{C_3}(t)}{dt} \\ \dfrac{dV_{C_4}(t)}{dt} \\ \dfrac{dV_{C_5}(t)}{dt} \end{bmatrix} = \begin{bmatrix} 1 & 0 & 0 & 0 & 0 \\ +1 & -1 & 0 & 0 & 0 \\ 0 & +1 & -1 & 0 & 0 \\ 0 & 0 & 1 & -1 & 0 \\ 0 & 0 & 0 & +1 & -1 \end{bmatrix}^{-1} \begin{bmatrix} -1 & -1 & -1 & -1 & -1 \\ 0 & 0 & 0 & 0 & 0 \\ 0 & 0 & 0 & 0 & 0 \\ 0 & 0 & 0 & 0 & 0 \\ 0 & 0 & 0 & 0 & 0 \end{bmatrix} \begin{bmatrix} V_{C_1}(t) \\ V_{C_2}(t) \\ V_{C_3}(t) \\ V_{C_4}(t) \\ V_{C_5}(t) \end{bmatrix}$$

$$+ \begin{bmatrix} 1 & 0 & 0 & 0 & 0 \\ +1 & -1 & 0 & 0 & 0 \\ 0 & +1 & -1 & 0 & 0 \\ 0 & 0 & 1 & -1 & 0 \\ 0 & 0 & 0 & +1 & -1 \end{bmatrix}^{-1} \begin{bmatrix} 1 \\ 0 \\ 0 \\ 0 \\ 0 \end{bmatrix} V_1(t) \Rightarrow \begin{bmatrix} \dfrac{dV_{C_1}(t)}{dt} \\ \dfrac{dV_{C_2}(t)}{dt} \\ \dfrac{dV_{C_3}(t)}{dt} \\ \dfrac{dV_{C_4}(t)}{dt} \\ \dfrac{dV_{C_5}(t)}{dt} \end{bmatrix}$$

$$= \begin{bmatrix} -1 & -1 & -1 & -1 & -1 \\ -1 & -1 & -1 & -1 & -1 \\ -1 & -1 & -1 & -1 & -1 \\ -1 & -1 & -1 & -1 & -1 \\ -1 & -1 & -1 & -1 & -1 \end{bmatrix} \begin{bmatrix} V_{C_1}(t) \\ V_{C_2}(t) \\ V_{C_3}(t) \\ V_{C_4}(t) \\ V_{C_5}(t) \end{bmatrix} + \begin{bmatrix} 1 \\ 1 \\ 1 \\ 1 \\ 1 \end{bmatrix} V_1(t)$$

$$v_R(t) = \begin{bmatrix} -1 & -1 & -1 & -1 & -1 \end{bmatrix} \begin{bmatrix} V_{C_1}(t) \\ V_{C_2}(t) \\ V_{C_3}(t) \\ V_{C_4}(t) \\ V_{C_5}(t) \end{bmatrix} + [1]V_1(t)$$

Therefore,

$$x(t) = \begin{bmatrix} V_{C_1}(t) \\ V_{C_2}(t) \\ V_{C_3}(t) \\ V_{C_4}(t) \\ V_{C_5}(t) \end{bmatrix}, y(t) = v_R(t)$$

$$A = \begin{bmatrix} -1 & -1 & -1 & -1 & -1 \\ -1 & -1 & -1 & -1 & -1 \\ -1 & -1 & -1 & -1 & -1 \\ -1 & -1 & -1 & -1 & -1 \\ -1 & -1 & -1 & -1 & -1 \end{bmatrix}$$

$$B = \begin{bmatrix} 1 \\ 1 \\ 1 \\ 1 \\ 1 \end{bmatrix}$$

$$C = \begin{bmatrix} -1 & -1 & -1 & -1 & -1 \end{bmatrix}$$

$$D = [1]$$

Calculations presented in this book are truncated to four decimal places. Consequently, minor discrepancies may arise between results obtained from different solution methodologies. These variations are solely attributable to accumulated rounding errors and do not reflect inaccuracies in the methodologies themselves.

4.2 Solved Examples

4.2.1 Example 4.1

Using MATLAB, determine the poles and zeros of the state-space system:

$$
\begin{cases}
\begin{bmatrix} \dfrac{di_L(t)}{dt} \\[2mm] \dfrac{dv_C(t)}{dt} \end{bmatrix} =
\begin{bmatrix} 0.5 & 0 \\[1mm] -\dfrac{1}{12} & -\dfrac{1}{3} \end{bmatrix}
\begin{bmatrix} i_L(t) \\[1mm] v_C(t) \end{bmatrix} +
\begin{bmatrix} 0 \\[1mm] \dfrac{1}{12} \end{bmatrix} V_1(t) \\[6mm]
v_o(t) = \begin{bmatrix} 0 & -1 \end{bmatrix}
\begin{bmatrix} i_L(t) \\[1mm] v_C(t) \end{bmatrix} + [1]V_1(t)
\end{cases}
$$

Solution:

$$
\begin{cases}
\begin{bmatrix} \dfrac{di_L(t)}{dt} \\[2mm] \dfrac{dv_C(t)}{dt} \end{bmatrix} =
\begin{bmatrix} 0.5 & 0 \\[1mm] -\dfrac{1}{12} & -\dfrac{1}{3} \end{bmatrix}
\begin{bmatrix} i_L(t) \\[1mm] v_C(t) \end{bmatrix} +
\begin{bmatrix} 0 \\[1mm] \dfrac{1}{12} \end{bmatrix} V_1(t) \\[6mm]
v_o(t) = \begin{bmatrix} 0 & -1 \end{bmatrix}
\begin{bmatrix} i_L(t) \\[1mm] v_C(t) \end{bmatrix} + [1]V_1(t)
\end{cases}
\Rightarrow
\begin{cases}
A = \begin{bmatrix} 0.5 & 0 \\[1mm] -\dfrac{1}{12} & -\dfrac{1}{3} \end{bmatrix} \\[6mm]
B = \begin{bmatrix} 0 \\[1mm] \dfrac{1}{12} \end{bmatrix} \\[4mm]
C = \begin{bmatrix} 0 & -1 \end{bmatrix} \\[2mm]
D = [1]
\end{cases}
$$

The code in Fig. 4.8 finds the poles and zeros.

Fig. 4.8 MATLAB code

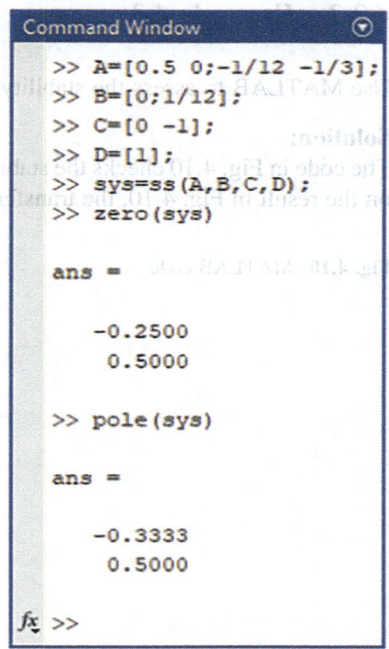

```
Command Window
>> A=[0.5 0;-1/12 -1/3];
>> B=[0;1/12];
>> C=[0 -1];
>> D=[1];
>> sys=ss(A,B,C,D);
>> zero(sys)

ans =

    -0.2500
     0.5000

>> pole(sys)

ans =

    -0.3333
     0.5000

fx >>
```

4.2.2 Example 4.2

Use MATLAB to assess the stability of the transfer function $\frac{s+4}{s^2 - 6s - 6}$.

Solution:
The code in Fig. 4.9 checks the stability of the given transfer function model. Based on the result in Fig. 4.9, the transfer function is unstable.

Fig. 4.9 MATLAB code

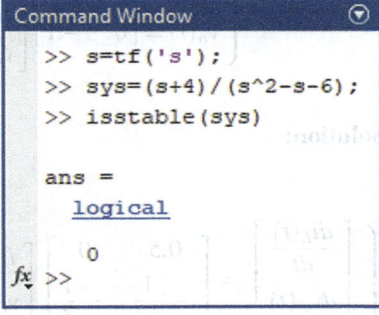

```
Command Window
>> s=tf('s');
>> sys=(s+4)/(s^2-s-6);
>> isstable(sys)

ans =
  logical

   0
fx >>
```

4.2.3   Example 4.3

Use **MATLAB** to assess the stability of the transfer function $\frac{s+4}{s^2+6s+6}$.

Solution:
The code in Fig. 4.10 checks the stability of the given transfer function model. Based on the result in Fig. 4.10, the transfer function is stable.

Fig. 4.10 MATLAB code

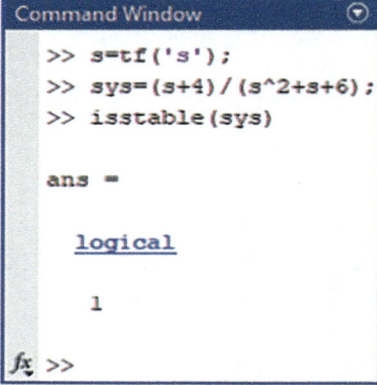

4.2.4   Example 4.4

Assess the stability of the following state-space model using **MATLAB**.

$$\begin{cases} \begin{bmatrix} \dfrac{di_L(t)}{dt} \\ \dfrac{dv_C(t)}{dt} \end{bmatrix} = \begin{bmatrix} 0.5 & 0 \\ -\dfrac{1}{12} & -\dfrac{1}{3} \end{bmatrix} \begin{bmatrix} i_L(t) \\ v_C(t) \end{bmatrix} + \begin{bmatrix} 0 \\ \dfrac{1}{12} \end{bmatrix} V_1(t) \\ v_o(t) = \begin{bmatrix} 0 & -1 \end{bmatrix} \begin{bmatrix} i_L(t) \\ v_C(t) \end{bmatrix} + [1]V_1(t) \end{cases}$$

Solution:

$$\begin{cases} \begin{bmatrix} \dfrac{di_L(t)}{dt} \\ \dfrac{dv_C(t)}{dt} \end{bmatrix} = \begin{bmatrix} 0.5 & 0 \\ -\dfrac{1}{12} & -\dfrac{1}{3} \end{bmatrix} \begin{bmatrix} i_L(t) \\ v_C(t) \end{bmatrix} + \begin{bmatrix} 0 \\ \dfrac{1}{12} \end{bmatrix} V_1(t) \\ v_o(t) = \begin{bmatrix} 0 & -1 \end{bmatrix} \begin{bmatrix} i_L(t) \\ v_C(t) \end{bmatrix} + [1]V_1(t) \end{cases} \Rightarrow \begin{cases} A = \begin{bmatrix} 0.5 & 0 \\ -\dfrac{1}{12} & -\dfrac{1}{3} \end{bmatrix} \\ B = \begin{bmatrix} 0 \\ \dfrac{1}{12} \end{bmatrix} \\ C = \begin{bmatrix} 0 & -1 \end{bmatrix} \\ D = [1] \end{cases}$$

The code in Fig. 4.11 checks the stability of the given state-space model. Based on the result in Fig. 4.11, the transfer function is unstable.

Fig. 4.11 MATLAB code

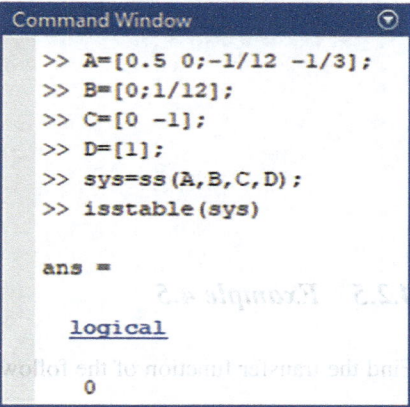

Let us proceed to calculate the poles of the given state-space model. Based on the results presented in Figs. 4.12 and 4.13, the system is determined to have a right half plane (RHP) pole, indicating instability.

Fig. 4.12 MATLAB code

Fig. 4.13 MATLAB code

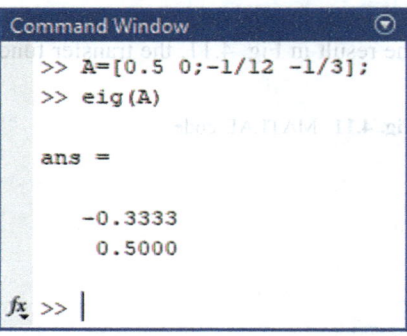

4.2.5   Example 4.5

Find the transfer function of the following state-space model using MATLAB.

$$
\begin{cases}
\begin{bmatrix} \dfrac{di_L(t)}{dt} \\[2mm] \dfrac{dv_C(t)}{dt} \end{bmatrix} =
\begin{bmatrix} 0 & 0.5 \\[1mm] -\dfrac{1}{12} & -\dfrac{1}{3} \end{bmatrix}
\begin{bmatrix} i_L(t) \\[1mm] v_C(t) \end{bmatrix} +
\begin{bmatrix} 0 \\[1mm] \dfrac{1}{12} \end{bmatrix} V_1(t) \\[6mm]
v_o(t) = \begin{bmatrix} 0 & -1 \end{bmatrix}
\begin{bmatrix} i_L(t) \\[1mm] v_C(t) \end{bmatrix} + [1]V_1(t)
\end{cases}
$$

Solution:

$$
\begin{cases}
\begin{bmatrix} \dfrac{di_L(t)}{dt} \\[2mm] \dfrac{dv_C(t)}{dt} \end{bmatrix} =
\begin{bmatrix} 0 & 0.5 \\[1mm] -\dfrac{1}{12} & -\dfrac{1}{3} \end{bmatrix}
\begin{bmatrix} i_L(t) \\[1mm] v_C(t) \end{bmatrix} +
\begin{bmatrix} 0 \\[1mm] \dfrac{1}{12} \end{bmatrix} V_1(t) \\[6mm]
v_o(t) = \begin{bmatrix} 0 & -1 \end{bmatrix}
\begin{bmatrix} i_L(t) \\[1mm] v_C(t) \end{bmatrix} + [1]V_1(t)
\end{cases}
\Rightarrow
\begin{cases}
A = \begin{bmatrix} 0 & 0.5 \\[1mm] -\dfrac{1}{12} & -\dfrac{1}{3} \end{bmatrix} \\[6mm]
B = \begin{bmatrix} 0 \\[1mm] \dfrac{1}{12} \end{bmatrix} \\[5mm]
C = \begin{bmatrix} 0 & -1 \end{bmatrix} \\[2mm]
D = [1]
\end{cases}
$$

Figures 4.14 and 4.15 contain code to determine the transfer function of the given state-space model.

Fig. 4.14 MATLAB code

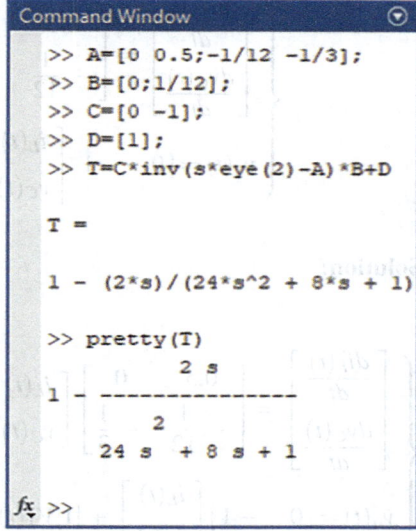

Fig. 4.15 MATLAB code

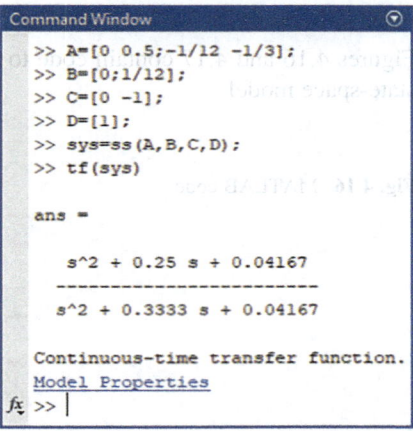

4.2.6 Example 4.6

Find the transfer function of the following state-space model using MATLAB.

$$\begin{cases} \begin{bmatrix} \dfrac{di_L(t)}{dt} \\ \dfrac{dv_C(t)}{dt} \end{bmatrix} = \begin{bmatrix} 0.5 & 0 \\ -\dfrac{1}{12} & -\dfrac{1}{3} \end{bmatrix} \begin{bmatrix} i_L(t) \\ v_C(t) \end{bmatrix} + \begin{bmatrix} 0 \\ \dfrac{1}{12} \end{bmatrix} V_1(t) \\ \\ v_o(t) = \begin{bmatrix} 0 & -1 \end{bmatrix} \begin{bmatrix} i_L(t) \\ v_C(t) \end{bmatrix} + [1]V_1(t) \end{cases}$$

Solution:

$$\begin{cases} \begin{bmatrix} \dfrac{di_L(t)}{dt} \\ \dfrac{dv_C(t)}{dt} \end{bmatrix} = \begin{bmatrix} 0.5 & 0 \\ -\dfrac{1}{12} & -\dfrac{1}{3} \end{bmatrix} \begin{bmatrix} i_L(t) \\ v_C(t) \end{bmatrix} + \begin{bmatrix} 0 \\ \dfrac{1}{12} \end{bmatrix} V_1(t) \\ \\ v_o(t) = \begin{bmatrix} 0 & -1 \end{bmatrix} \begin{bmatrix} i_L(t) \\ v_C(t) \end{bmatrix} + [1]V_1(t) \end{cases} \Rightarrow \begin{cases} A = \begin{bmatrix} 0.5 & 0 \\ -\dfrac{1}{12} & -\dfrac{1}{3} \end{bmatrix} \\ B = \begin{bmatrix} 0 \\ \dfrac{1}{12} \end{bmatrix} \\ C = \begin{bmatrix} 0 & -1 \end{bmatrix} \\ D = [1] \end{cases}$$

Figures 4.16 and 4.17 contain code to determine the transfer function of the given state-space model.

Fig. 4.16 MATLAB code

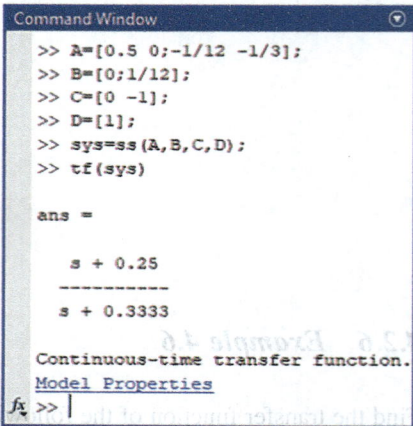

```
Command Window
>> A=[0.5 0;-1/12 -1/3];
>> B=[0;1/12];
>> C=[0 -1];
>> D=[1];
>> sys=ss(A,B,C,D);
>> tf(sys)

ans =

  s + 0.25
  ----------
  s + 0.3333

Continuous-time transfer function.
Model Properties
fx >> |
```

Fig. 4.17 MATLAB code

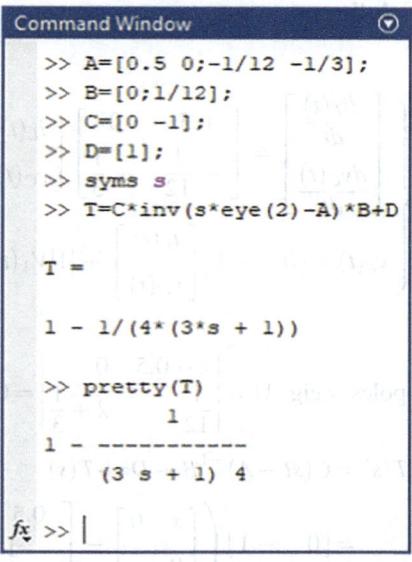

In Example 4.5, with a 2×2 A matrix, we obtained a second-order transfer function. However, in this example, despite having a 2×2 A matrix, the transfer function is first-order. This difference is due to pole-zero cancellation in this example, which was absent in Example 4.5.

4.2.7 Example 4.7

Determine the poles and transfer function of the following state-space model.

$$\begin{cases} \begin{bmatrix} \dfrac{di_L(t)}{dt} \\ \dfrac{dv_C(t)}{dt} \end{bmatrix} = \begin{bmatrix} 0.5 & 0 \\ -\dfrac{1}{12} & -\dfrac{1}{3} \end{bmatrix} \begin{bmatrix} i_L(t) \\ v_C(t) \end{bmatrix} + \begin{bmatrix} 0 \\ \dfrac{1}{12} \end{bmatrix} V_1(t) \\ v_o(t) = \begin{bmatrix} 0 & -1 \end{bmatrix} \begin{bmatrix} i_L(t) \\ v_C(t) \end{bmatrix} + [1]V_1(t) \end{cases}$$

Solution:

$$\begin{cases} \begin{bmatrix} \dfrac{di_L(t)}{dt} \\ \dfrac{dv_C(t)}{dt} \end{bmatrix} = \begin{bmatrix} 0.5 & 0 \\ -\dfrac{1}{12} & -\dfrac{1}{3} \end{bmatrix} \begin{bmatrix} i_L(t) \\ v_C(t) \end{bmatrix} + \begin{bmatrix} 0 \\ \dfrac{1}{12} \end{bmatrix} V_1(t) \\ \\ v_o(t) = \begin{bmatrix} 0 & -1 \end{bmatrix} \begin{bmatrix} i_L(t) \\ v_C(t) \end{bmatrix} + [1]V_1(t) \end{cases} \Rightarrow \begin{cases} A = \begin{bmatrix} 0.5 & 0 \\ -\dfrac{1}{12} & -\dfrac{1}{3} \end{bmatrix} \\ B = \begin{bmatrix} 0 \\ \dfrac{1}{12} \end{bmatrix} \\ C = \begin{bmatrix} 0 & -1 \end{bmatrix} \\ D = [1] \end{cases}$$

$$\text{poles} = \text{eig}(A) = \begin{vmatrix} \lambda - 0.5 & 0 \\ \dfrac{1}{12} & \lambda + \dfrac{1}{3} \end{vmatrix} = 0 \Rightarrow (\lambda - 0.5)\left(\lambda + \dfrac{1}{3}\right) = 0 \Rightarrow \lambda_{1,2} = 0.5, \ -\dfrac{1}{3}$$

$$T(s) = C(sI - A)^{-1}B + D \Rightarrow T(s)$$

$$= \begin{bmatrix} 0 & -1 \end{bmatrix} \left(\begin{bmatrix} s & 0 \\ 0 & s \end{bmatrix} - \begin{bmatrix} 0.5 & 0 \\ -\dfrac{1}{12} & -\dfrac{1}{3} \end{bmatrix} \right)^{-1} \begin{bmatrix} 0 \\ \dfrac{1}{12} \end{bmatrix} + [1] \Rightarrow T(s)$$

$$= \begin{bmatrix} 0 & -1 \end{bmatrix} \begin{bmatrix} s - 0.5 & 0 \\ \dfrac{1}{12} & s + \dfrac{1}{3} \end{bmatrix}^{-1} \begin{bmatrix} 0 \\ \dfrac{1}{12} \end{bmatrix} + [1] \Rightarrow T(s)$$

$$= \begin{bmatrix} 0 & -1 \end{bmatrix} \dfrac{1}{(s - 0.5)(s + \frac{1}{3}) - 0 \times \frac{1}{12}} \begin{bmatrix} s + \dfrac{1}{3} & 0 \\ -\dfrac{1}{12} & s - 0.5 \end{bmatrix} \begin{bmatrix} 0 \\ \dfrac{1}{12} \end{bmatrix} + [1] \Rightarrow T(s)$$

$$= \dfrac{1}{(s - 0.5)(s + \frac{1}{3})} \begin{bmatrix} 0 & -1 \end{bmatrix} \begin{bmatrix} s + \dfrac{1}{3} & 0 \\ -\dfrac{1}{12} & s - 0.5 \end{bmatrix} \begin{bmatrix} 0 \\ \dfrac{1}{12} \end{bmatrix} + [1] \Rightarrow T(s)$$

$$= \dfrac{1}{(s - 0.5)(s + \frac{1}{3})} \begin{bmatrix} \dfrac{1}{12} & 0.5 - s \end{bmatrix} \begin{bmatrix} 0 \\ \dfrac{1}{12} \end{bmatrix} + [1] \Rightarrow T(s)$$

$$= \dfrac{1}{(s - 0.5)(s + \frac{1}{3})} \dfrac{0.5 - s}{12} + [1] \Rightarrow T(s) = \dfrac{0.5 - s}{12(s - 0.5)(s + \frac{1}{3})} + 1 \Rightarrow T(s)$$

$$= \dfrac{0.5 - s}{12s^2 - 2s - 2} + 1 \Rightarrow T(s) = \dfrac{12s^2 - 3s - 1.5}{12s^2 - 2s - 2}$$

Poles and zeros of this transfer function are:

$$T(s) = \dfrac{12s^2 - 3s - 1.5}{12s^2 - 2s - 2} \Rightarrow \begin{cases} 12s^2 - 3s - 1.5 = 0 \Rightarrow s = 0.5, s = -0.25 \ \text{(zeros)} \\ 12s^2 - 2s - 2 = 0 \Rightarrow s = 0.5, s = -0.3333 \ \text{(poles)} \end{cases}$$

4.2.8 Example 4.8

Assess the stability of $T(s) = \frac{12s^2 - 3s - 1.5}{12s^2 - 2s - 2} = \frac{12(s-0.5)(s+0.25)}{12(s-0.5)(s+0.3333)} = \frac{(s-0.5)(s+0.25)}{(s-0.5)(s+0.3333)}$.

Solution:

$$T(s) = \frac{(s-0.5)(s+0.25)}{(s-0.5)(s+0.3333)} \Rightarrow \text{poles} = 0.5, \ -0.3333$$

The system is unstable due to the presence of an RHP pole. Note that $\frac{(s-0.5)(s+0.25)}{(s-0.5)(s+0.3333)}$ is not equivalent to $\frac{s+0.25}{s+0.3333}$.

4.2.9 Example 4.9

Determine the poles and transfer function of the following state-space model.

$$\begin{cases} \begin{bmatrix} \frac{di_L(t)}{dt} \\ \frac{dv_C(t)}{dt} \end{bmatrix} = \begin{bmatrix} 0 & 0.5 \\ -\frac{1}{12} & -\frac{1}{3} \end{bmatrix} \begin{bmatrix} i_L(t) \\ v_C(t) \end{bmatrix} + \begin{bmatrix} 0 \\ \frac{1}{12} \end{bmatrix} V_1(t) \\ v_o(t) = \begin{bmatrix} 0 & -1 \end{bmatrix} \begin{bmatrix} i_L(t) \\ v_C(t) \end{bmatrix} + [1]V_1(t) \end{cases}$$

Solution:

$$\begin{cases} \begin{bmatrix} \frac{di_L(t)}{dt} \\ \frac{dv_C(t)}{dt} \end{bmatrix} = \begin{bmatrix} 0 & 0.5 \\ -\frac{1}{12} & -\frac{1}{3} \end{bmatrix} \begin{bmatrix} i_L(t) \\ v_C(t) \end{bmatrix} + \begin{bmatrix} 0 \\ \frac{1}{12} \end{bmatrix} V_1(t) \\ v_o(t) = \begin{bmatrix} 0 & -1 \end{bmatrix} \begin{bmatrix} i_L(t) \\ v_C(t) \end{bmatrix} + [1]V_1(t) \end{cases} \Rightarrow \begin{cases} A = \begin{bmatrix} 0 & 0.5 \\ -\frac{1}{12} & -\frac{1}{3} \end{bmatrix} \\ B = \begin{bmatrix} 0 \\ \frac{1}{12} \end{bmatrix} \\ C = \begin{bmatrix} 0 & -1 \end{bmatrix} \\ D = [1] \end{cases}$$

$$\mathrm{eig}(A)=\begin{vmatrix} \lambda & -0.5 \\ \dfrac{1}{12} & \lambda+\dfrac{1}{3} \end{vmatrix}=0 \Rightarrow \lambda\left(\lambda+\dfrac{1}{3}\right)-(-0.5)\times \dfrac{1}{12}=0 \Rightarrow \lambda^2+\dfrac{1}{3}\lambda+\dfrac{1}{24}$$

$$=0 \Rightarrow \lambda_{1,2}=-0.1667\pm j0.1179$$

$$T(s)=C(sI-A)^{-1}B+D \Rightarrow T(s)$$

$$=[0 \quad -1]\left(\begin{bmatrix} s & 0 \\ 0 & s \end{bmatrix}-\begin{bmatrix} 0 & 0.5 \\ -\dfrac{1}{12} & -\dfrac{1}{3} \end{bmatrix}\right)^{-1}\begin{bmatrix} 0 \\ \dfrac{1}{12} \end{bmatrix}+[1] \Rightarrow T(s)$$

$$=[0 \quad -1]\begin{bmatrix} s & -0.5 \\ \dfrac{1}{12} & s+\dfrac{1}{3} \end{bmatrix}^{-1}\begin{bmatrix} 0 \\ \dfrac{1}{12} \end{bmatrix}+[1] \Rightarrow T(s)$$

$$=[0 \quad -1]\dfrac{1}{s\left(s+\frac{1}{3}\right)-(-0.5)\times \frac{1}{12}}\begin{bmatrix} s+\dfrac{1}{3} & 0.5 \\ -\dfrac{1}{12} & s \end{bmatrix}\begin{bmatrix} 0 \\ \dfrac{1}{12} \end{bmatrix}+[1] \Rightarrow T(s)$$

$$=\dfrac{1}{s^2+\frac{1}{3}s+\frac{1}{24}}[0 \quad -1]\begin{bmatrix} s+\dfrac{1}{3} & 0.5 \\ -\dfrac{1}{12} & s \end{bmatrix}\begin{bmatrix} 0 \\ \dfrac{1}{12} \end{bmatrix}+[1] \Rightarrow T(s)$$

$$=\dfrac{24}{24s^2+8s+1}\left[\dfrac{1}{12} \quad -s\right]\begin{bmatrix} 0 \\ \dfrac{1}{12} \end{bmatrix}+[1] \Rightarrow T(s)$$

$$=\dfrac{24}{24s^2+8s+1}\times \dfrac{-s}{12}+[1] \Rightarrow T(s)=\dfrac{-2s}{24s^2+8s+1}+[1] \Rightarrow T(s)$$

$$=1-\dfrac{2s}{24s^2+8s+1} \Rightarrow T(s)=\dfrac{24s^2+6s+1}{24s^2+8s+1} \Rightarrow T(s)=\dfrac{s^2+0.25s+0.0417}{s^2+0.3333s+0.0417}$$

$$T(s)=\dfrac{s^2+0.25s+0.0417}{s^2+0.3333s+0.0417} \Rightarrow$$

$$\begin{cases} s^2+0.25s+0.0417=0 \Rightarrow s_{1,2}=-0.1250\pm j0.1614 \ \text{(zeros)} \\ s^2+0.3333s+0.0417=0 \Rightarrow s_{1,2}=-0.1667\pm j0.1179 \ \text{(poles)} \end{cases}$$

4.2.10   Example 4.10

$A=\begin{bmatrix} 7 & -1 \\ 4 & 3 \end{bmatrix}$. Calculate e^{At}.

Solution:

$$e^{At} = \mathcal{L}^{-1}\left\{(sI_{n\times n} - A)^{-1}\right\} \Rightarrow e^{At} = \mathcal{L}^{-1}\left\{(sI_{2\times 2} - A)^{-1}\right\} \Rightarrow e^{At}$$

$$= \mathcal{L}^{-1}\left\{\left(\begin{bmatrix} s & 0 \\ 0 & s \end{bmatrix} - \begin{bmatrix} 7 & -1 \\ 4 & 3 \end{bmatrix}\right)^{-1}\right\} \Rightarrow e^{At}$$

$$= \mathcal{L}^{-1}\left\{\begin{bmatrix} s-7 & 1 \\ -4 & s-3 \end{bmatrix}^{-1}\right\} \Rightarrow e^{At}$$

$$= \mathcal{L}^{-1}\left\{\frac{1}{(s-7)(s-3)+4}\begin{bmatrix} s-3 & -1 \\ 4 & s-7 \end{bmatrix}^{-1}\right\} \Rightarrow e^{At}$$

$$= \mathcal{L}^{-1}\left\{\frac{1}{s^2 - 10s + 25}\begin{bmatrix} s-3 & -1 \\ 4 & s-7 \end{bmatrix}\right\} \Rightarrow e^{At}$$

$$= \mathcal{L}^{-1}\left\{\begin{bmatrix} \dfrac{s-3}{s^2 - 10s + 25} & -\dfrac{1}{s^2 - 10s + 25} \\ \dfrac{4}{s^2 - 10s + 25} & \dfrac{s-7}{s^2 - 10s + 25} \end{bmatrix}\right\} \Rightarrow e^{At}$$

$$= \begin{bmatrix} \mathcal{L}^{-1}\left\{\dfrac{s-3}{s^2 - 10s + 25}\right\} & \mathcal{L}^{-1}\left\{-\dfrac{1}{s^2 - 10s + 25}\right\} \\ \mathcal{L}^{-1}\left\{\dfrac{4}{s^2 - 10s + 25}\right\} & \mathcal{L}^{-1}\left\{\dfrac{s-7}{s^2 - 10s + 25}\right\} \end{bmatrix} \Rightarrow e^{At}$$

$$= \begin{bmatrix} \mathcal{L}^{-1}\left\{\dfrac{s-3}{(s-5)^2}\right\} & \mathcal{L}^{-1}\left\{-\dfrac{1}{(s-5)^2}\right\} \\ \mathcal{L}^{-1}\left\{\dfrac{4}{(s-5)^2}\right\} & \mathcal{L}^{-1}\left\{\dfrac{s-7}{(s-5)^2}\right\} \end{bmatrix} \Rightarrow e^{At}$$

$$= \begin{bmatrix} (1+2t)e^{5t} & -te^{5t} \\ 4te^{5t} & (1-2t)e^{5t} \end{bmatrix} \Rightarrow e^{At} = e^{5t}\begin{bmatrix} 1+2t & -t \\ 4t & 1-2t \end{bmatrix}$$

4.2.11 Example 4.11

$A = \begin{bmatrix} 7 & -1 \\ 4 & 3 \end{bmatrix}$. Calculate e^{At} with MATLAB.

Solution:
Figure 4.18 illustrates the necessary MATLAB code.

Fig. 4.18 MATLAB code

```
Command Window
>> syms t
>> A=[7 -1;4 3];
>> expm(A*t)

ans =

[exp(5*t) + 2*t*exp(5*t),          -t*exp(5*t)]
[             4*t*exp(5*t), -exp(5*t)*(2*t - 1)]

fx >> |
```

4.2.12   Example 4.12

Solve $\begin{cases} \dot{x}(t) = 3x(t) - 2y(t) \\ \dot{y}(t) = 4x(t) - y(t) \end{cases}$ with $\begin{bmatrix} x(0) = 1 \\ y(0) = 7 \end{bmatrix}$.

Solution:

$$\begin{cases} \dot{x}(t) = 3x(t) - 2y(t) \\ \dot{y}(t) = 4x(t) - y(t) \end{cases} \Rightarrow \begin{cases} L(\dot{x}(t)) = L(3x(t)) - L(2y(t)) \\ L(\dot{y}(t)) = L(4x(t)) - L(y(t)) \end{cases} \Rightarrow$$

$$\times \begin{cases} sX - x(0) = 3X - 2Y \\ sY - y(0) = 4X - Y \end{cases} \Rightarrow \begin{cases} sX - 1 = 3X - 2Y \\ sY - 7 = 4X - Y \end{cases} \Rightarrow$$

$$\times \begin{cases} (s-3)X + 2Y = 1 \\ -4X + (s+1)Y = 7 \end{cases} \Rightarrow \begin{bmatrix} s-3 & 2 \\ -4 & s+1 \end{bmatrix}\begin{bmatrix} X \\ Y \end{bmatrix} = \begin{bmatrix} 1 \\ 7 \end{bmatrix} \Rightarrow \begin{bmatrix} X \\ Y \end{bmatrix}$$

$$= \begin{bmatrix} s-3 & 2 \\ -4 & s+1 \end{bmatrix}^{-1}\begin{bmatrix} 1 \\ 7 \end{bmatrix} = \frac{1}{(s-3)(s+1) - (2)(-4)}\begin{bmatrix} s+1 & -2 \\ 4 & s-3 \end{bmatrix}\begin{bmatrix} 1 \\ 7 \end{bmatrix}$$

$$= \frac{1}{s^2 - 2s + 5}\begin{bmatrix} s+1-14 \\ 4+7(s-3) \end{bmatrix} = \frac{1}{s^2 - 2s + 5}\begin{bmatrix} s-13 \\ 7s-17 \end{bmatrix}$$

$$= \begin{bmatrix} \dfrac{s-13}{s^2 - 2s + 5} \\ \dfrac{7s-17}{s^2 - 2s + 5} \end{bmatrix} \Rightarrow \begin{cases} X = \dfrac{s-13}{s^2 - 2s + 5} \\ Y = \dfrac{7s-17}{s^2 - 2s + 5} \end{cases}$$

Therefore,

$$\begin{cases} x(t) = L^{-1}(X) = L^{-1}\left(\dfrac{s-13}{s^2-2s+5}\right) = L^{-1}\left(\dfrac{s-1}{(s-1)^2+2^2} - 6\dfrac{2}{(s-1)^2+2^2}\right) = e^t\cos(2t) - 6e^t\sin(2t) \\ y(t) = L^{-1}(Y) = L^{-1}\left(\dfrac{7s-17}{s^2-2s+5}\right) = L^{-1}\left(\dfrac{7(s-1)}{(s-1)^2+2^2} - 5\dfrac{2}{(s-1)^2+2^2}\right) = 7e^t\cos(2t) - 5e^t\sin(2t) \end{cases}$$

4.2.13 Example 4.13

Solve $\begin{cases} \dot{x}(t) = 7x(t) - y(t) \\ \dot{y}(t) = 4x(t) + 3y(t) \end{cases}$ with $\begin{bmatrix} x(0) = 1 \\ y(0) = 2 \end{bmatrix}$.

Solution:

$\begin{cases} \dot{x}(t) = 7x(t) - y(t) \\ \dot{y}(t) = 4x(t) + 3y(t) \end{cases} \Rightarrow \begin{cases} L(\dot{x}(t)) = L(7x(t)) - L(y(t)) \\ L(\dot{y}(t)) = L(4x(t)) + L(3y(t)) \end{cases} \Rightarrow$

$\times \begin{cases} sX - x(0) = 7X - Y \\ sY - y(0) = 4X + 3Y \end{cases} \Rightarrow \begin{cases} sX - 1 = 7X - Y \\ sY - 2 = 4X + 3Y \end{cases} \Rightarrow$

$\times \begin{cases} (s-7)X + Y = 1 \\ -4X + (s-3)Y = 2 \end{cases} \Rightarrow \begin{bmatrix} s-7 & 1 \\ -4 & s-3 \end{bmatrix} \begin{bmatrix} X \\ Y \end{bmatrix} = \begin{bmatrix} 1 \\ 2 \end{bmatrix} \Rightarrow \begin{bmatrix} X \\ Y \end{bmatrix}$

$= \begin{bmatrix} s-7 & 1 \\ -4 & s-3 \end{bmatrix}^{-1} \begin{bmatrix} 1 \\ 2 \end{bmatrix} = \dfrac{1}{(s-7)(s-3) - (1)(-4)} \begin{bmatrix} s-3 & -1 \\ 4 & s-7 \end{bmatrix} \begin{bmatrix} 1 \\ 2 \end{bmatrix}$

$= \dfrac{1}{s^2 - 10s + 25} \begin{bmatrix} s-5 \\ 2s-10 \end{bmatrix} = \dfrac{1}{(s-5)^2} \begin{bmatrix} s-5 \\ 2(s-5) \end{bmatrix} = \begin{bmatrix} \dfrac{1}{s-5} \\ \dfrac{2}{s-5} \end{bmatrix} \Rightarrow \begin{cases} X = \dfrac{1}{s-5} \\ Y = \dfrac{2}{s-5} \end{cases}$

Therefore,

$$\begin{cases} x(t) = L^{-1}(X) = L^{-1}\left(\dfrac{1}{s-5}\right) = e^{5t} \\ y(t) = L^{-1}(Y) = L^{-1}\left(\dfrac{2}{s-5}\right) = 2e^t \end{cases}$$

4.2.14 Example 4.14

Solve $\begin{cases} \dot{x}(t) = 7x(t) - y(t) \\ \dot{y}(t) = 4x(t) + 3y(t) \end{cases}$ with $\begin{bmatrix} x(0) = 1 \\ y(0) = 2 \end{bmatrix}$.

Solution:

$\begin{cases} \dot{x}(t) = 7x(t) - y(t) \\ \dot{y}(t) = 4x(t) + 3y(t) \end{cases} \Rightarrow \begin{bmatrix} \dot{x}(t) \\ \dot{y}(t) \end{bmatrix} = \begin{bmatrix} 7 & -1 \\ 4 & 3 \end{bmatrix} \begin{bmatrix} x(t) \\ y(t) \end{bmatrix} \Rightarrow A = \begin{bmatrix} 7 & -1 \\ 4 & 3 \end{bmatrix}$

$\begin{bmatrix} x(t) \\ y(t) \end{bmatrix} = e^{At}x_0 \Rightarrow \begin{bmatrix} x(t) \\ y(t) \end{bmatrix} = e^{5t} \begin{bmatrix} 1+2t & -t \\ 4t & 1-2t \end{bmatrix} \begin{bmatrix} 1 \\ 2 \end{bmatrix} \Rightarrow \begin{bmatrix} x(t) \\ y(t) \end{bmatrix}$

$= e^{5t} \begin{bmatrix} 1+2t-2t \\ 4t+2-4t \end{bmatrix} \Rightarrow \begin{bmatrix} x(t) \\ y(t) \end{bmatrix} = e^{5t} \begin{bmatrix} 1 \\ 2 \end{bmatrix} \Rightarrow \begin{bmatrix} x(t) \\ y(t) \end{bmatrix} = \begin{bmatrix} e^{5t} \\ 2e^{5t} \end{bmatrix}$

4.2.15   Example 4.15

Determine the state-space representation of the circuit in Fig. 4.19, with the resistor voltage as the output.

Fig. 4.19 Circuit for Example 4.15

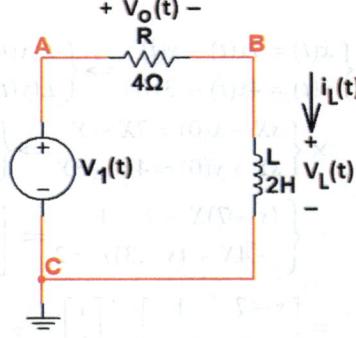

Solution:
From Fig. 4.19, we have

$$KVL@CABC : \ -V_1(t) + Ri_L(t) + L\frac{di_L(t)}{dt} = 0 \Rightarrow Ri_L(t) + L\frac{di_L(t)}{dt}$$

$$= V_1(t) \Rightarrow 4i_L(t) + 2\frac{di_L(t)}{dt} = V_1(t) \Rightarrow \frac{di_L(t)}{dt} = -2i_L(t) + \frac{1}{2}V_1(t)$$

$$V_o(t) = Ri_L(t) \Rightarrow V_o(t) = 4i_L(t) + 0V_1(t)$$

$$\begin{cases} \dfrac{di_L(t)}{dt} = -2i_L(t) + \dfrac{1}{2}V_1(t) \\ V_o(t) = 4i_L(t) + 0V_1(t) \end{cases} \Rightarrow \begin{cases} A = [-2] \\ B = \begin{bmatrix} \frac{1}{2} \end{bmatrix} \\ C = [4] \\ D = [0] \end{cases}$$

4.2.16   Example 4.16

Determine the state-space representation of the circuit in Fig. 4.20, with the resistor current as the output.

Fig. 4.20 Circuit for
Example 4.16

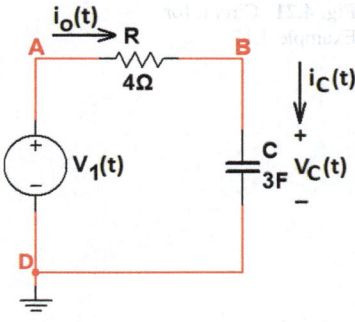

Solution:
From Fig. 4.20, we have

$$i_C(t) = C\frac{dv_C(t)}{dt} \Rightarrow i_C(t) = 3\frac{dv_C(t)}{dt}$$

$$KVL@DABD : -V_1(t) + Ri_C(t) + v_C(t) = 0 \Rightarrow RC\frac{dv_C(t)}{dt} + v_C(t)$$

$$= V_1(t) \Rightarrow \frac{dv_C(t)}{dt} = -\frac{1}{RC}v_C(t) + \frac{1}{RC}V_1(t) \Rightarrow \frac{dv_C(t)}{dt}$$

$$= -\frac{1}{12}v_C(t) + \frac{1}{12}V_1(t)$$

$$i_o(t) = \frac{v_A(t) - v_B(t)}{R} \Rightarrow i_o(t) = \frac{v_1(t) - v_C(t)}{R} \Rightarrow i_o(t) = \frac{-1}{4}v_C(t) + \frac{1}{4}v_1(t)$$

$$\begin{cases} \dfrac{dv_C(t)}{dt} = -\dfrac{1}{12}v_C(t) + \dfrac{1}{12}V_1(t) \\ i_o(t) = \dfrac{-1}{4}v_C(t) + \dfrac{1}{4}v_1(t) \end{cases} \Rightarrow \begin{cases} A = \left[-\dfrac{1}{12}\right] \\ B = \left[\dfrac{1}{12}\right] \\ C = \left[\dfrac{-1}{4}\right] \\ D = \left[\dfrac{1}{4}\right] \end{cases}$$

4.2.17 *Example 4.17*

Determine the state-space representation of the circuit in Fig. 4.21, with the resistor voltage as the output.

Fig. 4.21 Circuit for
Example 4.17

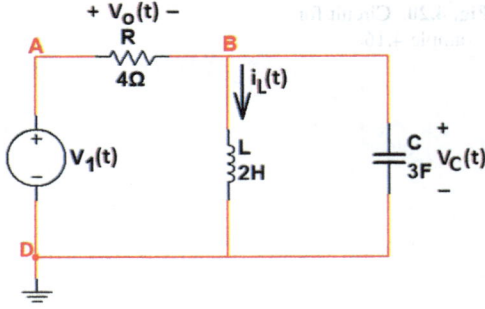

Solution:
The current flow in the distinct branches is depicted in Fig. 4.22.

Fig. 4.22 The current flow
in the distinct branches

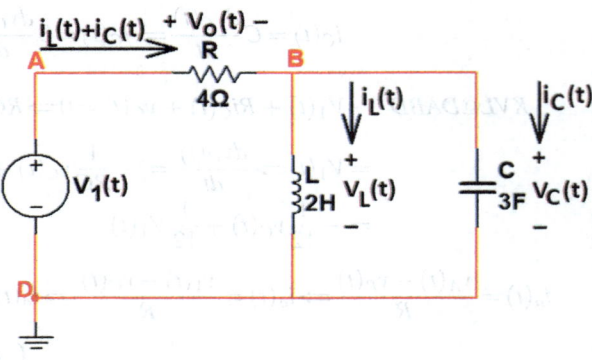

From Fig. 4.22, we have

$$KVL@DABD : \ -V_1(t) + R(i_L(t) + i_C(t)) + v_C(t) = 0 \Rightarrow -V_1(t)$$

$$+ R\left(i_L(t) + C\frac{dv_C(t)}{dt}\right) + v_C(t) = 0 \Rightarrow \frac{dv_C(t)}{dt} = -\frac{1}{RC}v_C(t)$$

$$-\frac{1}{C}i_L(t) + \frac{1}{RC}V_1(t) \Rightarrow \frac{dv_C(t)}{dt} = -\frac{1}{12}v_C(t) - \frac{1}{3}i_L(t) + \frac{1}{12}V_1(t)$$

$$v_L(t) = v_C(t) \Rightarrow L\frac{di_L(t)}{dt} = v_C(t) \Rightarrow \frac{di_L(t)}{dt} = \frac{1}{L}v_C(t) \Rightarrow \frac{di_L(t)}{dt} = 0.5v_C(t)$$

$$v_o(t) = v_A(t) - v_B(t) \Rightarrow v_o(t) = V_1(t) - v_C(t)$$

$$\begin{cases} \dfrac{dv_C(t)}{dt} = -\dfrac{1}{12}v_C(t) - \dfrac{1}{3}i_L(t) + \dfrac{1}{12}V_1(t) \\ \dfrac{di_L(t)}{dt} = 0.5v_C(t) \\ v_o(t) = V_1(t) - v_C(t) \end{cases} \Rightarrow \begin{cases} \begin{bmatrix} \dfrac{dv_C(t)}{dt} \\ \dfrac{di_L(t)}{dt} \end{bmatrix} = \begin{bmatrix} -\dfrac{1}{12} & -\dfrac{1}{3} \\ 0.5 & 0 \end{bmatrix} \begin{bmatrix} v_C(t) \\ i_L(t) \end{bmatrix} + \begin{bmatrix} \dfrac{1}{12} \\ 0 \end{bmatrix} V_1(t) \\ v_o(t) = \begin{bmatrix} -1 & 0 \end{bmatrix} \begin{bmatrix} v_C(t) \\ i_L(t) \end{bmatrix} + \begin{bmatrix} 1 \end{bmatrix} V_1(t) \end{cases}$$

$$\begin{cases} X = \begin{bmatrix} v_C(t) \\ i_L(t) \end{bmatrix} \\ A = \begin{bmatrix} -\dfrac{1}{12} & -\dfrac{1}{3} \\ 0.5 & 0 \end{bmatrix} \\ B = \begin{bmatrix} \dfrac{1}{12} \\ 0 \end{bmatrix} \\ C = \begin{bmatrix} -1 & 0 \end{bmatrix} \\ D = \begin{bmatrix} 1 \end{bmatrix} \end{cases}$$

Alternatively, the following state-space model can represent the given system.

$$\begin{cases} \dfrac{di_L(t)}{dt} = 0.5v_C(t) \\ \dfrac{dv_C(t)}{dt} = -\dfrac{1}{12}v_C(t) - \dfrac{1}{3}i_L(t) + \dfrac{1}{12}V_1(t) \\ v_o(t) = V_1(t) - v_C(t) \end{cases} \Rightarrow$$

$$\begin{cases} \begin{bmatrix} \dfrac{di_L(t)}{dt} \\ \dfrac{dv_C(t)}{dt} \end{bmatrix} = \begin{bmatrix} 0 & 0.5 \\ -\dfrac{1}{3} & -\dfrac{1}{12} \end{bmatrix} \begin{bmatrix} i_L(t) \\ v_C(t) \end{bmatrix} + \begin{bmatrix} 0 \\ \dfrac{1}{12} \end{bmatrix} V_1(t) \\ v_o(t) = \begin{bmatrix} 0 & -1 \end{bmatrix} \begin{bmatrix} i_L(t) \\ v_C(t) \end{bmatrix} + \begin{bmatrix} 1 \end{bmatrix} V_1(t) \end{cases}$$

$$\begin{cases} X = \begin{bmatrix} i_L(t) \\ v_C(t) \end{bmatrix} \\[2mm] A = \begin{bmatrix} 0 & 0.5 \\ -\dfrac{1}{3} & -\dfrac{1}{12} \end{bmatrix} \\[2mm] B = \begin{bmatrix} 0 \\ \dfrac{1}{12} \end{bmatrix} \\[2mm] C = \begin{bmatrix} 0 & -1 \end{bmatrix} \\[1mm] D = \begin{bmatrix} 1 \end{bmatrix} \end{cases}$$

4.2.18   Example 4.18

Determine the transfer function for the state-space model below, previously obtained
in the last example.

$$\begin{cases} \begin{bmatrix} \dfrac{di_L(t)}{dt} \\[3mm] \dfrac{dv_C(t)}{dt} \end{bmatrix} = \begin{bmatrix} 0 & 0.5 \\ -\dfrac{1}{3} & -\dfrac{1}{12} \end{bmatrix} \begin{bmatrix} i_L(t) \\ v_C(t) \end{bmatrix} + \begin{bmatrix} 0 \\ \dfrac{1}{12} \end{bmatrix} V_1(t) \\[6mm] v_o(t) = \begin{bmatrix} 0 & -1 \end{bmatrix} \begin{bmatrix} i_L(t) \\ v_C(t) \end{bmatrix} + [1] V_1(t) \end{cases}$$

Solution:

$$\begin{cases} \begin{bmatrix} \dfrac{di_L(t)}{dt} \\[3mm] \dfrac{dv_C(t)}{dt} \end{bmatrix} = \begin{bmatrix} 0 & 0.5 \\ -\dfrac{1}{3} & -\dfrac{1}{12} \end{bmatrix} \begin{bmatrix} i_L(t) \\ v_C(t) \end{bmatrix} + \begin{bmatrix} 0 \\ \dfrac{1}{12} \end{bmatrix} V_1(t) \\[6mm] v_o(t) = \begin{bmatrix} 0 & -1 \end{bmatrix} \begin{bmatrix} i_L(t) \\ v_C(t) \end{bmatrix} + [1] V_1(t) \end{cases} \Rightarrow \begin{cases} A = \begin{bmatrix} 0 & 0.5 \\ -\dfrac{1}{3} & -\dfrac{1}{12} \end{bmatrix} \\[4mm] B = \begin{bmatrix} 0 \\ \dfrac{1}{12} \end{bmatrix} \\[3mm] C = \begin{bmatrix} 0 & -1 \end{bmatrix} \\[1mm] D = [1] \end{cases}$$

$$T(s) = C(sI - A)^{-1}B + D \Rightarrow T(s)$$

$$= \begin{bmatrix} 0 & -1 \end{bmatrix} \left(\begin{bmatrix} s & 0 \\ 0 & s \end{bmatrix} - \begin{bmatrix} 0 & 0.5 \\ -\frac{1}{3} & -\frac{1}{12} \end{bmatrix} \right)^{-1} \begin{bmatrix} 0 \\ \frac{1}{12} \end{bmatrix} + [1] \Rightarrow T(s)$$

$$= \begin{bmatrix} 0 & -1 \end{bmatrix} \begin{bmatrix} s & -0.5 \\ \frac{1}{3} & s + \frac{1}{12} \end{bmatrix}^{-1} \begin{bmatrix} 0 \\ \frac{1}{12} \end{bmatrix} + [1] \Rightarrow T(s) = \begin{bmatrix} 0 & -1 \end{bmatrix}$$

$$\times \frac{1}{s(s + \frac{1}{12}) - (-0.5) \times \frac{1}{3}} \begin{bmatrix} s + \frac{1}{12} & 0.5 \\ -\frac{1}{3} & s \end{bmatrix} \begin{bmatrix} 0 \\ \frac{1}{12} \end{bmatrix} + [1] \Rightarrow T(s) = \frac{1}{s^2 + \frac{1}{12}s + \frac{1}{6}}$$

$$\times \begin{bmatrix} 0 & -1 \end{bmatrix} \begin{bmatrix} s + \frac{1}{12} & 0.5 \\ -\frac{1}{3} & s \end{bmatrix} \begin{bmatrix} 0 \\ \frac{1}{12} \end{bmatrix} + [1] \Rightarrow T(s) = \frac{12}{12s^2 + s + 2} \begin{bmatrix} \frac{1}{3} & -s \end{bmatrix}$$

$$\times \begin{bmatrix} 0 \\ \frac{1}{12} \end{bmatrix} + [1] \Rightarrow T(s) = \frac{12}{12s^2 + s + 2} \times \frac{-s}{12} + [1] \Rightarrow T(s) = \frac{-s}{12s^2 + s + 2}$$

$$+ [1] \Rightarrow T(s) = 1 - \frac{s}{12s^2 + s + 2} \Rightarrow T(s) = \frac{12s^2 + 2}{12s^2 + s + 2} \Rightarrow T(s)$$

$$= \frac{s^2 + 0.1667}{s^2 + 0.0833s + 0.1667}$$

We will calculate $\frac{V_o(s)}{V_1(s)}$ using the circuit in Fig. 4.23 to confirm the obtained result.

Fig. 4.23 Frequency domain equivalent of Fig. 4.21

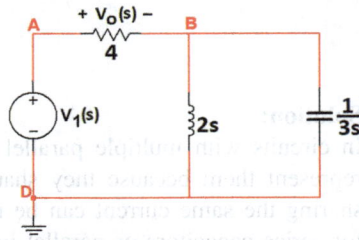

From Fig. 4.23, we have

$$V_o(s) = \frac{4}{4 + \frac{2s \times \frac{1}{3s}}{2s + \frac{1}{3s}}} V_1(s) \Rightarrow V_o(s) = \frac{4}{4 + \frac{\frac{2}{3}}{\frac{6s^2 + 1}{3s}}} V_1(s) \Rightarrow V_o(s)$$

$$= \frac{4}{4 + \frac{2s}{6s^2 + 1}} V_1(s) \Rightarrow V_o(s) = \frac{24s^2 + 4}{24s^2 + 2s + 4} V_1(s) \Rightarrow V_o(s)$$

$$= \frac{12s^2 + 2}{12s^2 + s + 2} V_1(s) \Rightarrow \frac{V_o(s)}{V_1(s)} = \frac{12s^2 + 2}{12s^2 + s + 2} \Rightarrow T(s)$$

$$= \frac{12s^2 + 2}{12s^2 + s + 2} \Rightarrow T(s) = \frac{s^2 + 0.1667}{s^2 + 0.0833s + 0.1667}$$

The current result validates the outcome of the previous method.

4.2.19 Example 4.19

Determine the state-space representation of the circuit in Fig. 4.24, with the $i_o(t)$ as the output.

Fig. 4.24 Circuit for Example 4.19

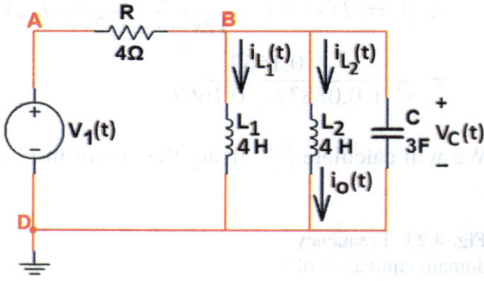

Solution:
In circuits with multiple parallel capacitors, a single equivalent capacitance can represent them because they share the same voltage. Similarly, series inductors sharing the same current can be replaced by an equivalent inductance. However, for series capacitors or parallel inductors, each component requires an individual state variable (voltage for capacitors, current for inductors) as they do not share these fundamental properties.

For the circuit depicted in Fig. 4.24, three state variables are necessary: $i_{L_1}(t), i_{L_2}(t)$, and $v_C(t)$. $V_{L_1}(t)$ and $V_{L_2}(t)$ are depicted in Fig. 4.25.

Fig. 4.25 $V_{L_1}(t)$ and $V_{L_2}(t)$

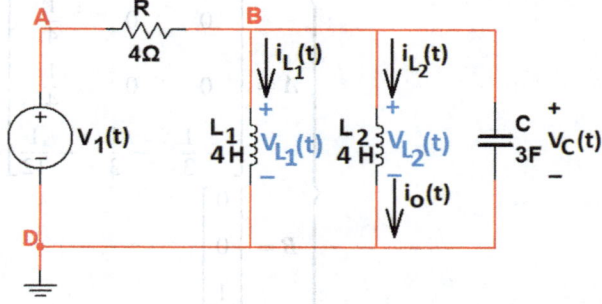

From Fig. 4.25, we have

$$v_{L_1}(t) = v_{L_2}(t) = v_C(t)$$

$$v_{L_1}(t) = v_C(t) \Rightarrow L_1 \frac{di_{L_1}(t)}{dt} = v_C(t) \Rightarrow \frac{di_{L_1}(t)}{dt} = \frac{1}{L_1} v_C(t) \Rightarrow \frac{di_{L_1}(t)}{dt} = \frac{1}{4} v_C(t)$$

$$v_{L_2}(t) = v_C(t) \Rightarrow L_2 \frac{di_{L_2}(t)}{dt} = v_C(t) \Rightarrow \frac{di_{L_2}(t)}{dt} = \frac{1}{L_2} v_C(t) \Rightarrow \frac{di_{L_2}(t)}{dt} = \frac{1}{4} v_C(t)$$

$$KCL@B : i_{L_1}(t) + i_{L_2}(t) + C\frac{dv_C(t)}{dt} + \frac{V_B - V_A}{R} = 0 \Rightarrow i_{L_1}(t) + i_{L_2}(t) + 3\frac{dv_C(t)}{dt}$$

$$+ \frac{v_C(t) - V_1(t)}{4} = 0 \Rightarrow \frac{dv_C(t)}{dt} = -\frac{1}{3} i_{L_1}(t) - \frac{1}{3} i_{L_2}(t) - \frac{v_C(t)}{12} + \frac{V_1(t)}{12}$$

$$\begin{cases} \dfrac{di_{L_1}(t)}{dt} = \dfrac{1}{4} v_C(t) \\[2mm] \dfrac{di_{L_2}(t)}{dt} = \dfrac{1}{4} v_C(t) \\[2mm] \dfrac{dv_C(t)}{dt} = -\dfrac{1}{3} i_{L_1}(t) - \dfrac{1}{3} i_{L_2}(t) - \dfrac{v_C(t)}{12} + \dfrac{V_1(t)}{12} \end{cases} \Rightarrow \begin{bmatrix} \dfrac{di_{L_1}(t)}{dt} \\[2mm] \dfrac{di_{L_2}(t)}{dt} \\[2mm] \dfrac{dv_C(t)}{dt} \end{bmatrix}$$

$$= \begin{bmatrix} 0 & 0 & \dfrac{1}{4} \\[2mm] 0 & 0 & \dfrac{1}{4} \\[2mm] -\dfrac{1}{3} & -\dfrac{1}{3} & -\dfrac{1}{12} \end{bmatrix} \begin{bmatrix} i_{L_1}(t) \\[2mm] i_{L_2}(t) \\[2mm] v_C(t) \end{bmatrix} + \begin{bmatrix} 0 \\[2mm] 0 \\[2mm] 1 \end{bmatrix} V_1(t)$$

$$i_o(t) = i_{L_2}(t) \Rightarrow i_o(t) = \begin{bmatrix} 0 & 1 & 0 \end{bmatrix} \begin{bmatrix} i_{L_1}(t) \\[2mm] i_{L_2}(t) \\[2mm] v_C(t) \end{bmatrix} + [0] V_1(t)$$

Therefore,

$$
\begin{cases}
A = \begin{bmatrix} 0 & 0 & \dfrac{1}{4} \\[2mm] 0 & 0 & \dfrac{1}{4} \\[2mm] -\dfrac{1}{3} & -\dfrac{1}{3} & -\dfrac{1}{12} \end{bmatrix} \\[10mm]
B = \begin{bmatrix} 0 \\ 0 \\ 1 \end{bmatrix} \\[6mm]
C = \begin{bmatrix} 0 & 1 & 0 \end{bmatrix} \\[2mm]
D = \begin{bmatrix} 0 \end{bmatrix}
\end{cases}
$$

4.2.20   Exercise 4.1

Determine the state-space representation of the circuits in Figs. 4.26 and 4.27, with the $i_o(t)$ as the output.

Fig. 4.26 Circuit for Exercise 4.1 (This circuit requires three state variables)

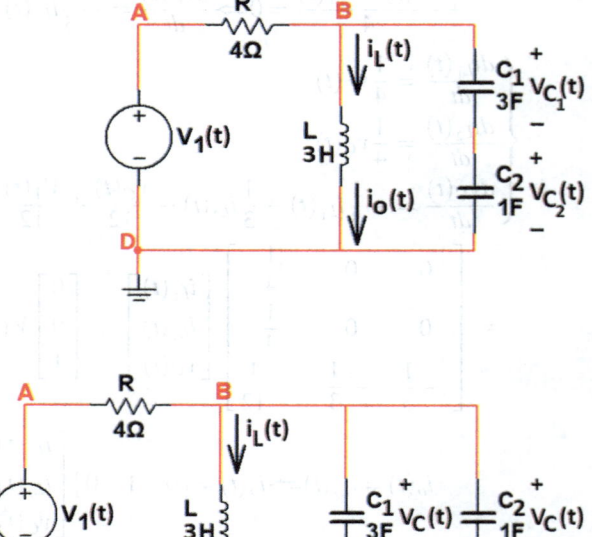

Fig. 4.27 Circuit for Exercise 4.1 (This circuit requires two state variables)

4.2.21 Example 4.20

Determine the state-space representation of the circuit in Fig. 4.28, with the $V_o(t)$ as the output.

Fig. 4.28 Circuit for Example 4.20

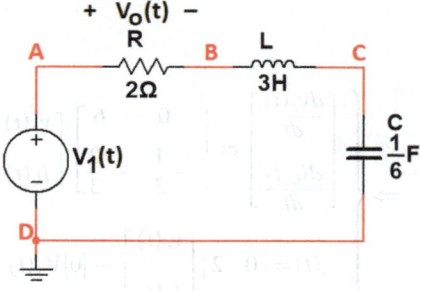

Solution:
The problem statement does not dictate the inductor current and capacitor voltage, so we will assign them (Fig. 4.29).

Fig. 4.29 $i_L(t)$ and $v_c(t)$

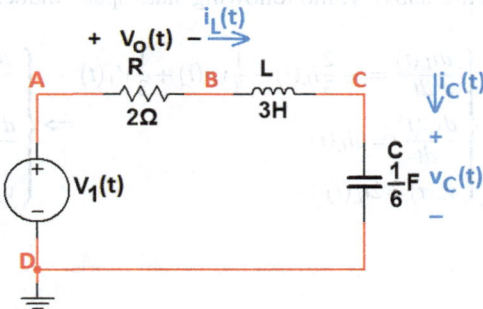

From Fig. 4.29, we have

$$KCL@C: i_L(t) = i_C(t) \Rightarrow i_L(t) = C\frac{dv_c(t)}{dt} \Rightarrow i_L(t) = \frac{1}{6}\frac{dv_c(t)}{dt} \Rightarrow \frac{dv_c(t)}{dt} = 6i_L(t)$$

$$KVL@DABCD: -V_1(t) + Ri_L(t) + L\frac{di_L(t)}{dt} + v_C(t) = 0 \Rightarrow -V_1(t) + 2i_L(t)$$

$$+ 3\frac{di_L(t)}{dt} + v_C(t) = 0 \Rightarrow 2i_L(t) + 3\frac{di_L(t)}{dt} + v_C(t)$$

$$= V_1(t) \Rightarrow \frac{di_L(t)}{dt} = -\frac{2}{3}i_L(t) - \frac{1}{3}v_C(t) + \frac{1}{3}V_1(t)$$

$$v_o(t) = Ri_L(t) \Rightarrow v_o(t) = 2i_L(t)$$

$$\begin{cases} \dfrac{dv_c(t)}{dt} = 6i_L(t) \\[2mm] \dfrac{di_L(t)}{dt} = -\dfrac{2}{3}i_L(t) - \dfrac{1}{3}v_C(t) + \dfrac{1}{3}V_1(t) \\[2mm] v_o(t) = 2i_L(t) \end{cases} \Rightarrow \begin{cases} \dfrac{dv_c(t)}{dt} = 6i_L(t) + 0v_C(t) + 0V_1(t) \\[2mm] \dfrac{di_L(t)}{dt} = -\dfrac{2}{3}i_L(t) - \dfrac{1}{3}v_C(t) + \dfrac{1}{3}V_1(t) \\[2mm] v_o(t) = 2i_L(t) + 0v_C(t) + 0V_1(t) \end{cases}$$

$$X = \begin{bmatrix} v_c(t) \\ i_L(t) \end{bmatrix}$$

$$\Rightarrow \begin{cases} \begin{bmatrix} \dfrac{dv_c(t)}{dt} \\[3mm] \dfrac{di_L(t)}{dt} \end{bmatrix} = \begin{bmatrix} 0 & 6 \\ -\dfrac{1}{3} & -\dfrac{2}{3} \end{bmatrix} \begin{bmatrix} v_c(t) \\ i_L(t) \end{bmatrix} + \begin{bmatrix} 0 \\ \dfrac{1}{3} \end{bmatrix} V_1(t) \\[8mm] v_o(t) = \begin{bmatrix} 0 & 2 \end{bmatrix} \begin{bmatrix} v_c(t) \\ i_L(t) \end{bmatrix} + [0]V_1(t) \end{cases} \Rightarrow \begin{cases} A = \begin{bmatrix} 0 & 6 \\ -\dfrac{1}{3} & -\dfrac{2}{3} \end{bmatrix} \\[8mm] B = \begin{bmatrix} 0 \\ 1 \\ \dfrac{1}{3} \end{bmatrix} \\[6mm] C = \begin{bmatrix} 0 & 2 \end{bmatrix} \\[2mm] D = [0] \end{cases}$$

Alternatively, the following state-space model can represent the given system.

$$\begin{cases} \dfrac{di_L(t)}{dt} = -\dfrac{2}{3}i_L(t) - \dfrac{1}{3}v_C(t) + \dfrac{1}{3}V_1(t) \\[2mm] \dfrac{dv_c(t)}{dt} = 6i_L(t) \\[2mm] v_o(t) = 2i_L(t) \end{cases} \Rightarrow \begin{cases} \dfrac{di_L(t)}{dt} = -\dfrac{2}{3}i_L(t) - \dfrac{1}{3}v_C(t) + \dfrac{1}{3}V_1(t) \\[2mm] \dfrac{dv_c(t)}{dt} = 6i_L(t) + 0v_C(t) + 0V_1(t) \\[2mm] v_o(t) = 2i_L(t) + 0v_C(t) + 0V_1(t) \end{cases}$$

$$X = \begin{bmatrix} i_L(t) \\ v_C(t) \end{bmatrix}$$

$$\Rightarrow \begin{cases} \begin{bmatrix} \dfrac{di_L(t)}{dt} \\[3mm] \dfrac{dv_C(t)}{dt} \end{bmatrix} = \begin{bmatrix} -\dfrac{2}{3} & -\dfrac{1}{3} \\ 6 & 0 \end{bmatrix} \begin{bmatrix} i_L(t) \\ v_C(t) \end{bmatrix} + \begin{bmatrix} \dfrac{1}{3} \\ 0 \end{bmatrix} V_1(t) \\[8mm] v_o(t) = \begin{bmatrix} 2 & 0 \end{bmatrix} \begin{bmatrix} i_L(t) \\ v_C(t) \end{bmatrix} + [0]V_1(t) \end{cases} \Rightarrow \begin{cases} A = \begin{bmatrix} -\dfrac{2}{3} & -\dfrac{1}{3} \\ 6 & 0 \end{bmatrix} \\[8mm] B = \begin{bmatrix} \dfrac{1}{3} \\ 0 \end{bmatrix} \\[6mm] C = \begin{bmatrix} 2 & 0 \end{bmatrix} \\[2mm] D = [0] \end{cases}$$

4.2.22 Example 4.21

Determine the state-space representation of the circuit in Fig. 4.30, with the $V_o(t)$ as the output.

Fig. 4.30 Circuit for Example 4.21

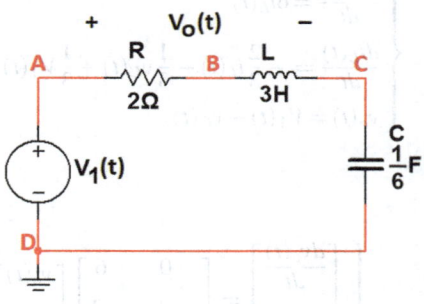

Solution:
The problem statement does not dictate the inductor current and capacitor voltage, so we will assign them (Fig. 4.31).

Fig. 4.31 $i_L(t)$ and $v_C(t)$

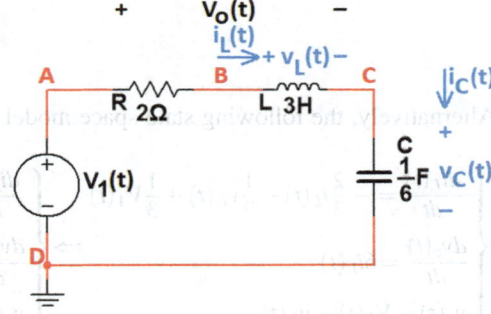

From Fig. 4.31, we have

$$KCL@C : i_L(t) = i_C(t) \Rightarrow i_L(t) = C\frac{dv_C(t)}{dt} \Rightarrow i_L(t) = \frac{1}{6}\frac{dv_C(t)}{dt} \Rightarrow \frac{dv_C(t)}{dt} = 6i_L(t)$$

$$KVL@DABCD : -V_1(t) + Ri_L(t) + L\frac{di_L(t)}{dt} + v_C(t) = 0 \Rightarrow -V_1(t) + 2i_L(t)$$

$$+ 3\frac{di_L(t)}{dt} + v_C(t) = 0 \Rightarrow 2i_L(t) + 3\frac{di_L(t)}{dt} + v_C(t)$$

$$= V_1(t) \Rightarrow \frac{di_L(t)}{dt} = -\frac{2}{3}i_L(t) - \frac{1}{3}v_C(t) + \frac{1}{3}V_1(t)$$

$$v_o(t) = v_R(t) + v_L(t) \Rightarrow v_o(t) = Ri_L(t) + v_L(t)$$

$$\textbf{KVL@DABCD}: \ -V_1(t) + Ri_L(t) + v_L(t) + v_C(t) = 0 \Rightarrow Ri_L(t) + v_L(t) = V_1(t) - v_C(t)$$

$$v_o(t) = Ri_L(t) + v_L(t) \Rightarrow v_o(t) = V_1(t) - v_C(t)$$

$$\begin{cases} \dfrac{dv_c(t)}{dt} = 6i_L(t) \\[2mm] \dfrac{di_L(t)}{dt} = -\dfrac{2}{3}i_L(t) - \dfrac{1}{3}v_C(t) + \dfrac{1}{3}V_1(t) \\[2mm] v_o(t) = V_1(t) - v_C(t) \end{cases} \Rightarrow \begin{cases} \dfrac{dv_c(t)}{dt} = 6i_L(t) + 0v_C(t) + 0V_1(t) \\[2mm] \dfrac{di_L(t)}{dt} = -\dfrac{2}{3}i_L(t) - \dfrac{1}{3}v_C(t) + \dfrac{1}{3}V_1(t) \\[2mm] v_o(t) = 0i_L(t) - v_C(t) + V_1(t) \end{cases}$$

$$\Rightarrow \begin{cases} \begin{bmatrix} \dfrac{dv_c(t)}{dt} \\[2mm] \dfrac{di_L(t)}{dt} \end{bmatrix} = \begin{bmatrix} 0 & 6 \\[1mm] -\dfrac{1}{3} & -\dfrac{2}{3} \end{bmatrix} \begin{bmatrix} v_c(t) \\[1mm] i_L(t) \end{bmatrix} + \begin{bmatrix} 0 \\[1mm] 1 \\[1mm] \dfrac{1}{3} \end{bmatrix} V_1(t) \\[6mm] v_o(t) = \begin{bmatrix} -1 & 0 \end{bmatrix} \begin{bmatrix} v_c(t) \\[1mm] i_L(t) \end{bmatrix} + [1]V_1(t) \end{cases}$$

$$\Rightarrow \begin{cases} X = \begin{bmatrix} v_c(t) \\[1mm] i_L(t) \end{bmatrix} \\[6mm] A = \begin{bmatrix} 0 & 6 \\[1mm] -\dfrac{1}{3} & -\dfrac{2}{3} \end{bmatrix} \\[6mm] B = \begin{bmatrix} 0 \\[1mm] 1 \\[1mm] \dfrac{1}{3} \end{bmatrix} \\[6mm] C = \begin{bmatrix} -1 & 0 \end{bmatrix} \\[2mm] D = [1] \end{cases}$$

Alternatively, the following state-space model can represent the given system.

$$\begin{cases} \dfrac{di_L(t)}{dt} = -\dfrac{2}{3}i_L(t) - \dfrac{1}{3}v_C(t) + \dfrac{1}{3}V_1(t) \\[2mm] \dfrac{dv_c(t)}{dt} = 6i_L(t) \\[2mm] v_o(t) = V_1(t) - v_C(t) \end{cases} \Rightarrow \begin{cases} \dfrac{di_L(t)}{dt} = -\dfrac{2}{3}i_L(t) - \dfrac{1}{3}v_C(t) + \dfrac{1}{3}V_1(t) \\[2mm] \dfrac{dv_c(t)}{dt} = 6i_L(t) + 0v_C(t) + 0V_1(t) \\[2mm] v_o(t) = 0i_L(t) - v_C(t) + V_1(t) \end{cases}$$

$$\Rightarrow \begin{cases} \begin{bmatrix} \dfrac{di_L(t)}{dt} \\[2mm] \dfrac{dv_C(t)}{dt} \end{bmatrix} = \begin{bmatrix} -\dfrac{2}{3} & -\dfrac{1}{3} \\[1mm] 6 & 0 \end{bmatrix} \begin{bmatrix} i_L(t) \\[1mm] v_C(t) \end{bmatrix} + \begin{bmatrix} \dfrac{1}{3} \\[1mm] 0 \end{bmatrix} V_1(t) \\[6mm] v_o(t) = \begin{bmatrix} 0 & -1 \end{bmatrix} \begin{bmatrix} i_L(t) \\[1mm] v_C(t) \end{bmatrix} + [1]V_1(t) \end{cases}$$

$$\Rightarrow \begin{cases} X = \begin{bmatrix} i_L(t) \\[1mm] v_C(t) \end{bmatrix} \\[6mm] A = \begin{bmatrix} -\dfrac{2}{3} & -\dfrac{1}{3} \\[1mm] 6 & 0 \end{bmatrix} \\[6mm] B = \begin{bmatrix} \dfrac{1}{3} \\[1mm] 0 \end{bmatrix} \\[6mm] C = \begin{bmatrix} 0 & -1 \end{bmatrix} \\[2mm] D = [1] \end{cases}$$

4.2.23 *Example 4.22*

Determine the state-space representation of the circuit in Fig. 4.32, with the $i_o(t)$ as the output.

Fig. 4.32 Circuit for
Example 4.22

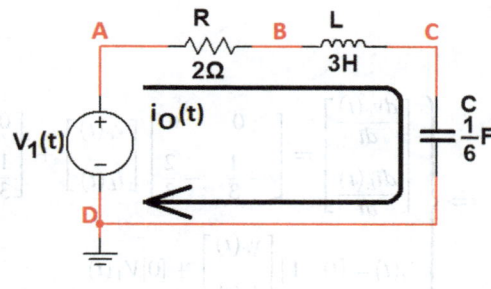

Solution:
The problem statement does not dictate the inductor current and capacitor voltage, so we will assign them (Fig. 4.33).

Fig. 4.33 $i_L(t)$ and $v_C(t)$

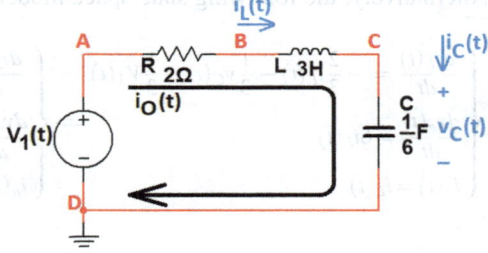

From Fig. 4.33, we have

$$KCL@C : i_L(t) = i_C(t) \Rightarrow i_L(t) = C\frac{dv_c(t)}{dt} \Rightarrow i_L(t) = \frac{1}{6}\frac{dv_c(t)}{dt} \Rightarrow \frac{dv_c(t)}{dt} = 6i_L(t)$$

$$KVL@DABCD : -V_1(t) + Ri_L(t) + L\frac{di_L(t)}{dt} + v_C(t) = 0 \Rightarrow -V_1(t) + 2i_L(t)$$

$$+ 3\frac{di_L(t)}{dt} + v_C(t) = 0 \Rightarrow 2i_L(t) + 3\frac{di_L(t)}{dt} + v_C(t)$$

$$= V_1(t) \Rightarrow \frac{di_L(t)}{dt} = -\frac{2}{3}i_L(t) - \frac{1}{3}v_C(t) + \frac{1}{3}V_1(t)$$

$$i_o(t) = i_L(t)$$

$$\begin{cases} \dfrac{dv_c(t)}{dt} = 6i_L(t) \\ \dfrac{di_L(t)}{dt} = -\dfrac{2}{3}i_L(t) - \dfrac{1}{3}v_C(t) + \dfrac{1}{3}V_1(t) \\ i_o(t) = i_L(t) \end{cases} \Rightarrow \begin{cases} \dfrac{dv_c(t)}{dt} = 6i_L(t) + 0v_C(t) + 0V_1(t) \\ \dfrac{di_L(t)}{dt} = -\dfrac{2}{3}i_L(t) - \dfrac{1}{3}v_C(t) + \dfrac{1}{3}V_1(t) \\ i_o(t) = i_L(t) + 0v_C(t) + 0V_1(t) \end{cases}$$

$$\Rightarrow \begin{cases} \begin{bmatrix} \dfrac{dv_c(t)}{dt} \\ \dfrac{di_L(t)}{dt} \end{bmatrix} = \begin{bmatrix} 0 & 6 \\ -\dfrac{1}{3} & -\dfrac{2}{3} \end{bmatrix} \begin{bmatrix} v_c(t) \\ i_L(t) \end{bmatrix} + \begin{bmatrix} 0 \\ \dfrac{1}{3} \end{bmatrix} V_1(t) \\ i_o(t) = \begin{bmatrix} 0 & 1 \end{bmatrix} \begin{bmatrix} v_c(t) \\ i_L(t) \end{bmatrix} + [0]V_1(t) \end{cases} \Rightarrow \begin{cases} X = \begin{bmatrix} v_c(t) \\ i_L(t) \end{bmatrix} \\ A = \begin{bmatrix} 0 & 6 \\ -\dfrac{1}{3} & -\dfrac{2}{3} \end{bmatrix} \\ B = \begin{bmatrix} 0 \\ \dfrac{1}{3} \end{bmatrix} \\ C = \begin{bmatrix} 0 & 1 \end{bmatrix} \\ D = [1] \end{cases}$$

Alternatively, the following state-space model can represent the given system.

$$\begin{cases} \dfrac{di_L(t)}{dt} = -\dfrac{2}{3}i_L(t) - \dfrac{1}{3}v_C(t) + \dfrac{1}{3}V_1(t) \\ \dfrac{dv_c(t)}{dt} = 6i_L(t) \\ i_o(t) = i_L(t) \end{cases} \Rightarrow \begin{cases} \dfrac{di_L(t)}{dt} = -\dfrac{2}{3}i_L(t) - \dfrac{1}{3}v_C(t) + \dfrac{1}{3}V_1(t) \\ \dfrac{dv_c(t)}{dt} = 6i_L(t) + 0v_C(t) + 0V_1(t) \\ i_o(t) = 1i_L(t) + 0v_C(t) + 0V_1(t) \end{cases}$$

$$\Rightarrow \begin{cases} \begin{bmatrix} \dfrac{di_L(t)}{dt} \\ \dfrac{dv_C(t)}{dt} \end{bmatrix} = \begin{bmatrix} -\dfrac{2}{3} & -\dfrac{1}{3} \\ 6 & 0 \end{bmatrix} \begin{bmatrix} i_L(t) \\ v_C(t) \end{bmatrix} + \begin{bmatrix} \dfrac{1}{3} \\ 0 \end{bmatrix} V_1(t) \\ i_o(t) = \begin{bmatrix} 1 & 0 \end{bmatrix} \begin{bmatrix} i_L(t) \\ v_C(t) \end{bmatrix} + [1]V_1(t) \end{cases} \Rightarrow \begin{cases} X = \begin{bmatrix} i_L(t) \\ v_C(t) \end{bmatrix} \\ A = \begin{bmatrix} -\dfrac{2}{3} & -\dfrac{1}{3} \\ 6 & 0 \end{bmatrix} \\ B = \begin{bmatrix} \dfrac{1}{3} \\ 0 \end{bmatrix} \\ C = \begin{bmatrix} 1 & 0 \end{bmatrix} \\ D = [1] \end{cases}$$

4.2.24 Example 4.23

Solve the
$$\begin{cases} \begin{bmatrix} \dfrac{di_L(t)}{dt} \\[2mm] \dfrac{dv_C(t)}{dt} \end{bmatrix} = \begin{bmatrix} -\dfrac{2}{3} & -\dfrac{1}{3} \\[2mm] 6 & 0 \end{bmatrix} \begin{bmatrix} i_L(t) \\[2mm] v_C(t) \end{bmatrix} + \begin{bmatrix} \dfrac{1}{3} \\[2mm] 0 \end{bmatrix} V_1(t) \\[6mm] i_o(t) = \begin{bmatrix} 1 & 0 \end{bmatrix} \begin{bmatrix} i_L(t) \\[2mm] v_C(t) \end{bmatrix} + [1]V_1(t) \end{cases}$$ with

$V_1(t) = H(t)$ and $x_0 = \begin{bmatrix} i_L(0) \\[1mm] v_C(0) \end{bmatrix} = \begin{bmatrix} 1 \\[1mm] 2 \end{bmatrix}$.

Solution:

$$\begin{cases} \begin{bmatrix} \dfrac{di_L(t)}{dt} \\[2mm] \dfrac{dv_C(t)}{dt} \end{bmatrix} = \begin{bmatrix} -\dfrac{2}{3} & -\dfrac{1}{3} \\[2mm] 6 & 0 \end{bmatrix} \begin{bmatrix} i_L(t) \\[2mm] v_C(t) \end{bmatrix} + \begin{bmatrix} \dfrac{1}{3} \\[2mm] 0 \end{bmatrix} H(t) \\[6mm] i_o(t) = \begin{bmatrix} 1 & 0 \end{bmatrix} \begin{bmatrix} i_L(t) \\[2mm] v_C(t) \end{bmatrix} + [1]H(t) \end{cases} \Rightarrow \begin{cases} \dfrac{di_L(t)}{dt} = -\dfrac{2}{3}i_L(t) - \dfrac{1}{3}v_C(t) + \dfrac{1}{3}H(t) \\[3mm] \dfrac{dv_c(t)}{dt} = 6i_L(t) \\[3mm] i_o(t) = i_L(t) + H(t) \end{cases}$$

$$\Rightarrow \begin{cases} \mathcal{L}\left\{\dfrac{di_L(t)}{dt}\right\} = \mathcal{L}\left\{-\dfrac{2}{3}i_L(t) - \dfrac{1}{3}v_C(t) + \dfrac{1}{3}H(t)\right\} \\[3mm] \mathcal{L}\left\{\dfrac{dv_c(t)}{dt}\right\} = \mathcal{L}\{6i_L(t)\} \\[3mm] \mathcal{L}\{i_o(t)\} = \mathcal{L}\{i_L(t) + H(t)\} \end{cases} \Rightarrow \begin{cases} sI_L(s) - i_L(0) = -\dfrac{2}{3}I_L(s) - \dfrac{1}{3}V_C(s) + \dfrac{1}{3s} \\[3mm] sV_C(s) - v_C(0) = 6I_L(s) \\[3mm] I_o(s) = I_L(s) + \dfrac{1}{s} \end{cases}$$

$$\Rightarrow \begin{cases} sI_L(s) - 1 = -\dfrac{2}{3}I_L(s) - \dfrac{1}{3}V_C(s) + \dfrac{1}{3s} \\[3mm] sV_C(s) - 2 = 6I_L(s) \\[3mm] I_o(s) = I_L(s) + \dfrac{1}{s} \end{cases} \Rightarrow \begin{cases} sI_L(s) = -\dfrac{2}{3}I_L(s) - \dfrac{1}{3}V_C(s) + \dfrac{1}{3s} + 1 \\[3mm] sV_C(s) = 6I_L(s) + 2 \\[3mm] I_o(s) = I_L(s) + \dfrac{1}{s} \end{cases}$$

$$\Rightarrow \begin{cases} \left(s + \dfrac{2}{3}\right)I_L(s) + \dfrac{1}{3}V_C(s) = \dfrac{3s+1}{3s} \\[3mm] 6I_L(s) - sV_C(s) = -2 \\[3mm] I_o(s) = I_L(s) + \dfrac{1}{s} \end{cases} \Rightarrow \begin{cases} \begin{bmatrix} s + \dfrac{2}{3} & \dfrac{1}{3} \\[2mm] 6 & -s \end{bmatrix} \begin{bmatrix} I_L(s) \\[2mm] V_C(s) \end{bmatrix} = \begin{bmatrix} \dfrac{3s+1}{3s} \\[2mm] -2 \end{bmatrix} \\[6mm] I_o(s) = I_L(s) + \dfrac{1}{s} \end{cases}$$

$$\Rightarrow \begin{cases} \begin{bmatrix} I_L(s) \\[2mm] V_C(s) \end{bmatrix} = \begin{bmatrix} s + \dfrac{2}{3} & \dfrac{1}{3} \\[2mm] 6 & -s \end{bmatrix}^{-1} \begin{bmatrix} \dfrac{3s+1}{3s} \\[2mm] -2 \end{bmatrix} \\[6mm] I_o(s) = I_L(s) + \dfrac{1}{s} \end{cases} \Rightarrow \begin{cases} \begin{bmatrix} I_L(s) \\[2mm] V_C(s) \end{bmatrix} = \dfrac{1}{-\left(s + \dfrac{2}{3}\right)s - 2} \begin{bmatrix} -s & -\dfrac{1}{3} \\[2mm] -6 & s + \dfrac{2}{3} \end{bmatrix} \begin{bmatrix} \dfrac{3s+1}{3s} \\[2mm] -2 \end{bmatrix} \\[6mm] I_o(s) = I_L(s) + \dfrac{1}{s} \end{cases}$$

$$
\Rightarrow \begin{cases} \begin{bmatrix} I_L(s) \\ V_C(s) \end{bmatrix} = -\dfrac{1}{s^2+\frac{2}{3}s+2} \begin{bmatrix} \dfrac{-3s-1}{3}+\dfrac{2}{3} \\ \dfrac{-6s-2}{s}-2s-\dfrac{4}{3} \end{bmatrix} \\ I_o(s)=I_L(s)+\dfrac{1}{s} \end{cases} \Rightarrow \begin{cases} \begin{bmatrix} I_L(s) \\ V_C(s) \end{bmatrix} = -\dfrac{1}{s^2+\frac{2}{3}s+2} \begin{bmatrix} -s+\dfrac{1}{3} \\ \dfrac{-2}{s}-2s-\dfrac{22}{3} \end{bmatrix} \\ I_o(s)=I_L(s)+\dfrac{1}{s} \end{cases}
$$

$$
\Rightarrow \begin{cases} \begin{bmatrix} I_L(s) \\ V_C(s) \end{bmatrix} = \dfrac{1}{s^2+\frac{2}{3}s+2} \begin{bmatrix} s-\dfrac{1}{3} \\ \dfrac{2}{s}+2s+\dfrac{22}{3} \end{bmatrix} \\ I_o(s)=I_L(s)+\dfrac{1}{s} \end{cases} \Rightarrow \begin{cases} \begin{bmatrix} I_L(s) \\ V_C(s) \end{bmatrix} = \begin{bmatrix} \dfrac{s-\dfrac{1}{3}}{s^2+\frac{2}{3}s+2} \\ \dfrac{\dfrac{2}{s}+2s+\dfrac{22}{3}}{s^2+\frac{2}{3}s+2} \end{bmatrix} \\ I_o(s)=I_L(s)+\dfrac{1}{s} \end{cases}
$$

$$
\Rightarrow \begin{cases} \begin{bmatrix} I_L(s) \\ V_C(s) \end{bmatrix} = \begin{bmatrix} \dfrac{3s-1}{3s^2+2s+6} \\ \dfrac{6s^2+22s+6}{(3s^2+2s+6)s} \end{bmatrix} \\ I_o(s)=\dfrac{3s-1}{3s^2+2s+6}+\dfrac{1}{s} \end{cases} \Rightarrow \begin{cases} \begin{bmatrix} i_L(t) \\ v_C(t) \end{bmatrix} = \mathcal{L}^{-1} \left\{ \begin{bmatrix} \dfrac{3s-1}{3s^2+2s+6} \\ \dfrac{6s^2+22s+6}{(3s^2+2s+6)s} \end{bmatrix} \right\} \\ i_o(t)=\mathcal{L}^{-1}\left\{\dfrac{3s-1}{3s^2+2s+6}+\dfrac{1}{s}\right\} \end{cases}
$$

$$
\Rightarrow \begin{cases} i_L(t)=\mathcal{L}^{-1}\left\{\dfrac{3s-1}{3s^2+2s+6}\right\} \\ v_C(t)=\mathcal{L}^{-1}\left\{\dfrac{6s^2+22s+6}{(3s^2+2s+6)s}\right\} \\ i_o(t)=\mathcal{L}^{-1}\left\{\dfrac{3s-1}{3s^2+2s+6}+\dfrac{1}{s}\right\} \end{cases}
$$

You can use MATLAB for inverse Laplace calculations (Figs. 4.34, 4.35, and 4.36).

```
Command Window
>> syms s
>> ilaplace((3*s-1)/(3*s^2+2*s+6))

ans =

exp(-t/3)*(cos((17^(1/2)*t)/3) - (2*17^(1/2)*sin((17^(1/2)*t)/3))/17)

>> vpa(ans,4)

ans =

exp(-0.3333*t)*(cos(1.374*t) - 0.4851*sin(1.374*t))

fx >>
```

Fig. 4.34   MATLAB code

```
Command Window                                                    ⊙

  >> syms s
  >> ilaplace((6*s^2+22*s+6)/s/(3*s^2+2*s+6))

  ans =

  exp(-t/3)*(cos((17^(1/2)*t)/3) + (19*17^(1/2)*sin((17^(1/2)*t)/3))/17) + 1

  >> vpa(ans,4)

  ans =

  exp(-0.3333*t)*(cos(1.374*t) + 4.608*sin(1.374*t)) + 1.0

fx >>
```

Fig. 4.35 MATLAB code

```
Command Window                                                    ⊙

  >> syms s
  >> ilaplace((3*s-1)/(3*s^2+2*s+6)+1/s)

  ans =

  exp(-t/3)*(cos((17^(1/2)*t)/3) - (2*17^(1/2)*sin((17^(1/2)*t)/3))/17) + 1

  >> vpa(ans,4)

  ans =

  exp(-0.3333*t)*(cos(1.374*t) - 0.4851*sin(1.374*t)) + 1.0

fx >>
```

Fig. 4.36 MATLAB code

As shown in Figs. 4.34, 4.35, and 4.36, we have

$$i_L(t) = e^{-0.3333t}(\cos(1.374t) - 0.4851\sin(1.374t))$$
$$v_C(t) = e^{-0.3333t}(\cos(1.374t) + 4.608\sin(1.374t)) + 1$$
$$i_o(t) = e^{-0.3333t}(\cos(1.374t) - 0.4851\sin(1.374t)) + 1$$

4.2.25   *Example 4.24*

Use MATLAB to find solutions for
$$\begin{cases} \begin{bmatrix} \dfrac{di_L(t)}{dt} \\ \dfrac{dv_C(t)}{dt} \end{bmatrix} = \begin{bmatrix} -\dfrac{2}{3} & -\dfrac{1}{3} \\ 6 & 0 \end{bmatrix} \begin{bmatrix} i_L(t) \\ v_C(t) \end{bmatrix} + \begin{bmatrix} \dfrac{1}{3} \\ 0 \end{bmatrix} V_1(t) \\[4ex] i_o(t) = [1\ 0] \begin{bmatrix} i_L(t) \\ v_C(t) \end{bmatrix} + [1] V_1(t) \end{cases}$$

with $V_1(t) = H(t)$ and $x_0 = \begin{bmatrix} i_L(0) \\ v_C(0) \end{bmatrix} = \begin{bmatrix} 1 \\ 2 \end{bmatrix}$.

Solution:

$$\begin{cases} \dot{x}(t) = Ax(t) + Bu(t) \\ y(t) = Cx(t) + Du(t) \end{cases} \Rightarrow \begin{cases} sX(s) - x_0 = AX(s) + BU(s) \\ Y(s) = CX(s) + DU(s) \end{cases}$$

$$\Rightarrow \begin{cases} sI_{n \times n}X(s) - AX(s) = x_0 + BU(s) \\ Y(s) = CX(s) + DU(s) \end{cases} \Rightarrow \begin{cases} (sI_{n \times n} - A)X(s) = x_0 + BU(s) \\ Y(s) = CX(s) + DU(s) \end{cases}$$

$$\Rightarrow \begin{cases} X(s) = (sI_{n \times n} - A)^{-1}x_0 + (sI_{n \times n} - A)^{-1}BU(s) \\ Y(s) = CX(s) + DU(s) \end{cases}$$

$$\Rightarrow \begin{cases} X(s) = (sI_{n \times n} - A)^{-1}(x_0 + BU(s)) \\ Y(s) = CX(s) + DU(s) \end{cases}$$

$$\Rightarrow \begin{cases} x(t) = \mathcal{L}^{-1}\left\{ (sI_{n \times n} - A)^{-1}(x_0 + BU(s)) \right\} \\ y(t) = C\mathcal{L}^{-1}\left\{ (sI_{n \times n} - A)^{-1}(x_0 + BU(s)) \right\} + Du(t) \end{cases}$$

Figure 4.37 displays the MATLAB code that solves the equation.

```
Command Window                                                          ⊙
  >> A=[-2/3 -1/3;6 0]; n=2;
  >> B=[1/3; 0];
  >> C=[1 0];
  >> D=1;
  >> x0=[1;2];
  >> syms s
  >> U=1/s;
  >> x_t=ilaplace(inv(s*eye(n)-A)*(x0+B*U))

  x_t =

       exp(-t/3)*(cos((17^(1/2)*t)/3) - (2*17^(1/2)*sin((17^(1/2)*t)/3))/17)
  exp(-t/3)*(cos((17^(1/2)*t)/3) + (19*17^(1/2)*sin((17^(1/2)*t)/3))/17) + 1

  >> vpa(x_t,4)

  ans =

       exp(-0.3333*t)*(cos(1.374*t) - 0.4851*sin(1.374*t))
  exp(-0.3333*t)*(cos(1.374*t) + 4.608*sin(1.374*t)) + 1.0

  >> iO_t=C*x_t+D

  iO_t =

  1/s + exp(-t/3)*(cos((17^(1/2)*t)/3) - (2*17^(1/2)*sin((17^(1/2)*t)/3))/17)

  >> vpa(iO_t,4)

  ans =

  exp(-0.3333*t)*(cos(1.374*t) - 0.4851*sin(1.374*t)) + 1

fx >>
```

Fig. 4.37 MATLAB code

4.2.26 Example 4.25

Solve the
$$
\begin{cases}
\begin{bmatrix} \dfrac{di_L(t)}{dt} \\[2mm] \dfrac{dv_C(t)}{dt} \end{bmatrix} = \begin{bmatrix} -\dfrac{2}{3} & -\dfrac{1}{3} \\[2mm] 6 & 0 \end{bmatrix} \begin{bmatrix} i_L(t) \\[1mm] v_C(t) \end{bmatrix} + \begin{bmatrix} \dfrac{1}{3} \\[1mm] 0 \end{bmatrix} V_1(t) \\[6mm]
i_o(t) = \begin{bmatrix} 1 & 0 \end{bmatrix} \begin{bmatrix} i_L(t) \\[1mm] v_C(t) \end{bmatrix} + [1]V_1(t)
\end{cases}
$$
with

$V_1(t) = 0$ and $x_0 = \begin{bmatrix} i_L(0) \\ v_C(0) \end{bmatrix} = \begin{bmatrix} 1 \\ 2 \end{bmatrix}.$

Solution:

$$
\begin{cases}
\begin{bmatrix} \dfrac{di_L(t)}{dt} \\[3mm] \dfrac{dv_C(t)}{dt} \end{bmatrix} = \begin{bmatrix} -\dfrac{2}{3} & -\dfrac{1}{3} \\[2mm] 6 & 0 \end{bmatrix} \begin{bmatrix} i_L(t) \\[1mm] v_C(t) \end{bmatrix} + \begin{bmatrix} \dfrac{1}{3} \\[1mm] 0 \end{bmatrix} 0 \\[8mm]
i_o(t) = \begin{bmatrix} 1 & 0 \end{bmatrix} \begin{bmatrix} i_L(t) \\[1mm] v_C(t) \end{bmatrix} + [1]0
\end{cases}
$$

$$
\Rightarrow
\begin{cases}
\dfrac{di_L(t)}{dt} = -\dfrac{2}{3}i_L(t) - \dfrac{1}{3}v_C(t) \\[3mm]
\dfrac{dv_c(t)}{dt} = 6i_L(t) \\[3mm]
i_o(t) = i_L(t)
\end{cases}
\Rightarrow
\begin{cases}
L\left\{\dfrac{di_L(t)}{dt}\right\} = L\left\{ -\dfrac{2}{3}i_L(t) - \dfrac{1}{3}v_C(t)\right\} \\[3mm]
L\left\{\dfrac{dv_c(t)}{dt}\right\} = L\{6i_L(t)\} \\[3mm]
L\{i_o(t)\} = L\{i_L(t)\}
\end{cases}
$$

$$
\Rightarrow
\begin{cases}
sI_L(s) - i_L(0) = -\dfrac{2}{3}I_L(s) - \dfrac{1}{3}V_C(s) \\[2mm]
sV_C(s) - v_C(0) = 6I_L(s) \\[2mm]
I_o(s) = I_L(s)
\end{cases}
\Rightarrow
\begin{cases}
sI_L(s) - 1 = -\dfrac{2}{3}I_L(s) - \dfrac{1}{3}V_C(s) \\[2mm]
sV_C(s) - 2 = 6I_L(s) \\[2mm]
I_o(s) = I_L(s)
\end{cases}
$$

$$
\Rightarrow
\begin{cases}
sI_L(s) = -\dfrac{2}{3}I_L(s) - \dfrac{1}{3}V_C(s) + 1 \\[2mm]
sV_C(s) = 6I_L(s) + 2 \\[2mm]
I_o(s) = I_L(s)
\end{cases}
\Rightarrow
\begin{cases}
\left(s + \dfrac{2}{3}\right)I_L(s) + \dfrac{1}{3}V_C(s) = 1 \\[2mm]
6I_L(s) - sV_C(s) = -2 \\[2mm]
I_o(s) = I_L(s)
\end{cases}
$$

$$
\Rightarrow
\begin{cases}
\begin{bmatrix} s + \dfrac{2}{3} & \dfrac{1}{3} \\[2mm] 6 & -s \end{bmatrix} \begin{bmatrix} I_L(s) \\[1mm] V_C(s) \end{bmatrix} = \begin{bmatrix} 1 \\[1mm] -2 \end{bmatrix} \\[8mm]
I_o(s) = I_L(s)
\end{cases}
\Rightarrow
\begin{cases}
\begin{bmatrix} I_L(s) \\[1mm] V_C(s) \end{bmatrix} = \begin{bmatrix} s + \dfrac{2}{3} & \dfrac{1}{3} \\[2mm] 6 & -s \end{bmatrix}^{-1} \begin{bmatrix} 1 \\[1mm] -2 \end{bmatrix} \\[8mm]
I_o(s) = I_L(s)
\end{cases}
$$

$$
\Rightarrow
\begin{cases}
\begin{bmatrix} I_L(s) \\[1mm] V_C(s) \end{bmatrix} = \dfrac{1}{-\left(s + \dfrac{2}{3}\right)s - 2} \begin{bmatrix} -s & -\dfrac{1}{3} \\[2mm] -6 & s + \dfrac{2}{3} \end{bmatrix} \begin{bmatrix} 1 \\[1mm] -2 \end{bmatrix} \\[10mm]
I_o(s) = I_L(s)
\end{cases}
$$

$$
\Rightarrow
\begin{cases}
\begin{bmatrix} I_L(s) \\[1mm] V_C(s) \end{bmatrix} = -\dfrac{1}{s^2 + \dfrac{2}{3}s + 2} \begin{bmatrix} -s + \dfrac{2}{3} \\[2mm] -6 - 2s - \dfrac{4}{3} \end{bmatrix} \\[10mm]
I_o(s) = I_L(s)
\end{cases}
$$

$$\Rightarrow \begin{cases} \begin{bmatrix} I_L(s) \\ V_C(s) \end{bmatrix} = -\dfrac{1}{s^2 + \dfrac{2}{3}s + 2} \begin{bmatrix} \dfrac{2-3s}{3} \\ \dfrac{-6s-22}{3} \end{bmatrix} \\ I_o(s) = I_L(s) \end{cases} \Rightarrow \begin{cases} \begin{bmatrix} I_L(s) \\ V_C(s) \end{bmatrix} = \begin{bmatrix} \dfrac{3s-2}{3s^2+2s+6} \\ \dfrac{6s+22}{3s^2+2s+6} \end{bmatrix} \\ I_o(s) = I_L(s) \end{cases}$$

$$\Rightarrow \begin{cases} \begin{bmatrix} i_L(t) \\ v_C(t) \end{bmatrix} = \mathcal{L}^{-1} \left\{ \begin{bmatrix} \dfrac{3s-2}{3s^2+2s+6} \\ \dfrac{6s+22}{3s^2+2s+6} \end{bmatrix} \right\} \\ i_o(t) = \mathcal{L}^{-1} \left\{ \dfrac{3s-2}{3s^2+2s+6} \right\} \end{cases}$$

You can use MATLAB for inverse Laplace calculations (Figs. 4.38 and 4.39).

```
Command Window
>> syms s
>> ilaplace((3*s-2)/(3*s^2+2*s+6))

ans =

exp(-t/3)*(cos((17^(1/2)*t)/3) - (3*17^(1/2)*sin((17^(1/2)*t)/3))/17)

>> vpa(ans,4)

ans =

exp(-0.3333*t)*(cos(1.374*t) - 0.7276*sin(1.374*t))

fx >> |
```

Fig. 4.38 MATLAB code

```
Command Window
>> syms s
>> ilaplace((6*s+22)/(3*s^2+2*s+6))

ans =

2*exp(-t/3)*(cos((17^(1/2)*t)/3) + (10*17^(1/2)*sin((17^(1/2)*t)/3))/17)

>> vpa(ans,4)

ans =

2.0*exp(-0.3333*t)*(cos(1.374*t) + 2.425*sin(1.374*t))

fx >> |
```

Fig. 4.39 MATLAB code

As shown in Figs. 4.38 and 4.39, we have

$$i_L(t) = i_o(t) = e^{-0.3333t}\cos(1.374t) - 0.7276\sin(1.374t)$$

$$v_C(t) = 2e^{-0.3333t}(\cos(1.374t) + 2.425\sin(1.374t))$$

4.2.27   Example 4.26

Determine the state-space representation of the circuit in Fig. 4.40, with the $i_o(t)$ as the output.

Fig. 4.40 Circuit for
Example 4.26

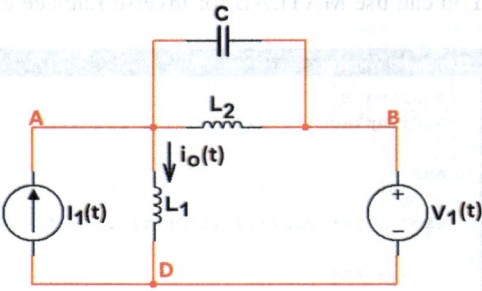

Solution:
The problem did not specify voltage polarities or current directions, so we will define them (Fig. 4.41).

Fig. 4.41 The circuit's
voltages and currents

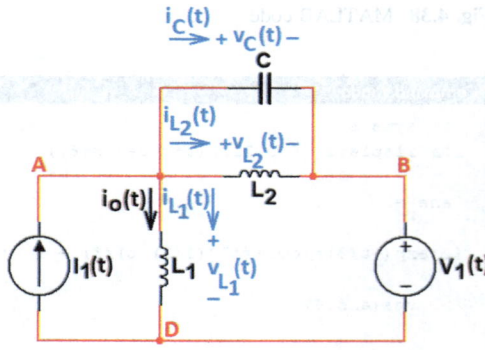

From Fig. 4.41, we have

$$KVL@DABD: -v_{L_1}(t) + v_{L_2}(t) + V_1(t) = 0 \Rightarrow -v_{L_1}(t) + v_C(t) + V_1(t) = 0$$

$$\Rightarrow -L_1\frac{di_{L_1}(t)}{dt} + v_C(t) + V_1(t) = 0 \Rightarrow \frac{di_{L_1}(t)}{dt} = \frac{1}{L_1}v_C(t) + \frac{1}{L_1}V_1(t)$$

$$v_{L_2}(t) = v_C(t) \Rightarrow L_2\frac{di_{L_2}(t)}{dt} = v_C(t) \Rightarrow \frac{di_{L_2}(t)}{dt} = \frac{1}{L_2}v_C(t)$$

$$KCL@A: I_1(t) = i_{L_1}(t) + i_{L_2}(t) + i_C(t) \Rightarrow I_1(t) = i_{L_1}(t) + i_{L_2}(t)$$

$$+ C\frac{dv_C(t)}{dt} \Rightarrow \frac{dv_C(t)}{dt} = -\frac{1}{C}i_{L_1}(t) - \frac{1}{C}i_{L_2}(t) + \frac{1}{C}I_1(t)$$

$$b\begin{cases} \dfrac{di_{L_1}(t)}{dt} = \dfrac{1}{L_1}v_C(t) + \dfrac{1}{L_1}V_1(t) \\[2mm] \dfrac{di_{L_2}(t)}{dt} = \dfrac{1}{L_2}v_C(t) \\[2mm] \dfrac{dv_C(t)}{dt} = -\dfrac{1}{C}i_{L_1}(t) - \dfrac{1}{C}i_{L_2}(t) + \dfrac{1}{C}I_1(t) \\[2mm] i_o(t) = i_{L_1}(t) \end{cases}$$

$$\Rightarrow \begin{cases} \begin{bmatrix} \dfrac{di_{L_1}(t)}{dt} \\[2mm] \dfrac{di_{L_2}(t)}{dt} \\[2mm] \dfrac{dv_C(t)}{dt} \end{bmatrix} = \begin{bmatrix} 0 & 0 & \dfrac{1}{L_1} \\[2mm] 0 & 0 & \dfrac{1}{L_2} \\[2mm] -\dfrac{1}{C} & -\dfrac{1}{C} & 0 \end{bmatrix} \begin{bmatrix} i_{L_1}(t) \\[2mm] i_{L_1}(t) \\[2mm] v_C(t) \end{bmatrix} + \begin{bmatrix} \dfrac{1}{L_1} \\[2mm] 0 \\[2mm] 0 \end{bmatrix} V_1(t) + \begin{bmatrix} 0 \\[2mm] 0 \\[2mm] \dfrac{1}{C} \end{bmatrix} I_1(t) \\[6mm] [i_o(t)] = [1 \quad 0 \quad 0]\begin{bmatrix} i_{L_1}(t) \\[2mm] i_{L_1}(t) \\[2mm] v_C(t) \end{bmatrix} + 0[V_1(t)] + 0[I_1(t)] \end{cases}$$

$$\Rightarrow \begin{cases} \begin{bmatrix} \dfrac{di_{L_1}(t)}{dt} \\[2mm] \dfrac{di_{L_2}(t)}{dt} \\[2mm] \dfrac{dv_C(t)}{dt} \end{bmatrix} = \begin{bmatrix} 0 & 0 & \dfrac{1}{L_1} \\[2mm] 0 & 0 & \dfrac{1}{L_2} \\[2mm] -\dfrac{1}{C} & -\dfrac{1}{C} & 0 \end{bmatrix} \begin{bmatrix} i_{L_1}(t) \\[2mm] i_{L_1}(t) \\[2mm] v_C(t) \end{bmatrix} + \begin{bmatrix} 1/L_1 & 0 \\[2mm] 0 & 0 \\[2mm] 0 & 1/C_1 \end{bmatrix} \begin{bmatrix} V_1(t) \\[2mm] I_1(t) \end{bmatrix} \\[6mm] [i_o(t)] = [1 \quad 0 \quad 0]\begin{bmatrix} i_{L_1}(t) \\[2mm] i_{L_1}(t) \\[2mm] v_C(t) \end{bmatrix} + [0 \quad 0]\begin{bmatrix} V_1(t) \\[2mm] I_1(t) \end{bmatrix} \end{cases}$$

$$\begin{cases} A = \begin{bmatrix} 0 & 0 & \dfrac{1}{L_1} \\[2mm] 0 & 0 & \dfrac{1}{L_2} \\[2mm] -\dfrac{1}{C} & -\dfrac{1}{C} & 0 \end{bmatrix} \\[10mm] B = \begin{bmatrix} 1/L_1 & 0 \\[2mm] 0 & 0 \\[2mm] 0 & 1/C_1 \end{bmatrix} \\[6mm] C = \begin{bmatrix} 1 & 0 & 0 \end{bmatrix} \\[2mm] D = \begin{bmatrix} 0 & 0 \end{bmatrix} \end{cases}$$

4.2.28 Example 4.27

Determine the state-space representation of the circuit in Fig. 4.42, with the $i_o(t)$ as the output.

Fig. 4.42 Circuit for Example 4.27

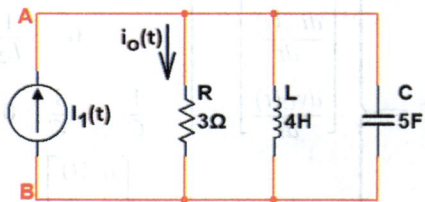

Solution:
The problem did not specify voltage polarities or current directions, so we will define them (Fig. 4.43).

Fig. 4.43 The circuit's voltages and currents

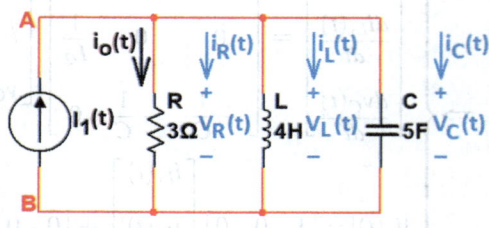

From Fig. 4.43, we have

$$v_R(t) = v_L(t) = v_C(t)$$

$$i_R(t) = \frac{v_R(t)}{R} \Rightarrow i_R(t) = \frac{v_C(t)}{R}$$

$$KCL@A : i_R(t) + i_L(t) + i_C(t) = i_1(t) \Rightarrow \frac{v_C(t)}{R} + i_L(t) + C\frac{dv_C(t)}{dt} = i_1(t) \Rightarrow \frac{v_C(t)}{3}$$

$$+ i_L(t) + 5\frac{dv_C(t)}{dt} = i_1(t) \Rightarrow \frac{dv_C(t)}{dt} = -\frac{1}{15}v_C(t) - \frac{1}{5}i_L(t) + i_1(t)$$

$$v_L(t) = v_C(t) \Rightarrow L\frac{di_L(t)}{dt} = v_C(t) \Rightarrow \frac{di_L(t)}{dt} = \frac{1}{L}v_C(t) \Rightarrow \frac{di_L(t)}{dt} = \frac{1}{4}v_C(t)$$

$$\begin{cases} \dfrac{dv_C(t)}{dt} = -\dfrac{1}{15}v_C(t) - \dfrac{1}{5}i_L(t) + i_1(t) \\ \dfrac{di_L(t)}{dt} = \dfrac{1}{4}v_C(t) \end{cases} \Rightarrow \begin{bmatrix} \dfrac{dv_C(t)}{dt} \\ \dfrac{di_L(t)}{dt} \end{bmatrix} = \begin{bmatrix} -\dfrac{1}{15} & -\dfrac{1}{5} \\ \dfrac{1}{4} & 0 \end{bmatrix}$$

$$\times \begin{bmatrix} v_C(t) \\ i_L(t) \end{bmatrix} + \begin{bmatrix} 1 \\ 0 \end{bmatrix} i_1(t)$$

$$i_o(t) = i_R(t) \Rightarrow i_o(t) = \frac{1}{R}v_R(t) \Rightarrow i_o(t) = \frac{1}{R}v_C(t) \Rightarrow i_o(t) = \frac{1}{3}v_C(t) \Rightarrow i_o(t)$$

$$= \begin{bmatrix} \frac{1}{3} & 0 \end{bmatrix} \begin{bmatrix} v_C(t) \\ i_L(t) \end{bmatrix} + [0]i_1(t)$$

$$\begin{cases} A = \begin{bmatrix} -\dfrac{1}{15} & -\dfrac{1}{5} \\ \dfrac{1}{4} & 0 \end{bmatrix} \\[6pt] B = \begin{bmatrix} 1 \\ 0 \end{bmatrix} \\[6pt] C = \begin{bmatrix} \frac{1}{3} & 0 \end{bmatrix} \\[6pt] D = [0] \end{cases}$$

4.2.29 Example 4.28

Determine the state-space representation of the circuit in Fig. 4.44, with the $V_o(t)$ as the output.

Fig. 4.44 Circuit for
Example 4.28

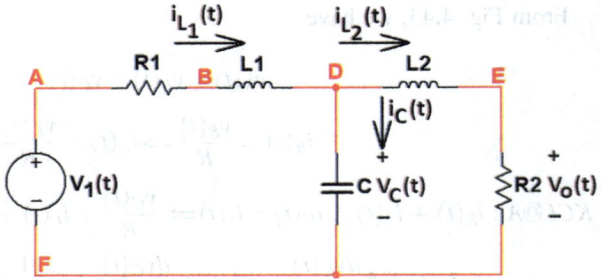

Solution:
From Fig. 4.44, we have

$$KVL@FABDF : \; -V_1(t) + R_1 i_{L_1}(t) + L_1 \frac{di_{L_1}(t)}{dt} + v_C(t) = 0 \Rightarrow \frac{di_{L_1}(t)}{dt}$$

$$= \frac{-R_1}{L_1} i_{L_1}(t) - \frac{1}{L_1} v_C(t) + \frac{1}{L_1} V_1(t)$$

$$KVL@DEFD : L_2 \frac{di_{L_2}(t)}{dt} + R_2 i_{L_2}(t) - v_C(t) = 0 \Rightarrow \frac{di_{L_2}(t)}{dt} = -\frac{R_2}{L_2} i_{L_2}(t) + \frac{1}{L_2} v_C(t)$$

$$KCL@D : i_{L_1}(t) = i_{L_2}(t) + i_C(t) \Rightarrow i_{L_1}(t) = i_{L_2}(t) + C \frac{dv_C(t)}{dt} \Rightarrow \frac{dv_C(t)}{dt}$$

$$= \frac{1}{C} i_{L_1}(t) - \frac{1}{C} i_{L_2}(t)$$

$$v_o(t) = R_2 i_{L_2}$$

$$\begin{cases} \dfrac{di_{L_1}(t)}{dt} = \dfrac{-R_1}{L_1} i_{L_1}(t) - \dfrac{1}{L_1} v_C(t) + \dfrac{1}{L_1} V_1(t) \\[2mm] \dfrac{di_{L_2}(t)}{dt} = -\dfrac{R_2}{L_2} i_{L_2}(t) + \dfrac{1}{L_2} v_C(t) \\[2mm] \dfrac{dv_C(t)}{dt} = \dfrac{1}{C} i_{L_1}(t) - \dfrac{1}{C} i_{L_2}(t) \\[2mm] v_o(t) = R_2 i_{L_2} \end{cases}$$

$$\Rightarrow \begin{cases} \begin{bmatrix} \dfrac{di_{L_1}(t)}{dt} \\[3mm] \dfrac{di_{L_2}(t)}{dt} \\[3mm] \dfrac{dv_C(t)}{dt} \end{bmatrix} = \begin{bmatrix} \dfrac{-R_1}{L_1} & 0 & \dfrac{-1}{L_1} \\[3mm] 0 & -\dfrac{R_2}{L_2} & \dfrac{1}{L_2} \\[3mm] \dfrac{1}{C} & -\dfrac{1}{C} & 0 \end{bmatrix} \begin{bmatrix} i_{L_1}(t) \\[3mm] i_{L_2}(t) \\[3mm] v_C(t) \end{bmatrix} + \begin{bmatrix} \dfrac{1}{L_1} \\[3mm] 0 \\[3mm] 0 \end{bmatrix} V_1(t) \\[10mm] v_o(t) = \begin{bmatrix} 0 & R_2 & 0 \end{bmatrix} \begin{bmatrix} i_{L_1}(t) \\[3mm] i_{L_2}(t) \\[3mm] v_C(t) \end{bmatrix} + 0 V_1(t) \end{cases}$$

$$
\begin{cases}
A = \begin{bmatrix} \dfrac{-R_1}{L_1} & 0 & \dfrac{-1}{L_1} \\[3mm] 0 & -\dfrac{R_2}{L_2} & \dfrac{1}{L_2} \\[3mm] \dfrac{1}{C} & -\dfrac{1}{C} & 0 \end{bmatrix} \\[12mm]
B = \begin{bmatrix} \dfrac{1}{L_1} \\[2mm] 0 \\[2mm] 0 \end{bmatrix} \\[10mm]
C = \begin{bmatrix} 0 & R_2 & 0 \end{bmatrix} \\[2mm]
D = \begin{bmatrix} 0 \end{bmatrix}
\end{cases}
$$

4.2.30 Example 4.29

Determine the state-space representation of the circuit in Fig. 4.45, with the $i_o(t)$ as the output.

Fig. 4.45 Circuit for Example 4.29

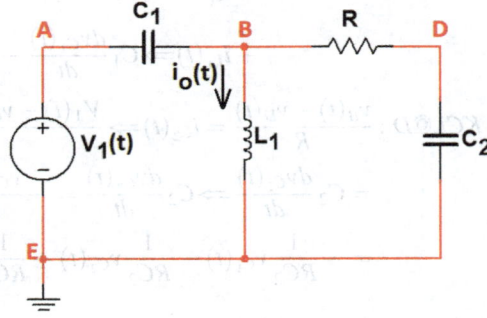

Solution:
The problem did not specify voltage polarities or current directions, so we will define them (Fig. 4.46).

Fig. 4.46 The circuit's voltages and currents

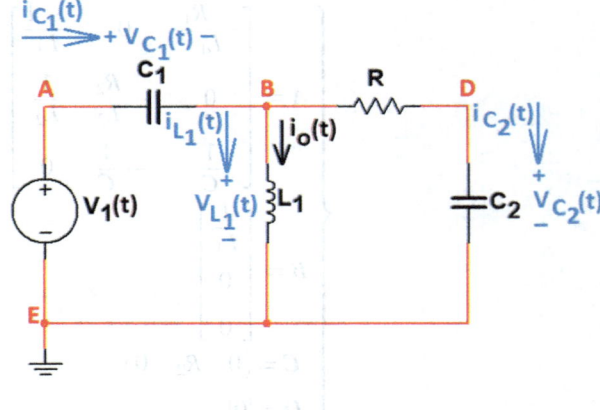

From Fig. 4.46, we have

$$v_E = 0\ V$$

$$v_A(t) = V_1(t)$$

$$v_B(t) = V_1(t) - v_{C_1}(t)$$

$$v_D(t) = v_{C_2}(t)$$

$$KCL@B : i_{C_1}(t) = i_{C_2}(t) + i_{L_1}(t) \Rightarrow C_1 \frac{dv_{C_1}(t)}{dt} = C_2 \frac{dv_{C_2}(t)}{dt}$$

$$+ i_{L_1}(t) \Rightarrow C_1 \frac{dv_{C_1}(t)}{dt} - C_2 \frac{dv_{C_2}(t)}{dt} = i_{L_1}(t)$$

$$KCL@D : \frac{v_B(t) - v_D(t)}{R} = i_{C_2}(t) \Rightarrow \frac{V_1(t) - v_{C_1}(t) - v_{C_2}(t)}{R}$$

$$= C_2 \frac{dv_{C_2}(t)}{dt} \Rightarrow C_2 \frac{dv_{C_2}(t)}{dt} = -\frac{v_{C_1}(t)}{R} - \frac{v_{C_2}(t)}{R} + \frac{V_1(t)}{R} \Rightarrow \frac{dv_{C_2}(t)}{dt}$$

$$= -\frac{1}{RC_2} v_{C_1}(t) - \frac{1}{RC_2} v_{C_2}(t) + \frac{1}{RC_2} V_1(t)$$

$$KVL@EABE : -V_1(t) + v_{C_1}(t) + v_L(t) = 0 \Rightarrow -V_1(t) + v_{C_1}(t) + L \frac{di_L(t)}{dt}$$

$$= 0 \Rightarrow L \frac{di_L(t)}{dt} = -v_{C_1}(t) + V_1(t) \Rightarrow \frac{di_L(t)}{dt} = -\frac{1}{L} v_{C_1}(t) + \frac{1}{L} V_1(t)$$

$$\begin{cases} C_1 \dfrac{dv_{C_1}(t)}{dt} - C_2 \dfrac{dv_{C_2}(t)}{dt} = i_{L_1}(t) \\[2mm] \dfrac{dv_{C_2}(t)}{dt} = -\dfrac{1}{RC_2}v_{C_1}(t) - \dfrac{1}{RC_2}v_{C_2}(t) + \dfrac{1}{RC_2}V_1(t) \\[2mm] \dfrac{di_L(t)}{dt} = -\dfrac{1}{L}v_{C_1}(t) + \dfrac{1}{L}V_1(t) \end{cases}$$

$$\Rightarrow \begin{cases} C_1 \dfrac{dv_{C_1}(t)}{dt} + \dfrac{1}{R}v_{C_1}(t) + \dfrac{1}{R}v_{C_2}(t) - \dfrac{1}{R}V_1(t) = i_{L_1}(t) \\[2mm] \dfrac{dv_{C_2}(t)}{dt} = -\dfrac{1}{RC_2}v_{C_1}(t) - \dfrac{1}{RC_2}v_{C_2}(t) + \dfrac{1}{RC_2}V_1(t) \\[2mm] \dfrac{di_L(t)}{dt} = -\dfrac{1}{L}v_{C_1}(t) + \dfrac{1}{L}V_1(t) \end{cases}$$

$$\Rightarrow \begin{cases} \dfrac{dv_{C_1}(t)}{dt} = -\dfrac{1}{RC_1}v_{C_1}(t) - \dfrac{1}{RC_1}v_{C_2}(t) + \dfrac{1}{RC_1}V_1(t) + \dfrac{1}{C_1}i_{L_1}(t) \\[2mm] \dfrac{dv_{C_2}(t)}{dt} = -\dfrac{1}{RC_2}v_{C_1}(t) - \dfrac{1}{RC_2}v_{C_2}(t) + \dfrac{1}{RC_2}V_1(t) \\[2mm] \dfrac{di_L(t)}{dt} = -\dfrac{1}{L}v_{C_1}(t) + \dfrac{1}{L}V_1(t) \end{cases}$$

$$\Rightarrow \begin{bmatrix} \dfrac{dv_{C_1}(t)}{dt} \\[3mm] \dfrac{dv_{C_2}(t)}{dt} \\[3mm] \dfrac{di_L(t)}{dt} \end{bmatrix} = \begin{bmatrix} -\dfrac{1}{RC_1} & -\dfrac{1}{RC_1} & \dfrac{1}{C_1} \\[3mm] -\dfrac{1}{RC_2} & -\dfrac{1}{RC_2} & 0 \\[3mm] -\dfrac{1}{L} & 0 & 0 \end{bmatrix} \begin{bmatrix} v_{C_1}(t) \\[3mm] v_{C_2}(t) \\[3mm] i_L(t) \end{bmatrix} + \begin{bmatrix} \dfrac{1}{RC_1} \\[3mm] \dfrac{1}{RC_2} \\[3mm] \dfrac{1}{L} \end{bmatrix} V_1(t)$$

$$i_o(t) = i_L(t) = \begin{bmatrix} 0 & 0 & 1 \end{bmatrix} \begin{bmatrix} v_{C_1}(t) \\ v_{C_2}(t) \\ i_L(t) \end{bmatrix} + [0]V_1(t)$$

$$\begin{cases} A = \begin{bmatrix} -\dfrac{1}{RC_1} & -\dfrac{1}{RC_1} & \dfrac{1}{C_1} \\[3mm] -\dfrac{1}{RC_2} & -\dfrac{1}{RC_2} & 0 \\[3mm] -\dfrac{1}{L} & 0 & 0 \end{bmatrix} \\[10mm] B = \begin{bmatrix} \dfrac{1}{RC_1} \\[3mm] \dfrac{1}{RC_2} \\[3mm] \dfrac{1}{L} \end{bmatrix} \\[10mm] C = \begin{bmatrix} 0 & 0 & 1 \end{bmatrix} \\[2mm] D = [0] \end{cases}$$

4.2.31   Example 4.30

Determine the state-space representation of the circuit in Fig. 4.47, with the $V_o(t)$ as the output.

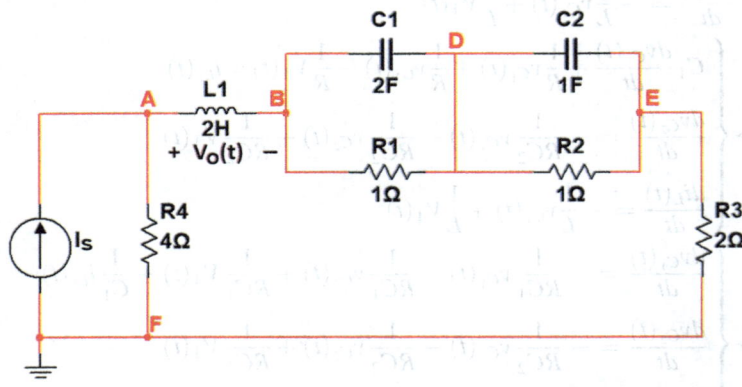

Fig. 4.47 Circuit for Example 4.30

Solution:
The circuit shown in Fig. 4.48 was obtained by employing the current-to-voltage source transformation technique.

Fig. 4.48 Equivalent circuit
for Fig. 4.47

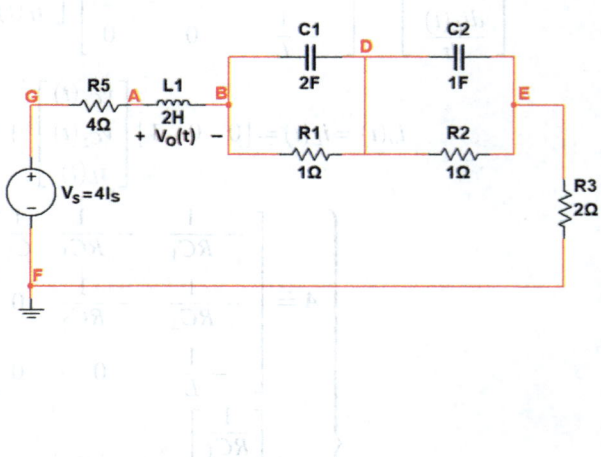

The current flow in the distinct branches is depicted in Fig. 4.49.

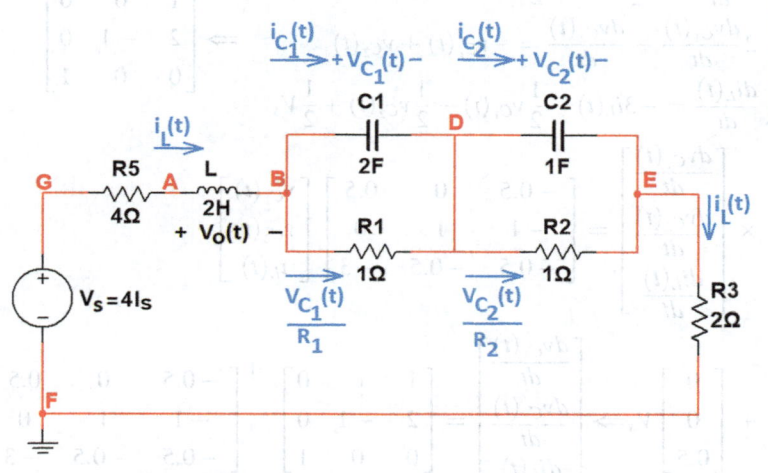

Fig. 4.49 The current flow in the distinct branches

From Fig. 4.49, we have

$$KCL@B : i_L(t) = i_{C_1}(t) + \frac{v_{C_1}(t)}{R_1} \Rightarrow i_L(t) = C_1 \frac{dv_{C_1}(t)}{dt} + \frac{v_{C_1}(t)}{R_1} \Rightarrow i_L(t) = 2 \frac{dv_{C_1}(t)}{dt}$$

$$+ v_{C_1}(t) \Rightarrow \frac{dv_{C_1}(t)}{dt} = \frac{1}{2} i_L(t) - \frac{1}{2} v_{C_1}(t)$$

$$KCL@D : i_{C_1}(t) + \frac{v_{C_1}(t)}{R_1} = i_{C_2}(t) + \frac{v_{C_2}(t)}{R_2} \Rightarrow C_1 \frac{dv_{C_1}(t)}{dt} + \frac{v_{C_1}(t)}{R_1} = C_2 \frac{dv_{C_2}(t)}{dt}$$

$$+ \frac{v_{C_2}(t)}{R_2} \Rightarrow 2 \frac{dv_{C_1}(t)}{dt} - \frac{dv_{C_2}(t)}{dt} = -\frac{v_{C_1}(t)}{1} + \frac{v_{C_2}(t)}{1} \Rightarrow 2 \frac{dv_{C_1}(t)}{dt}$$

$$- \frac{dv_{C_2}(t)}{dt} = -v_{C_1}(t) + v_{C_2}(t)$$

$$KVL@FGABDEF : -V_s + R_5 i_L(t) + L \frac{di_L(t)}{dt} + v_{C_1}(t) + v_{C_2}(t) + R_3 i_L(t)$$

$$= 0 \Rightarrow \frac{di_L(t)}{dt} = -\frac{R_3 + R_5}{L} i_L(t) - \frac{1}{L} v_{C_1}(t) - \frac{1}{L} v_{C_2}(t)$$

$$+ \frac{1}{L} V_s \Rightarrow \frac{di_L(t)}{dt} = -\frac{2 + 4}{2} i_L(t) - \frac{1}{2} v_{C_1}(t) - \frac{1}{2} v_{C_2}(t)$$

$$+ \frac{1}{2} V_s \Rightarrow \frac{di_L(t)}{dt} = -3 i_L(t) - \frac{1}{2} v_{C_1}(t) - \frac{1}{2} v_{C_2}(t) + \frac{1}{2} V_s$$

$$\begin{cases} \dfrac{dv_{C_1}(t)}{dt} = \dfrac{1}{2}i_L(t) - \dfrac{1}{2}v_{C_1}(t) \\[2mm] 2\dfrac{dv_{C_1}(t)}{dt} - \dfrac{dv_{C_2}(t)}{dt} = -v_{C_1}(t) + v_{C_2}(t) \\[2mm] \dfrac{di_L(t)}{dt} = -3i_L(t) - \dfrac{1}{2}v_{C_1}(t) - \dfrac{1}{2}v_{C_2}(t) + \dfrac{1}{2}V_s \end{cases} \implies \begin{bmatrix} 1 & 0 & 0 \\ 2 & -1 & 0 \\ 0 & 0 & 1 \end{bmatrix}$$

$$\times \begin{bmatrix} \dfrac{dv_{C_1}(t)}{dt} \\[2mm] \dfrac{dv_{C_2}(t)}{dt} \\[2mm] \dfrac{di_L(t)}{dt} \end{bmatrix} = \begin{bmatrix} -0.5 & 0 & 0.5 \\ -1 & 1 & 0 \\ -0.5 & -0.5 & -3 \end{bmatrix} \begin{bmatrix} v_{C_1}(t) \\ v_{C_2}(t) \\ i_L(t) \end{bmatrix}$$

$$+ \begin{bmatrix} 0 \\ 0 \\ 0.5 \end{bmatrix} V_s \implies \begin{bmatrix} \dfrac{dv_{C_1}(t)}{dt} \\[2mm] \dfrac{dv_{C_2}(t)}{dt} \\[2mm] \dfrac{di_L(t)}{dt} \end{bmatrix} = \begin{bmatrix} 1 & 0 & 0 \\ 2 & -1 & 0 \\ 0 & 0 & 1 \end{bmatrix}^{-1} \begin{bmatrix} -0.5 & 0 & 0.5 \\ -1 & 1 & 0 \\ -0.5 & -0.5 & -3 \end{bmatrix}$$

$$\times \begin{bmatrix} v_{C_1}(t) \\ v_{C_2}(t) \\ i_L(t) \end{bmatrix} + \begin{bmatrix} 1 & 0 & 0 \\ 2 & -1 & 0 \\ 0 & 0 & 1 \end{bmatrix}^{-1} \begin{bmatrix} 0 \\ 0 \\ 0.5 \end{bmatrix} V_s \implies \begin{bmatrix} \dfrac{dv_{C_1}(t)}{dt} \\[2mm] \dfrac{dv_{C_2}(t)}{dt} \\[2mm] \dfrac{di_L(t)}{dt} \end{bmatrix}$$

$$= \begin{bmatrix} -0.5 & 0 & 0.5 \\ 0 & -1 & 1 \\ -0.5 & -0.5 & -3 \end{bmatrix} \begin{bmatrix} v_{C_1}(t) \\ v_{C_2}(t) \\ i_L(t) \end{bmatrix} + \begin{bmatrix} 0 \\ 0 \\ 0.5 \end{bmatrix} V_s \implies \begin{bmatrix} \dfrac{dv_{C_1}(t)}{dt} \\[2mm] \dfrac{dv_{C_2}(t)}{dt} \\[2mm] \dfrac{di_L(t)}{dt} \end{bmatrix}$$

$$= \begin{bmatrix} -0.5 & 0 & 0.5 \\ 0 & -1 & 1 \\ -0.5 & -0.5 & -3 \end{bmatrix} \begin{bmatrix} v_{C_1}(t) \\ v_{C_2}(t) \\ i_L(t) \end{bmatrix} + \begin{bmatrix} 0 \\ 0 \\ 0.5 \end{bmatrix} 4I_s \implies \begin{bmatrix} \dfrac{dv_{C_1}(t)}{dt} \\[2mm] \dfrac{dv_{C_2}(t)}{dt} \\[2mm] \dfrac{di_L(t)}{dt} \end{bmatrix}$$

$$= \begin{bmatrix} -0.5 & 0 & 0.5 \\ 0 & -1 & 1 \\ -0.5 & -0.5 & -3 \end{bmatrix} \begin{bmatrix} v_{C_1}(t) \\ v_{C_2}(t) \\ i_L(t) \end{bmatrix} + \begin{bmatrix} 0 \\ 0 \\ 2 \end{bmatrix} I_s$$

$$KVL@FGABDEF : -V_s + R_5 i_L(t) + v_o(t) + v_{C_1}(t) + v_{C_2}(t) + R_3 i_L(t)$$
$$= 0 \Rightarrow v_o(t) = -v_{C_1}(t) - v_{C_2}(t) - (R_3 + R_5) i_L(t)$$
$$+ V_s \Rightarrow v_o(t) = -v_{C_1}(t) - v_{C_2}(t) - 6 i_L(t) + 4 I_s \Rightarrow v_o(t)$$

$$= [-1 \quad -1 \quad -6] \begin{bmatrix} v_{C_1}(t) \\ v_{C_2}(t) \\ i_L(t) \end{bmatrix} + [4] I_s$$

$$\begin{cases} A = \begin{bmatrix} -0.5 & 0 & 0.5 \\ 0 & -1 & 1 \\ -0.5 & -0.5 & -3 \end{bmatrix} \\ B = \begin{bmatrix} 0 \\ 0 \\ 2 \end{bmatrix} \\ C = [-1 \quad -1 \quad -6] \\ D = [4] \end{cases}$$

4.2.32 Example 4.31

Determine the state-space representation of the circuit in Fig. 4.50, with the $i_o(t)$ as the output.

Fig. 4.50 Circuit for Example 4.31

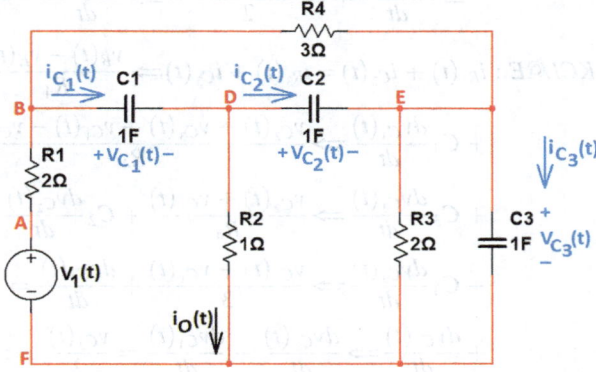

Solution:
The problem did not specify voltage polarities or current directions, so we will define them (Fig. 4.51).

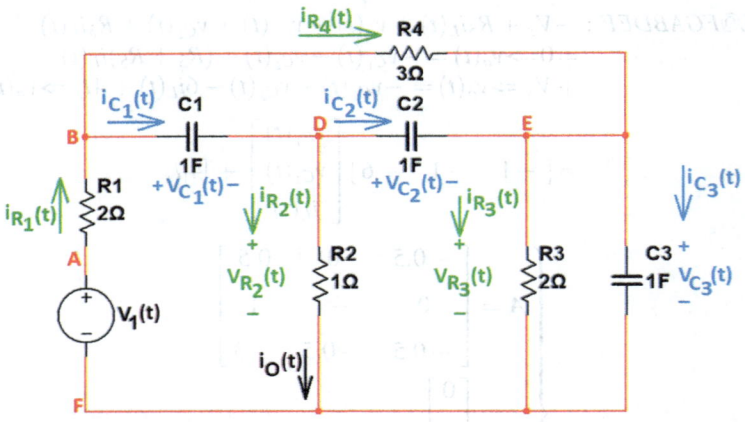

Fig. 4.51 The circuit's voltages and currents

From Fig. 4.51, we have

$$KVL@DEFD : v_{C_2}(t) + v_{C_3}(t) - v_{R_2}(t) = 0 \Rightarrow v_{R_2}(t) = v_{C_2}(t) + v_{C_3}(t)$$

$$KVL@EFE : v_{C_3}(t) - v_{R_3}(t) = 0 \Rightarrow v_{R_3}(t) = v_{C_3}(t)$$

$$v_B(t) = v_D(t) + v_{C_1}(t) \Rightarrow v_B(t) = v_{R_2}(t) + v_{C_1}(t) \Rightarrow v_B(t) = v_{C_2}(t) + v_{C_3}(t) + v_{C_1}(t)$$

$$KCL@D : i_{C_1}(t) = i_{C_2}(t) + i_{R_2}(t) \Rightarrow C_1 \frac{dv_{C_1}(t)}{dt} = C_2 \frac{dv_{C_2}(t)}{dt}$$

$$+ \frac{v_{R_2}(t)}{R_2} \Rightarrow C_1 \frac{dv_{C_1}(t)}{dt} = C_2 \frac{dv_{C_2}(t)}{dt} + \frac{v_{C_2}(t) + v_{C_3}(t)}{R_2} \Rightarrow \frac{dv_{C_1}(t)}{dt}$$

$$= \frac{dv_{C_2}(t)}{dt} + \frac{v_{C_2}(t) + v_{C_3}(t)}{2} \Rightarrow \frac{dv_{C_1}(t)}{dt} - \frac{dv_{C_2}(t)}{dt} = \frac{v_{C_2}(t)}{2} + \frac{v_{C_3}(t)}{2}$$

$$KCL@E : i_{R_4}(t) + i_{C_2}(t) = i_{R_3}(t) + i_{C_3}(t) \Rightarrow \frac{v_B(t) - v_E(t)}{R_4} + C_2 \frac{dv_{C_2}(t)}{dt} = \frac{v_{R_3}(t)}{R_3}$$

$$+ C_3 \frac{dv_{C_3}(t)}{dt} \Rightarrow \frac{v_{C_2}(t) + v_{C_3}(t) + v_{C_1}(t) - v_{C_3}(t)}{R_4} + C_2 \frac{dv_{C_2}(t)}{dt} = \frac{v_{C_3}(t)}{R_3}$$

$$+ C_3 \frac{dv_{C_3}(t)}{dt} \Rightarrow \frac{v_{C_2}(t) + v_{C_1}(t)}{R_4} + C_2 \frac{dv_{C_2}(t)}{dt} = \frac{v_{C_3}(t)}{R_3}$$

$$+ C_3 \frac{dv_{C_3}(t)}{dt} \Rightarrow \frac{v_{C_2}(t) + v_{C_1}(t)}{3} + \frac{dv_{C_2}(t)}{dt} = \frac{v_{C_3}(t)}{2}$$

$$+ \frac{dv_{C_3}(t)}{dt} \Rightarrow \frac{dv_{C_2}(t)}{dt} - \frac{dv_{C_3}(t)}{dt} = \frac{v_{C_3}(t)}{2}$$

$$- \frac{v_{C_2}(t) + v_{C_1}(t)}{3} \Rightarrow \frac{dv_{C_2}(t)}{dt} - \frac{dv_{C_3}(t)}{dt} = -\frac{v_{C_1}(t)}{3} - \frac{v_{C_2}(t)}{3} + \frac{v_{C_3}(t)}{2}$$

$$KCL@B : i_{R_1}(t) = i_{C_1}(t) + i_{R_4}(t) \Rightarrow \frac{v_A(t) - v_B(t)}{R_1} = C_1 \frac{dv_{C_1}(t)}{dt}$$

$$+ \frac{v_B(t) - v_E(t)}{R_4} \Rightarrow \frac{v_1(t) - (v_{C_1}(t) + v_{C_2}(t) + v_{C_3}(t))}{R_1} = C_1 \frac{dv_{C_1}(t)}{dt}$$

$$+ \frac{v_{C_1}(t) + v_{C_2}(t) + v_{C_3}(t) - v_{C_3}(t)}{R_4} \Rightarrow \frac{v_1(t) - v_{C_1}(t) - v_{C_2}(t) - v_{C_3}(t)}{R_1}$$

$$= C_1 \frac{dv_{C_1}(t)}{dt} + \frac{v_{C_1}(t) + v_{C_2}(t)}{R_4} \Rightarrow \frac{dv_{C_1}(t)}{dt}$$

$$= \frac{v_1(t) - v_{C_1}(t) - v_{C_2}(t) - v_{C_3}(t)}{2} + \frac{- v_{C_1}(t) - v_{C_2}(t)}{3} \Rightarrow \frac{dv_{C_1}(t)}{dt} =$$

$$- \frac{5}{6} v_{C_1}(t) - \frac{5}{6} v_{C_2}(t) - \frac{v_{C_3}(t)}{2} + \frac{v_1(t)}{2}$$

$$\begin{cases} \dfrac{dv_{C_1}(t)}{dt} - \dfrac{dv_{C_2}(t)}{dt} = \dfrac{v_{C_2}(t)}{2} + \dfrac{v_{C_3}(t)}{2} \\[2mm] \dfrac{dv_{C_2}(t)}{dt} - \dfrac{dv_{C_3}(t)}{dt} = -\dfrac{v_{C_1}(t)}{3} - \dfrac{v_{C_2}(t)}{3} + \dfrac{v_{C_3}(t)}{2} \\[2mm] \dfrac{dv_{C_1}(t)}{dt} = -\dfrac{5}{6} v_{C_1}(t) - \dfrac{5}{6} v_{C_2}(t) - \dfrac{v_{C_3}(t)}{2} + \dfrac{v_1(t)}{2} \end{cases} \Rightarrow \begin{bmatrix} 1 & -1 & 0 \\ 0 & 1 & -1 \\ 1 & 0 & 0 \end{bmatrix}$$

$$\times \begin{bmatrix} \dfrac{dv_{C_1}(t)}{dt} \\[2mm] \dfrac{dv_{C_2}(t)}{dt} \\[2mm] \dfrac{dv_{C_3}(t)}{dt} \end{bmatrix} = \begin{bmatrix} 0 & \dfrac{1}{2} & \dfrac{1}{2} \\[2mm] -\dfrac{1}{3} & -\dfrac{1}{3} & \dfrac{1}{2} \\[2mm] -\dfrac{5}{6} & -\dfrac{5}{6} & -\dfrac{1}{2} \end{bmatrix} \begin{bmatrix} v_{C_1}(t) \\[2mm] v_{C_2}(t) \\[2mm] v_{C_3}(t) \end{bmatrix} + \begin{bmatrix} 0 \\[1mm] 0 \\[1mm] \dfrac{1}{2} \end{bmatrix} v_1(t) \Rightarrow \begin{bmatrix} \dfrac{dv_{C_1}(t)}{dt} \\[2mm] \dfrac{dv_{C_2}(t)}{dt} \\[2mm] \dfrac{dv_{C_3}(t)}{dt} \end{bmatrix}$$

$$= \begin{bmatrix} 1 & -1 & 0 \\ 0 & 1 & -1 \\ 1 & 0 & 0 \end{bmatrix}^{-1} \begin{bmatrix} 0 & \dfrac{1}{2} & \dfrac{1}{2} \\[2mm] -\dfrac{1}{3} & -\dfrac{1}{3} & \dfrac{1}{2} \\[2mm] -\dfrac{5}{6} & -\dfrac{5}{6} & -\dfrac{1}{2} \end{bmatrix} \begin{bmatrix} v_{C_1}(t) \\[2mm] v_{C_2}(t) \\[2mm] v_{C_3}(t) \end{bmatrix}$$

$$+ \begin{bmatrix} 1 & -1 & 0 \\ 0 & 1 & -1 \\ 1 & 0 & 0 \end{bmatrix}^{-1} \begin{bmatrix} 0 \\[1mm] 0 \\[1mm] \dfrac{1}{2} \end{bmatrix} v_1(t) \Rightarrow \begin{bmatrix} \dfrac{dv_{C_1}(t)}{dt} \\[2mm] \dfrac{dv_{C_2}(t)}{dt} \\[2mm] \dfrac{dv_{C_3}(t)}{dt} \end{bmatrix}$$

$$= \begin{bmatrix} -0.8333 & -0.8333 & -0.5 \\ -0.8333 & -1.3333 & -1 \\ -0.5 & -1 & -1.5 \end{bmatrix} \begin{bmatrix} v_{C_1}(t) \\ v_{C_2}(t) \\ v_{C_3}(t) \end{bmatrix} + \begin{bmatrix} 0.5 \\ 0.5 \\ 0.5 \end{bmatrix} v_1(t)$$

$$i_O(t) = i_{R_2}(t) \Rightarrow i_O(t) = \frac{v_{R_2}(t)}{R_2} \Rightarrow i_O(t) = \frac{v_{C_2}(t) + v_{C_3}(t)}{R_2} \Rightarrow i_O(t)$$

$$= \frac{v_{C_2}(t) + v_{C_3}(t)}{1} \Rightarrow i_O(t) = v_{C_2}(t) + v_{C_3}(t) \Rightarrow i_O(t) = \begin{bmatrix} 0 & 1 & 1 \end{bmatrix} \begin{bmatrix} v_{C_1}(t) \\ v_{C_2}(t) \\ v_{C_3}(t) \end{bmatrix}$$

$$+ [0]v_1(t)$$

$$\begin{cases} A = \begin{bmatrix} -0.8333 & -0.8333 & -0.5 \\ -0.8333 & -1.3333 & -1 \\ -0.5 & -1 & -1.5 \end{bmatrix} \\ B = \begin{bmatrix} 0.5 \\ 0.5 \\ 0.5 \end{bmatrix} \\ C = \begin{bmatrix} 0 & 1 & 1 \end{bmatrix} \\ D = [0] \end{cases}$$

4.2.33   Example 4.32

Determine the state-space representation of the circuit in Fig. 4.52, with the $v_o(t)$ as the output.

Fig. 4.52 Circuit for
Example 4.32

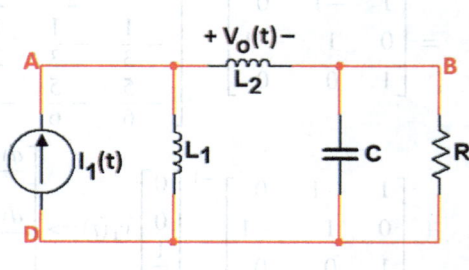

Solution:
Initially, one might expect three state variables in this circuit, given its two inductors and one capacitor. However, due to a node consisting solely of inductors and current sources, the actual number of state variables is reduced to two.

The problem did not specify voltage polarities or current directions, so we will define them (Fig. 4.53).

Fig. 4.53 The circuit's
voltages and currents

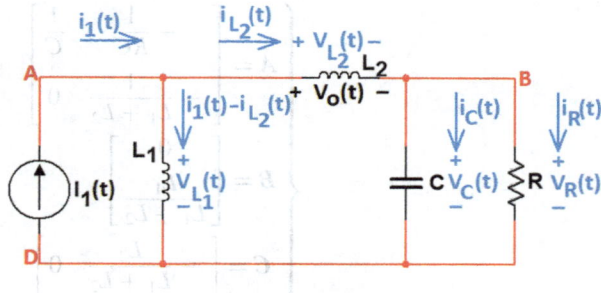

From Fig. 4.53, we have

$$v_C(t) = v_R(t)$$

$$KCL@B : i_{L_2}(t) = i_C(t) + i_R(t) \Rightarrow i_{L_2}(t) = C\frac{dv_C(t)}{dt} + \frac{v_R(t)}{R} \Rightarrow i_{L_2}(t) = C\frac{dv_C(t)}{dt}$$

$$+ \frac{v_C(t)}{R} \Rightarrow \frac{dv_C(t)}{dt} = \frac{1}{C}i_{L_2}(t) - \frac{1}{RC}v_C(t)$$

$$KVL@DABD : -v_{L_1}(t) + v_{L_2}(t) + v_C(t) = 0 \Rightarrow -L_1\frac{d}{dt}(i_1(t) - i_{L_2}(t)) + L_2\frac{di_{L_2}(t)}{dt}$$

$$+ v_C(t) = 0 \Rightarrow -L_1\frac{di_1(t)}{dt} + L_1\frac{di_{L_2}(t)}{dt} + L_2\frac{di_{L_2}(t)}{dt} + v_C(t)$$

$$= 0 \Rightarrow (L_1 + L_2)\frac{di_{L_2}(t)}{dt} = -v_C(t) + L_1\frac{di_1(t)}{dt} \Rightarrow \frac{di_{L_2}(t)}{dt} =$$

$$-\frac{1}{L_1 + L_2}v_C(t) + \frac{L_1}{L_1 + L_2}\frac{di_1(t)}{dt}$$

$$\begin{cases} \dfrac{dv_C(t)}{dt} = \dfrac{1}{C}i_{L_2}(t) - \dfrac{1}{RC}v_C(t) \\ \dfrac{di_{L_2}(t)}{dt} = -\dfrac{1}{L_1 + L_2}v_C(t) + \dfrac{L_1}{L_1 + L_2}\dfrac{di_1(t)}{dt} \end{cases} \Rightarrow \begin{bmatrix} \dfrac{dv_C(t)}{dt} \\ \dfrac{di_{L_2}(t)}{dt} \end{bmatrix}$$

$$= \begin{bmatrix} -\dfrac{1}{RC} & \dfrac{1}{C} \\ -\dfrac{1}{L_1 + L_2} & 0 \end{bmatrix} \begin{bmatrix} v_C(t) \\ i_{L_2}(t) \end{bmatrix} + \begin{bmatrix} 0 \\ \dfrac{L_1}{L_1 + L_2} \end{bmatrix} \dfrac{di_1(t)}{dt}$$

$$v_o(t) = v_{L_2}(t) \Rightarrow v_o(t) = L_2\frac{di_{L_2}(t)}{dt} \Rightarrow v_o(t)$$

$$= L_2\left(-\frac{1}{L_1 + L_2}v_C(t) + \frac{L_1}{L_1 + L_2}\frac{di_1(t)}{dt}\right) \Rightarrow v_o(t) = -\frac{L_2}{L_1 + L_2}v_C(t)$$

$$+ \frac{L_1 L_2}{L_1 + L_2}\frac{di_1(t)}{dt} \Rightarrow v_o(t) = \begin{bmatrix} -\dfrac{L_2}{L_1 + L_2} & 0 \end{bmatrix} \begin{bmatrix} v_C(t) \\ i_{L_2}(t) \end{bmatrix} + \begin{bmatrix} \dfrac{L_1 L_2}{L_1 + L_2} \end{bmatrix} \dfrac{di_1(t)}{dt}$$

$$x(t) = \begin{bmatrix} v_C(t) \\ i_{L_2}(t) \end{bmatrix}, u(t) = \frac{di_1(t)}{dt}$$

$$
\begin{cases}
A = \begin{bmatrix} -\dfrac{1}{RC} & \dfrac{1}{C} \\[2ex] -\dfrac{1}{L_1 + L_2} & 0 \end{bmatrix} \\[5ex]
B = \begin{bmatrix} 0 \\[1ex] \dfrac{L_1}{L_1 + L_2} \end{bmatrix} \\[5ex]
C = \begin{bmatrix} -\dfrac{L_2}{L_1 + L_2} & 0 \end{bmatrix} \\[4ex]
D = \begin{bmatrix} \dfrac{L_1 L_2}{L_1 + L_2} \end{bmatrix}
\end{cases}
$$

4.2.34   Example 4.33

Determine the state-space representation of the circuit in Fig. 4.54, with the $v_o(t)$ as the output.

Fig. 4.54 Circuit for
Example 4.33

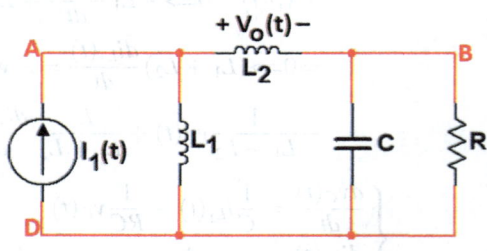

Solution:

We previously solved this problem using $x(t) = \begin{bmatrix} v_C(t) \\ i_{L_2}(t) \end{bmatrix}$. Now, we will solve it using $x(t) = \begin{bmatrix} v_C(t) \\ i_{L_1}(t) \end{bmatrix}$ Fig. 4.55.

Fig. 4.55 The circuit's
voltages and currents

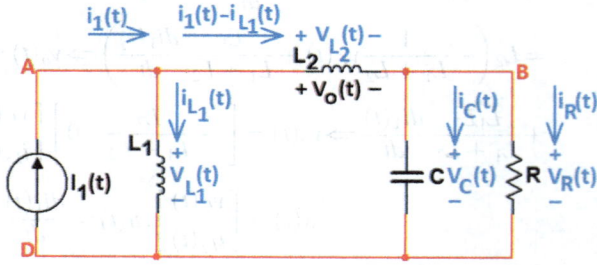

From Fig. 4.55, we have

$$v_C(t) = v_R(t)$$

$$KCL@B : i_1(t) - i_{L_1}(t) = i_C(t) + i_R(t) \Rightarrow i_1(t) - i_{L_1}(t) = C\frac{dv_C(t)}{dt}$$

$$+ \frac{v_R(t)}{R} \Rightarrow C\frac{dv_C(t)}{dt} = i_1(t) - i_{L_1}(t) - \frac{v_C(t)}{R} \Rightarrow \frac{dv_C(t)}{dt} = -\frac{1}{C}i_{L_1}(t)$$

$$- \frac{1}{RC}v_C(t) + \frac{1}{C}i_1(t)$$

$$KVL@DABD : -v_{L_1}(t) + v_{L_2}(t) + v_C(t) = 0 \Rightarrow -L_1\frac{di_{L_1}(t)}{dt} + L_2\frac{d}{dt}(i_1(t) - i_{L_1}(t))$$

$$+ v_C(t) = 0 \Rightarrow -L_1\frac{di_{L_1}(t)}{dt} - L_2\frac{di_{L_1}(t)}{dt} + L_2\frac{di_1(t)}{dt} + v_C(t)$$

$$= 0 \Rightarrow (L_1 + L_2)\frac{di_{L_1}(t)}{dt} = v_C(t) + L_2\frac{di_1(t)}{dt} \Rightarrow \frac{di_{L_1}(t)}{dt}$$

$$= \frac{1}{L_1 + L_2}v_C(t) + \frac{L_2}{L_1 + L_2}\frac{di_1(t)}{dt}$$

$$\begin{cases} \dfrac{dv_C(t)}{dt} = -\dfrac{1}{C}i_{L_1}(t) - \dfrac{1}{RC}v_C(t) + \dfrac{1}{C}i_1(t) \\ \dfrac{di_{L_1}(t)}{dt} = \dfrac{1}{L_1 + L_2}v_C(t) + \dfrac{L_2}{L_1 + L_2}\dfrac{di_1(t)}{dt} \end{cases} \Rightarrow \begin{bmatrix} \dfrac{dv_C(t)}{dt} \\ \dfrac{di_{L_1}(t)}{dt} \end{bmatrix} = \begin{bmatrix} -\dfrac{1}{RC} & -\dfrac{1}{C} \\ \dfrac{1}{L_1 + L_2} & 0 \end{bmatrix}$$

$$\times \begin{bmatrix} v_C(t) \\ i_{L_1}(t) \end{bmatrix} + \begin{bmatrix} \dfrac{1}{C} \\ 0 \end{bmatrix} i_1(t) + \begin{bmatrix} 0 \\ \dfrac{L_2}{L_1 + L_2} \end{bmatrix} \frac{di_1(t)}{dt} \Rightarrow \begin{bmatrix} \dfrac{dv_C(t)}{dt} \\ \dfrac{di_{L_1}(t)}{dt} \end{bmatrix}$$

$$= \begin{bmatrix} -\dfrac{1}{RC} & -\dfrac{1}{C} \\ \dfrac{1}{L_1 + L_2} & 0 \end{bmatrix} \begin{bmatrix} v_C(t) \\ i_{L_1}(t) \end{bmatrix} + \begin{bmatrix} \dfrac{1}{C} & 0 \\ 0 & \dfrac{L_2}{L_1 + L_2} \end{bmatrix} \begin{bmatrix} i_1(t) \\ \dfrac{di_1(t)}{dt} \end{bmatrix}$$

$$v_o(t) = v_A(t) - v_B(t) \Rightarrow v_o(t) = L_1\frac{di_{L_1}(t)}{dt} - v_C(t) \Rightarrow v_o(t) = \frac{L_1}{L_1 + L_2}v_C(t)$$

$$+ \frac{L_1 L_2}{L_1 + L_2}\frac{di_1(t)}{dt} - v_C(t) \Rightarrow v_o(t) = -\frac{L_2}{L_1 + L_2}v_C(t)$$

$$+ \frac{L_1 L_2}{L_1 + L_2}\frac{di_1(t)}{dt} \Rightarrow v_o(t) = \begin{bmatrix} -\dfrac{L_2}{L_1 + L_2} & 0 \end{bmatrix} \begin{bmatrix} v_C(t) \\ i_{L_1}(t) \end{bmatrix} + [0]i_1(t) + \begin{bmatrix} \dfrac{L_1 L_2}{L_1 + L_2} \end{bmatrix}$$

$$\times \frac{di_1(t)}{dt} \Rightarrow v_o(t) = \begin{bmatrix} -\dfrac{L_2}{L_1 + L_2} & 0 \end{bmatrix} \begin{bmatrix} v_C(t) \\ i_{L_1}(t) \end{bmatrix} + \begin{bmatrix} 0 & \dfrac{L_1 L_2}{L_1 + L_2} \end{bmatrix} \begin{bmatrix} i_1(t) \\ \dfrac{di_1(t)}{dt} \end{bmatrix}$$

$$x(t) = \begin{bmatrix} v_C(t) \\ i_{L_1}(t) \end{bmatrix}, u(t) = \begin{bmatrix} i_1(t) \\ \dfrac{di_1(t)}{dt} \end{bmatrix}$$

$$
\begin{cases}
A = \begin{bmatrix} -\dfrac{1}{RC} & -\dfrac{1}{C} \\[2ex] \dfrac{1}{L_1 + L_2} & 0 \end{bmatrix} \\[5ex]
B = \begin{bmatrix} \dfrac{1}{C} & 0 \\[2ex] 0 & \dfrac{L_2}{L_1 + L_2} \end{bmatrix} \\[5ex]
C = \begin{bmatrix} -\dfrac{L_2}{L_1 + L_2} & 0 \end{bmatrix} \\[4ex]
D = \begin{bmatrix} 0 & \dfrac{L_1 L_2}{L_1 + L_2} \end{bmatrix}
\end{cases}
$$

4.2.35   Example 4.34

Determine the state-space representation of the circuit in Fig. 4.56, with the $i_o(t)$ as the output.

Fig. 4.56 Circuit for Example 4.34

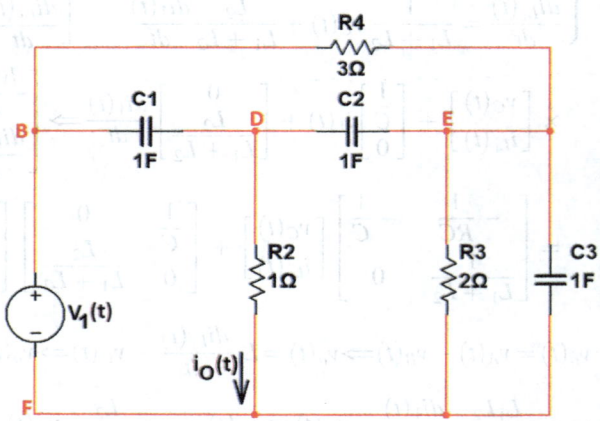

Solution:
This circuit contains three capacitors, initially suggesting three state variables. However, the loop formed by $V_1(t)$, C_1, C_2, and C_3 introduces a dependency, reducing the number of independent state variables to two (Compare this example to Example 4.31). For our analysis, we select the voltages across capacitors C_1 and C_2 to serve as the independent variables.

The voltages and currents are assigned as shown in Fig. 4.57.

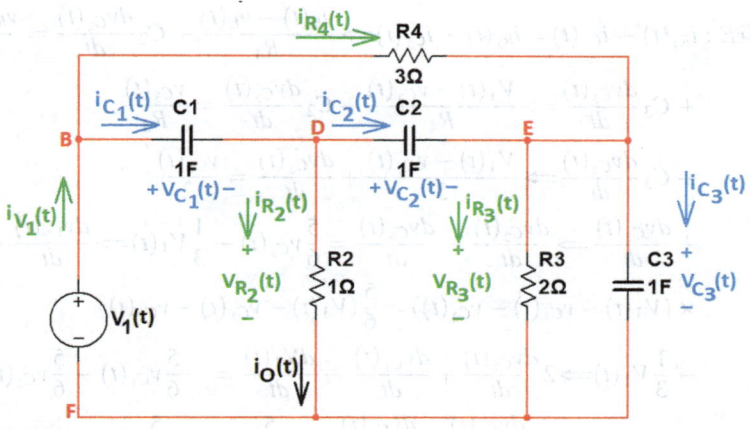

Fig. 4.57 The circuit's voltages and currents

From Fig. 4.57, we have

$$KVL@FBDEF : -V_1(t) + v_{C_1}(t) + v_{C_2}(t) + v_{C_3}(t) = 0 \Rightarrow v_{C_3}(t) = V_1(t)$$
$$- v_{C_1}(t) - v_{C_2}(t)$$

$$KVL@DEFD : v_{C_2}(t) + v_{C_3}(t) - v_{R_2}(t) = 0 \Rightarrow v_{R_2}(t) = v_{C_2}(t) + v_{C_3}(t)$$

$$v_D(t) = v_{R_2}(t) \Rightarrow v_D(t) = v_{C_2}(t) + v_{C_3}(t)$$

$$KVL@EFE : v_{C_3}(t) - v_{R_3}(t) = 0 \Rightarrow v_{R_3}(t) = v_{C_3}(t)$$

$$v_B(t) = V_1(t)$$

$$KCL@D : i_{C_1}(t) = i_{C_2}(t) + i_{R_2}(t) \Rightarrow C_1 \frac{dv_{C_1}(t)}{dt} = C_2 \frac{dv_{C_2}(t)}{dt}$$

$$+ \frac{v_{R_2}(t)}{R_2} \Rightarrow C_1 \frac{dv_{C_1}(t)}{dt} = C_2 \frac{dv_{C_2}(t)}{dt} + \frac{v_{C_2}(t) + v_{C_3}(t)}{R_2} \Rightarrow \frac{dv_{C_1}(t)}{dt}$$

$$= \frac{dv_{C_2}(t)}{dt} + \frac{v_{C_2}(t) + v_{C_3}(t)}{1} \Rightarrow \frac{dv_{C_1}(t)}{dt} - \frac{dv_{C_2}(t)}{dt} = v_{C_2}(t)$$

$$+ v_{C_3}(t) \Rightarrow \frac{dv_{C_1}(t)}{dt} - \frac{dv_{C_2}(t)}{dt} = v_{C_2}(t) + V_1(t) - v_{C_1}(t)$$

$$- v_{C_2}(t) \Rightarrow \frac{dv_{C_1}(t)}{dt} - \frac{dv_{C_2}(t)}{dt} = -v_{C_1}(t) + V_1(t)$$

$$KCL@E : i_{R_4}(t) + i_{C_2}(t) = i_{R_3}(t) + i_{C_3}(t) \Rightarrow \frac{v_B(t) - v_E(t)}{R_4} + C_2 \frac{dv_{C_2}(t)}{dt} = \frac{v_{R_3}(t)}{R_3}$$

$$+ C_3 \frac{dv_{C_3}(t)}{dt} \Rightarrow \frac{V_1(t) - v_{C_3}(t)}{R_4} + C_2 \frac{dv_{C_2}(t)}{dt} = \frac{v_{C_3}(t)}{R_3}$$

$$+ C_3 \frac{dv_{C_3}(t)}{dt} \Rightarrow \frac{V_1(t) - v_{C_3}(t)}{3} + \frac{dv_{C_2}(t)}{dt} = \frac{v_{C_3}(t)}{2}$$

$$+ \frac{dv_{C_3}(t)}{dt} \Rightarrow \frac{dv_{C_2}(t)}{dt} - \frac{dv_{C_3}(t)}{dt} = \frac{5}{6} v_{C_3}(t) - \frac{1}{3} V_1(t) \Rightarrow \frac{dv_{C_2}(t)}{dt} - \frac{d}{dt}$$

$$\times (V_1(t) - v_{C_1}(t) - v_{C_2}(t)) = \frac{5}{6}(V_1(t) - v_{C_1}(t) - v_{C_2}(t))$$

$$- \frac{1}{3} V_1(t) \Rightarrow 2 \frac{dv_{C_2}(t)}{dt} + \frac{dv_{C_1}(t)}{dt} - \frac{dV_1(t)}{dt} = -\frac{5}{6} v_{C_1}(t) - \frac{5}{6} v_{C_2}(t)$$

$$+ 0.5 V_1(t) \Rightarrow 2 \frac{dv_{C_2}(t)}{dt} + \frac{dv_{C_1}(t)}{dt} = -\frac{5}{6} v_{C_1}(t) - \frac{5}{6} v_{C_2}(t) + 0.5 V_1(t)$$

$$+ \frac{dV_1(t)}{dt}$$

$$\begin{cases} \dfrac{dv_{C_1}(t)}{dt} - \dfrac{dv_{C_2}(t)}{dt} = -v_{C_1}(t) + V_1(t) \\ 2 \dfrac{dv_{C_2}(t)}{dt} + \dfrac{dv_{C_1}(t)}{dt} = -\dfrac{5}{6} v_{C_1}(t) - \dfrac{5}{6} v_{C_2}(t) + 0.5 V_1(t) + \dfrac{dV_1(t)}{dt} \end{cases} \Rightarrow \begin{bmatrix} 1 & -1 \\ 1 & 2 \end{bmatrix}$$

$$\times \begin{bmatrix} \dfrac{dv_{C_1}(t)}{dt} \\ \dfrac{dv_{C_2}(t)}{dt} \end{bmatrix} = \begin{bmatrix} -1 & 0 \\ -\dfrac{5}{6} & -\dfrac{5}{6} \end{bmatrix} \begin{bmatrix} v_{C_1}(t) \\ v_{C_2}(t) \end{bmatrix} + \begin{bmatrix} 1 \\ 0.5 \end{bmatrix} V_1(t) + \begin{bmatrix} 0 \\ 1 \end{bmatrix}$$

$$\times \frac{dV_1(t)}{dt} \Rightarrow \begin{bmatrix} \dfrac{dv_{C_1}(t)}{dt} \\ \dfrac{dv_{C_2}(t)}{dt} \end{bmatrix} = \begin{bmatrix} 1 & -1 \\ 1 & 2 \end{bmatrix}^{-1} \begin{bmatrix} -1 & 0 \\ -\dfrac{5}{6} & -\dfrac{5}{6} \end{bmatrix} \begin{bmatrix} v_{C_1}(t) \\ v_{C_2}(t) \end{bmatrix}$$

$$+ \begin{bmatrix} 1 & -1 \\ 1 & 2 \end{bmatrix}^{-1} \begin{bmatrix} 1 \\ 0.5 \end{bmatrix} V_1(t) + \begin{bmatrix} 1 & -1 \\ 1 & 2 \end{bmatrix}^{-1} \begin{bmatrix} 0 \\ 1 \end{bmatrix} \frac{dV_1(t)}{dt} \Rightarrow \begin{bmatrix} \dfrac{dv_{C_1}(t)}{dt} \\ \dfrac{dv_{C_2}(t)}{dt} \end{bmatrix}$$

$$= \begin{bmatrix} -0.9444 & -0.2778 \\ 0.0556 & -0.2778 \end{bmatrix} \begin{bmatrix} v_{C_1}(t) \\ v_{C_2}(t) \end{bmatrix} + \begin{bmatrix} 0.8333 \\ -0.1667 \end{bmatrix} V_1(t) + \begin{bmatrix} 0.3333 \\ 0.3333 \end{bmatrix}$$

$$\times \frac{dV_1(t)}{dt} \Rightarrow \begin{bmatrix} \dfrac{dv_{C_1}(t)}{dt} \\ \dfrac{dv_{C_2}(t)}{dt} \end{bmatrix} = \begin{bmatrix} -0.9444 & -0.2778 \\ 0.0556 & -0.2778 \end{bmatrix} \begin{bmatrix} v_{C_1}(t) \\ v_{C_2}(t) \end{bmatrix}$$

$$+ \begin{bmatrix} 0.8333 & 0.3333 \\ -0.1667 & 0.3333 \end{bmatrix} \begin{bmatrix} V_1(t) \\ \dfrac{dV_1(t)}{dt} \end{bmatrix}$$

$$i_o(t) = \frac{v_{R_2}(t)}{R_2} \Rightarrow i_o(t) = \frac{v_D(t)}{R_2} \Rightarrow i_o(t) = \frac{v_{C_2}(t) + v_{C_3}(t)}{R_2} \Rightarrow i_o(t)$$

$$= \frac{v_{C_2}(t) + V_1(t) - v_{C_1}(t) - v_{C_2}(t)}{R_2} \Rightarrow i_o(t) = \frac{V_1(t) - v_{C_1}(t)}{R_2} \Rightarrow i_o(t)$$

$$= \frac{V_1(t) - v_{C_1}(t)}{1} \Rightarrow i_o(t) = V_1(t) - v_{C_1}(t) \Rightarrow i_o(t) = -v_{C_1}(t) + V_1(t)$$

$$+ 0\frac{dV_1(t)}{dt} \Rightarrow i_o(t) = \begin{bmatrix} -1 & 0 \end{bmatrix} \begin{bmatrix} v_{C_1}(t) \\ v_{C_2}(t) \end{bmatrix} + \begin{bmatrix} 1 & 0 \end{bmatrix} \begin{bmatrix} V_1(t) \\ \dfrac{dV_1(t)}{dt} \end{bmatrix}$$

$$x = \begin{bmatrix} v_{C_1}(t) \\ v_{C_2}(t) \end{bmatrix}, u = \begin{bmatrix} V_1(t) \\ \dfrac{dV_1(t)}{dt} \end{bmatrix}$$

$$\begin{cases} A = \begin{bmatrix} -0.9444 & -0.2778 \\ 0.0556 & -0.2778 \end{bmatrix} \\ B = \begin{bmatrix} 0.5 & 0.3333 \\ 0 & 0.3333 \end{bmatrix} \\ C = \begin{bmatrix} -1 & 0 \end{bmatrix} \\ D = \begin{bmatrix} 1 & 0 \end{bmatrix} \end{cases}$$

4.2.36 Example 4.35

State-space model of the circuit shown in Fig. 4.58 is given as

$$\begin{cases} \begin{bmatrix} \dfrac{dv_{C_1}(t)}{dt} \\ \dfrac{dv_{C_2}(t)}{dt} \end{bmatrix} = \begin{bmatrix} -0.9444 & -0.2778 \\ 0.0556 & -0.2778 \end{bmatrix} \begin{bmatrix} v_{C_1}(t) \\ v_{C_2}(t) \end{bmatrix} + \begin{bmatrix} 0.8333 & 0.3333 \\ -0.1667 & 0.3333 \end{bmatrix} \begin{bmatrix} V_1(t) \\ \dfrac{dV_1(t)}{dt} \end{bmatrix} \\ i_o(t) = \begin{bmatrix} -1 & 0 \end{bmatrix} \begin{bmatrix} v_{C_1}(t) \\ v_{C_2}(t) \end{bmatrix} + \begin{bmatrix} 1 & 0 \end{bmatrix} \begin{bmatrix} V_1(t) \\ \dfrac{dV_1(t)}{dt} \end{bmatrix} \end{cases}$$

Determine the voltages across capacitors C_1 and C_2, as well as current through resistor R_2 for an input voltage $V_1(t) = 5$ V (Fig. 4.59).

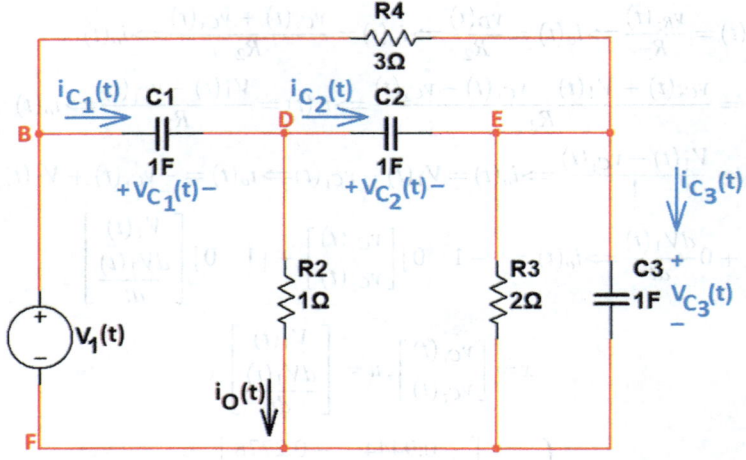

Fig. 4.58 Circuit for Example 4.35

Fig. 4.59 Circuit of
Fig. 4.58 with $V_1(t) = 5V$

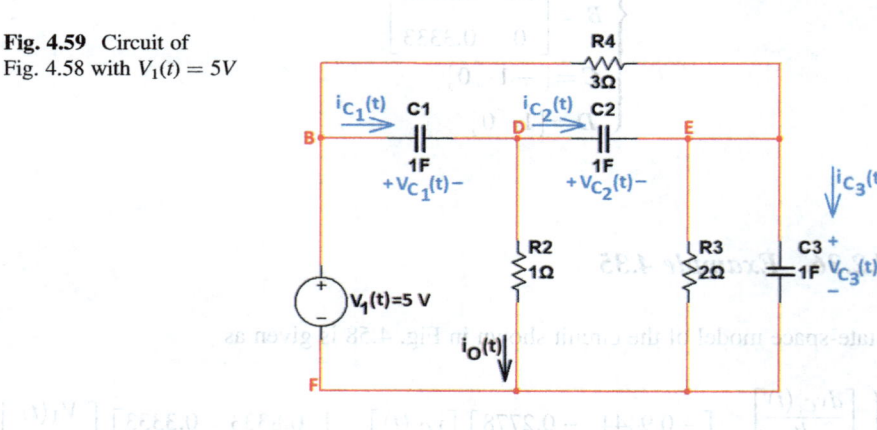

Solution:

$$\begin{bmatrix} V_1(t) \\ \dfrac{dV_1(t)}{dt} \end{bmatrix} = \begin{bmatrix} 5 \\ \dfrac{d5}{dt} \end{bmatrix} = \begin{bmatrix} 5 \\ 0 \end{bmatrix}$$

$$\begin{cases} \begin{bmatrix} \dfrac{dv_{C_1}(t)}{dt} \\ \dfrac{dv_{C_2}(t)}{dt} \end{bmatrix} = \begin{bmatrix} -0.9444 & -0.2778 \\ 0.0556 & -0.2778 \end{bmatrix} \begin{bmatrix} v_{C_1}(t) \\ v_{C_2}(t) \end{bmatrix} + \begin{bmatrix} 0.8333 & 0.3333 \\ -0.1667 & 0.3333 \end{bmatrix} \begin{bmatrix} V_1(t) \\ \dfrac{dV_1(t)}{dt} \end{bmatrix} \\[2em] i_o(t) = \begin{bmatrix} -1 & 0 \end{bmatrix} \begin{bmatrix} v_{C_1}(t) \\ v_{C_2}(t) \end{bmatrix} + \begin{bmatrix} 1 & 0 \end{bmatrix} \begin{bmatrix} V_1(t) \\ \dfrac{dV_1(t)}{dt} \end{bmatrix} \end{cases}$$

$$\Rightarrow \begin{cases} \begin{bmatrix} \dfrac{dv_{C_1}(t)}{dt} \\ \dfrac{dv_{C_2}(t)}{dt} \end{bmatrix} = \begin{bmatrix} -0.9444 & -0.2778 \\ 0.0556 & -0.2778 \end{bmatrix} \begin{bmatrix} v_{C_1}(t) \\ v_{C_2}(t) \end{bmatrix} + \begin{bmatrix} 0.8333 & 0.3333 \\ -0.1667 & 0.3333 \end{bmatrix} \begin{bmatrix} 5 \\ 0 \end{bmatrix} \\[2em] i_o(t) = \begin{bmatrix} -1 & 0 \end{bmatrix} \begin{bmatrix} v_{C_1}(t) \\ v_{C_2}(t) \end{bmatrix} + \begin{bmatrix} 1 & 0 \end{bmatrix} \begin{bmatrix} 5 \\ 0 \end{bmatrix} \end{cases}$$

$$\Rightarrow \begin{cases} \begin{bmatrix} \dfrac{dv_{C_1}(t)}{dt} \\ \dfrac{dv_{C_2}(t)}{dt} \end{bmatrix} = \begin{bmatrix} -0.9444 & -0.2778 \\ 0.0556 & -0.2778 \end{bmatrix} \begin{bmatrix} v_{C_1}(t) \\ v_{C_2}(t) \end{bmatrix} + \begin{bmatrix} 4.1665 \\ -0.8335 \end{bmatrix} \\[2em] i_o(t) = -v_{C_1}(t) + 5 \end{cases}$$

$$\overset{\mathcal{L}}{\Rightarrow} \begin{cases} \begin{bmatrix} s & 0 \\ 0 & s \end{bmatrix} \begin{bmatrix} V_{C_1}(s) \\ V_{C_2}(s) \end{bmatrix} = \begin{bmatrix} -0.9444 & -0.2778 \\ 0.0556 & -0.2778 \end{bmatrix} \begin{bmatrix} V_{C_1}(s) \\ V_{C_2}(s) \end{bmatrix} + \dfrac{1}{s}\begin{bmatrix} 4.1665 \\ -0.8335 \end{bmatrix} \\[2em] I_o(s) = -V_{C_1}(s) + \dfrac{5}{s} \end{cases}$$

$$\Rightarrow \begin{cases} \begin{bmatrix} s+0.9444 & 0.2778 \\ -0.0556 & s+0.2778 \end{bmatrix} \begin{bmatrix} V_{C_1}(s) \\ V_{C_2}(s) \end{bmatrix} = \dfrac{1}{s}\begin{bmatrix} 4.1665 \\ -0.8335 \end{bmatrix} \\[2em] I_o(s) = -V_{C_1}(s) + \dfrac{5}{s} \end{cases}$$

$$\Rightarrow \begin{cases} \begin{bmatrix} V_{C_1}(s) \\ V_{C_2}(s) \end{bmatrix} = \dfrac{1}{s}\begin{bmatrix} s+0.9444 & 0.2778 \\ -0.0556 & s+0.2778 \end{bmatrix}^{-1}\begin{bmatrix} 4.1665 \\ -0.8335 \end{bmatrix} \\[2em] I_o(s) = -V_{C_1}(s) + \dfrac{5}{s} \end{cases}$$

$$\Rightarrow \begin{cases} \begin{bmatrix} V_{C_1}(s) \\ V_{C_2}(s) \end{bmatrix} = \dfrac{1}{s} \times \dfrac{1}{(s+0.9444)(s+0.2778)+0.2778\times0.0556}\begin{bmatrix} s+0.2778 & -0.2778 \\ 0.0556 & s+0.9444 \end{bmatrix}\begin{bmatrix} 4.1665 \\ -0.8335 \end{bmatrix} \\[2em] I_o(s) = -V_{C_1}(s) + \dfrac{5}{s} \end{cases}$$

$$\Rightarrow \begin{cases} \begin{bmatrix} V_{C_1}(s) \\ V_{C_2}(s) \end{bmatrix} = \dfrac{1}{s(s^2+1.2222s+0.2778)}\begin{bmatrix} 4.167s+1.389 \\ -0.8335s-0.5555 \end{bmatrix} \\[2em] I_o(s) = -V_{C_1}(s) + \dfrac{5}{s} \end{cases}$$

$$\overset{\mathcal{L}^{-1}}{\Rightarrow} \begin{cases} \begin{bmatrix} v_{C_1}(t) \\ v_{C_2}(t) \end{bmatrix} = \mathcal{L}^{-1}\left\{ \dfrac{1}{s(s^2+1.2222s+0.2778)}\begin{bmatrix} 4.167s+1.389 \\ -0.8335s-0.5555 \end{bmatrix}\right\} \\[2em] i_o(t) = \mathcal{L}^{-1}\left\{-V_{C_1}(s) + \dfrac{5}{s}\right\} = -\mathcal{L}^{-1}\{V_{C_1}(s)\} + 5 \end{cases}$$

You can use MATLAB for inverse Laplace calculations (Figs. 4.60 and 4.61).

```
Command Window
>> syms s
>> ilaplace((4.167*s+1.389)/s/(s^2+1.2222*s+0.2778))

ans =

5 - 5*exp(-(6111*t)/10000)*(cosh((9564321^(1/2)*t)/10000) - (57*9564321^(1/2)*sinh((9564321^(1/2)*t)/10000))/245239)

>> vpa(ans,4)

ans =

5.0 - 5.0*exp(-0.6111*t)*(cosh(0.3093*t) - 0.7188*sinh(0.3093*t))

>>
```

Fig. 4.60 MATLAB code

```
Command Window
>> syms s
>> ilaplace((-0.8335*s-0.5555)/s/(s^2+1.2222*s+0.2778))

ans =

(5555*exp(-(6111*t)/10000)*(cosh((9564321^(1/2)*t)/10000) + (719465*9564321^(1/2)*sinh((9564321^(1/2)*t)/10000))/3541986877))/2778 - 5555/2778

>> vpa(ans,4)

ans =

2.0*exp(-0.6111*t)*(cosh(0.3093*t) + 0.6282*sinh(0.3093*t)) - 2.0

>>
```

Fig. 4.61 MATLAB code

As shown in Figs. 4.60 and 4.61, we have

$$\begin{cases} v_{C_1}(t) = 5 - 5e^{-0.6111t}(\cos h(0.3093t) - 0.7188 \sin h(0.3093t)) \\ v_{C_2}(t) = 2e^{-0.6111t}(\cos h(0.3093t) + 0.6282 \sin h(0.3093t)) - 2 \end{cases}$$

Therefore,

$$i_o(t) = -v_{C_1}(t) + 5 = 5e^{-0.6111t}(\cos h(0.3093t) - 0.7188 \sin h(0.3093t))$$

$$v_{C_3}(t) = V_1(t) - v_{C_1}(t) - v_{C_2}(t) \Longrightarrow v_{C_3}(t) = 5 - 5$$
$$+ 5e^{-0.6111t}(\cos h(0.3093t) - 0.7188 \sin h(0.3093t))$$
$$- 2e^{-0.6111t}(\cos h(0.3093t) + 0.6282 \sin h(0.3093t)) + 2 \Longrightarrow v_{C_3}(t) = 2$$
$$+ e^{-0.6111t}(5\cos h(0.3093t) - 3.5940 \sin h(0.3093t))$$
$$+ e^{-0.6111t}(-2\cos h(0.3093t) - 1.2564 \sin h(0.3093t)) \Longrightarrow v_{C_3}(t) = 2$$
$$+ e^{-0.6111t}(3\cos h(0.3093t) - 4.8504 \sin h(0.3093t))$$

4.2.37 Example 4.36

The results of Example 4.35 yielded the following expressions for $v_{C_1}(t), v_{C_2}(t), v_{C_3}(t)$ and $i_o(t)$ for the circuit illustrated in Fig. 4.62.

$$v_{C_1}(t) = 5 - 5e^{-0.6111t}(\cos h(0.3093t) - 0.7188 \sin h(0.3093t))$$

$$v_{C_2}(t) = 2e^{-0.6111t}(\cos h(0.3093t) + 0.6282 \sin h(0.3093t)) - 2$$

$$v_{C_3}(t) = 2 + e^{-0.6111t}(3\cos h(0.3093t) - 4.8504 \sin h(0.3093t))$$

$$i_o(t) = -v_{C_1}(t) + 5 = 5e^{-0.6111t}(\cos h(0.3093t) - 0.7188 \sin h(0.3093t))$$

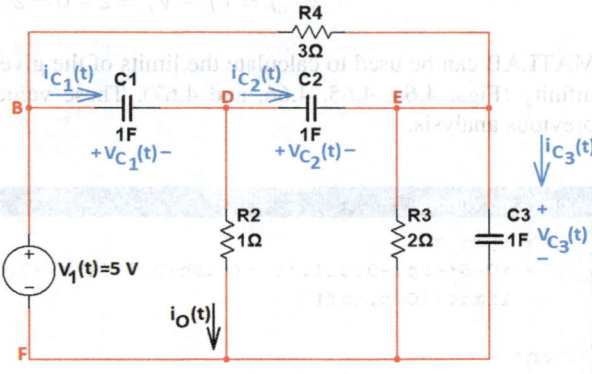

Fig. 4.62 Circuit for Example 4.36

The obtained results will be examined to determine if the derived expressions converge to the expected values. The DC steady-state equivalent of Fig. 4.62 is depicted in Fig. 4.63.

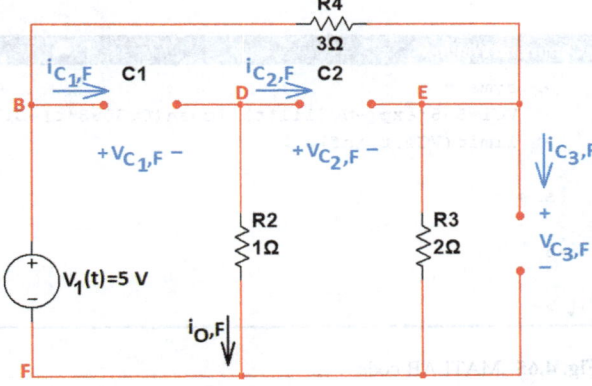

Fig. 4.63 The DC steady-state equivalent of Fig. 4.62

From Fig. 4.63, we have

$$V_F = 0 \text{ V}$$

$$V_B = 5 \text{ V}$$

$$V_E = \frac{R_3}{R_3 + R_4} V_1 = \frac{2}{2+3} 5 = 2 \text{ V}$$

$$V_D = V_F = 0 \text{ V}$$

$$i_{O,F} = 0 \text{ A}$$

$$V_{C_1,F} = V_B - V_D = 5 - 0 = 5 \text{ V}$$

$$V_{C_2,F} = V_D - V_E = 0 - 2 = -2 \text{ V}$$

$$V_{C_3,F} = V_E - V_F = 2 - 0 = 2 \text{ V}$$

MATLAB can be used to calculate the limits of the given functions as t approaches infinity (Figs. 4.64, 4.65, 4.66, and 4.67). These values match the results of the previous analysis.

```
Command Window                                                    ⊙
  >> syms t
  >> iO=5*exp(-0.6111*t)*(cosh(0.3093*t)-0.7188*sinh(0.3093*t));
  >> limit(iO,t,inf)

  ans =

  0
fx >> |
```

Fig. 4.64 MATLAB code

```
Command Window                                                    ⊙
  >> syms t
  >> VC1=5-5*exp(-0.6111*t)*(cosh(0.3093*t)-0.7188*sinh(0.3093*-t));
  >> limit(VC1,t,inf)

  ans =

  5
fx >>
```

Fig. 4.65 MATLAB code

```
Command Window                                                          ⊙
  >> syms t
  >> VC2=2*exp  (-0.6111*t)*(cosh(0.3093*t)+0.6282*sinh(0.3093*t))-2;
  >> limit(VC2,t,inf)

  ans =

  -2

fx >> |
```

Fig. 4.66 MATLAB code

```
Command Window                                                          ⊙
  >> syms t
  >> VC3=2+exp(-0.6111*t)*(3*cosh(0.3093*t)-4.8504*sinh(0.3093*t));
  >> limit(VC3,t,inf)

  ans =

  2

fx >>
```

Fig. 4.67 MATLAB code

4.2.38 *Exercise 4.2*

Determine the state-space representation of the circuit in Fig. 4.68, with the $v_o(t)$ as the output.

Fig. 4.68 Circuit for
Exercise 4.2

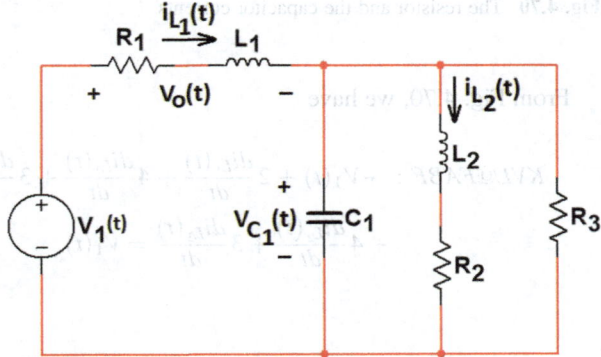

4.2.39 Example 4.37

Determine the state-space representation of the circuit in Fig. 4.69, with the $v_C(t)$ as the output.

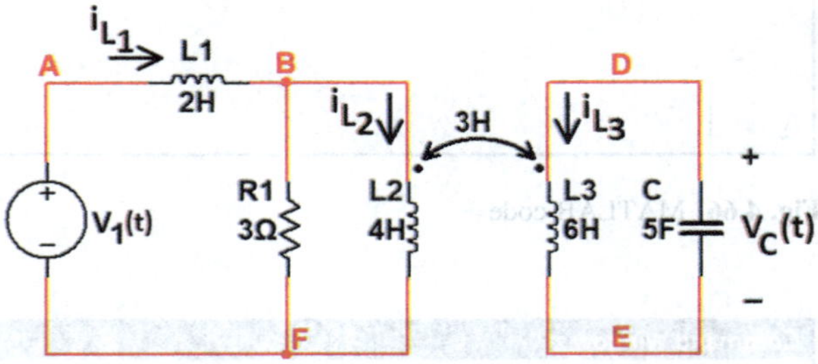

Fig. 4.69 Circuit for Example 4.37

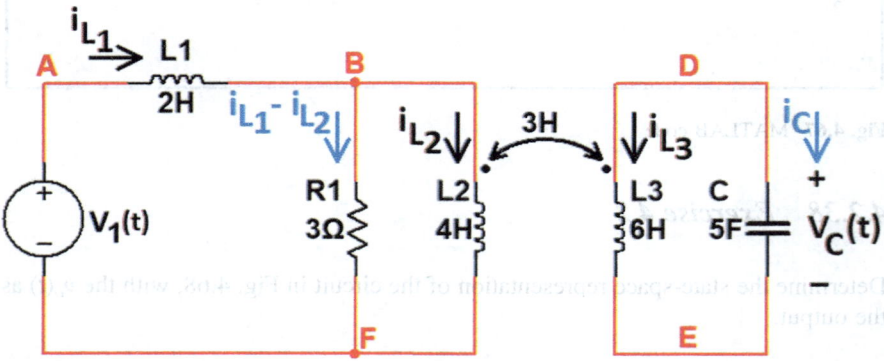

Solution:
The resistor and the capacitor currents are depicted in Fig. 4.70.

Fig. 4.70 The resistor and the capacitor currents

From Fig. 4.70, we have

$$KVL@FABF : -V_1(t) + 2\frac{di_{L_1}(t)}{dt} + 4\frac{di_{L_2}(t)}{dt} + 3\frac{di_{L_3}(t)}{dt} = 0 \Longrightarrow 2\frac{di_{L_1}(t)}{dt}$$
$$+ 4\frac{di_{L_2}(t)}{dt} + 3\frac{di_{L_3}(t)}{dt} = V_1(t)$$

$$KVL@FABF: -V_1(t) + 2\frac{di_{L_1}(t)}{dt} + 3(i_{L1}(t) - i_{L2}(t)) = 0 \Rightarrow 2\frac{di_{L_1}(t)}{dt} = -3i_{L1}(t)$$
$$+ 3i_{L2}(t) + V_1(t)$$

$$KVL@DED: 6\frac{di_{L_3}(t)}{dt} + 3\frac{di_{L_2}(t)}{dt} - V_C(t) = 0 \Rightarrow 6\frac{di_{L_3}(t)}{dt} + 3\frac{di_{L_2}(t)}{dt} = V_C(t)$$

$$KCL@D: i_C(t) + i_{L3}(t) = 0 \Rightarrow 5\frac{dv_C(t)}{dt} + i_{L3}(t) = 0 \Rightarrow 5\frac{dv_C(t)}{dt} = -i_{L3}(t)$$

$$\begin{cases} 2\dfrac{di_{L_1}(t)}{dt} + 4\dfrac{di_{L_2}(t)}{dt} + 3\dfrac{di_{L_3}(t)}{dt} = V_1(t) \\[2mm] 2\dfrac{di_{L_1}(t)}{dt} = -3i_{L1}(t) + 3i_{L2}(t) + V_1(t) \\[2mm] 6\dfrac{di_{L_3}(t)}{dt} + 3\dfrac{di_{L_2}(t)}{dt} = V_C(t) \\[2mm] 5\dfrac{dv_C(t)}{dt} = -i_{L3}(t) \end{cases} \Rightarrow \begin{bmatrix} 2 & 4 & 3 & 0 \\ 2 & 0 & 0 & 0 \\ 0 & 3 & 6 & 0 \\ 0 & 0 & 0 & 5 \end{bmatrix}$$

$$\times \begin{bmatrix} \dfrac{di_{L_1}(t)}{dt} \\[2mm] \dfrac{di_{L_2}(t)}{dt} \\[2mm] \dfrac{di_{L_3}(t)}{dt} \\[2mm] \dfrac{dv_C(t)}{dt} \end{bmatrix} = \begin{bmatrix} 0 & 0 & 0 & 0 \\ -3 & 3 & 0 & 0 \\ 0 & 0 & 0 & 1 \\ 0 & 0 & -1 & 0 \end{bmatrix} \begin{bmatrix} i_{L_1}(t) \\ i_{L_2}(t) \\ i_{L_3}(t) \\ v_C(t) \end{bmatrix}$$

$$+ \begin{bmatrix} 1 \\ 1 \\ 0 \\ 0 \end{bmatrix} V_1(t) \Rightarrow \begin{bmatrix} \dfrac{di_{L_1}(t)}{dt} \\[2mm] \dfrac{di_{L_2}(t)}{dt} \\[2mm] \dfrac{di_{L_3}(t)}{dt} \\[2mm] \dfrac{dv_C(t)}{dt} \end{bmatrix} = \begin{bmatrix} 2 & 4 & 3 & 0 \\ 2 & 0 & 0 & 0 \\ 0 & 3 & 6 & 0 \\ 0 & 0 & 0 & 5 \end{bmatrix}^{-1} \begin{bmatrix} 0 & 0 & 0 & 0 \\ -3 & 3 & 0 & 0 \\ 0 & 0 & 0 & 1 \\ 0 & 0 & -1 & 0 \end{bmatrix}$$

$$\times \begin{bmatrix} i_{L_1}(t) \\ i_{L_2}(t) \\ i_{L_3}(t) \\ v_C(t) \end{bmatrix} + \begin{bmatrix} 2 & 4 & 3 & 0 \\ 2 & 0 & 0 & 0 \\ 0 & 3 & 6 & 0 \\ 0 & 0 & 0 & 5 \end{bmatrix}^{-1} \begin{bmatrix} 1 \\ 1 \\ 0 \\ 0 \end{bmatrix} V_1(t)$$

MATLAB can be utilized to perform the calculations.

```
Command Window                                                              ⊙
>> inv([2 4 3 0;2 0 0 0;0 3 6 0;0 0 0 5])*[0 0 0 0;-3 3 0 0;0 0 0 1;0 0 -1 0]

ans =

   -1.5000    1.5000         0         0
    1.2000   -1.2000         0   -0.2000
   -0.6000    0.6000         0    0.2667
         0         0   -0.2000         0

>> inv([2 4 3 0;2 0 0 0;0 3 6 0;0 0 0 5])*[1;1;0;0]

ans =

    0.5000
         0
         0
         0

fx >> |
```

Fig. 4.71 MATLAB code

From Fig. 4.71, we have

$$
\begin{bmatrix} \dfrac{di_{L_1}(t)}{dt} \\[2mm] \dfrac{di_{L_2}(t)}{dt} \\[2mm] \dfrac{di_{L_3}(t)}{dt} \\[2mm] \dfrac{dv_C(t)}{dt} \end{bmatrix} = \begin{bmatrix} 2 & 4 & 3 & 0 \\ 2 & 0 & 0 & 0 \\ 0 & 3 & 6 & 0 \\ 0 & 0 & 0 & 5 \end{bmatrix}^{-1} \begin{bmatrix} 0 & 0 & 0 & 0 \\ -3 & 3 & 0 & 0 \\ 0 & 0 & 0 & 1 \\ 0 & 0 & -1 & 0 \end{bmatrix} \begin{bmatrix} i_{L_1}(t) \\ i_{L_2}(t) \\ i_{L_3}(t) \\ v_C(t) \end{bmatrix}
$$

$$
+ \begin{bmatrix} 2 & 4 & 3 & 0 \\ 2 & 0 & 0 & 0 \\ 0 & 3 & 6 & 0 \\ 0 & 0 & 0 & 5 \end{bmatrix}^{-1} \begin{bmatrix} 1 \\ 1 \\ 0 \\ 0 \end{bmatrix} V_1(t) \Rightarrow \begin{bmatrix} \dfrac{di_{L_1}(t)}{dt} \\[2mm] \dfrac{di_{L_2}(t)}{dt} \\[2mm] \dfrac{di_{L_3}(t)}{dt} \\[2mm] \dfrac{dv_C(t)}{dt} \end{bmatrix}
$$

$$
= \begin{bmatrix} -1.5 & 1.5 & 0 & 0 \\ 1.2 & -1.2 & 0 & -0.2 \\ -0.6 & 0.6 & 0 & 0.2667 \\ 0 & 0 & -0.2 & 0 \end{bmatrix} \begin{bmatrix} i_{L_1}(t) \\ i_{L_2}(t) \\ i_{L_3}(t) \\ v_C(t) \end{bmatrix} + \begin{bmatrix} 0.5 \\ 0 \\ 0 \\ 0 \end{bmatrix} V_1(t)
$$

$$V_O(t) = v_C(t) \Rightarrow V_O(t) = \begin{bmatrix} 0 & 0 & 0 & 1 \end{bmatrix} \begin{bmatrix} i_{L_1}(t) \\ i_{L_2}(t) \\ i_{L_3}(t) \\ v_C(t) \end{bmatrix} + 0V_1(t)$$

Therefore,

$$\begin{cases} \begin{bmatrix} \dfrac{di_{L_1}(t)}{dt} \\ \dfrac{di_{L_2}(t)}{dt} \\ \dfrac{di_{L_3}(t)}{dt} \\ \dfrac{dv_C(t)}{dt} \end{bmatrix} = \begin{bmatrix} -1.5 & 1.5 & 0 & 0 \\ 1.2 & -1.2 & 0 & -0.2 \\ -0.6 & 0.6 & 0 & 0.2667 \\ 0 & 0 & -0.2 & 0 \end{bmatrix} \begin{bmatrix} i_{L_1}(t) \\ i_{L_2}(t) \\ i_{L_3}(t) \\ v_C(t) \end{bmatrix} + \begin{bmatrix} 0.5 \\ 0 \\ 0 \\ 0 \end{bmatrix} V_1(t) \\ \\ V_O(t) = v_C(t) \Rightarrow V_O(t) = \begin{bmatrix} 0 & 0 & 0 & 1 \end{bmatrix} \begin{bmatrix} i_{L_1}(t) \\ i_{L_2}(t) \\ i_{L_3}(t) \\ v_C(t) \end{bmatrix} + 0V_1(t) \end{cases}$$

The matrices A, B, C, and D are

$$\begin{cases} A = \begin{bmatrix} -1.5 & 1.5 & 0 & 0 \\ 1.2 & -1.2 & 0 & -0.2 \\ -0.6 & 0.6 & 0 & 0.2667 \\ 0 & 0 & -0.2 & 0 \end{bmatrix} \\ \\ B = \begin{bmatrix} 0.5 \\ 0 \\ 0 \\ 0 \end{bmatrix} \\ \\ C = \begin{bmatrix} 0 & 0 & 0 & 1 \end{bmatrix} \\ D = [0] \end{cases}$$

Further Reading

1. Asadi F (2022) Essential circuit analysis using NI Multisim™ and MATLAB®, Springer. https://doi.org/10.1007/978-3-030-89850-2
2. Asadi F (2022) Essential circuit analysis using Proteus®, Springer. https://doi.org/10.1007/978-981-19-4353-9
3. Asadi F (2022) Essential circuit analysis using LTspice®, Springer. https://doi.org/10.1007/978-3-031-09853-6
4. Asadi F (2022) Electric circuit analysis with EasyEDA, Springer. https://doi.org/10.1007/978-3-031-00292-2
5. Asadi F, Eguchi K (2022) Electric and electronic circuit simulation using TINA-TI®, River Publishers. https://doi.org/10.13052/rp-9788770226851
6. Asadi F (2023) Analog electronic circuits laboratory manual, Springer. https://doi.org/10.1007/978-3-031-25122-1

Chapter 5
The Laplace Transform

5.1 Introduction

The Laplace transform is a mathematical tool that converts a function of a real variable (often representing time) into a function of a complex variable (often denoted as s). This transformation is particularly useful for solving differential equations, as it transforms differential equations into algebraic equations, which are often easier to solve.

This chapter begins with a review of the Laplace transform and its applications. The second part demonstrates MATLAB's application to the problems discussed.

Calculations presented in this book are truncated to four decimal places. Consequently, minor discrepancies may arise between results obtained from different solution methodologies. These variations are solely attributable to accumulated rounding errors and do not reflect inaccuracies in the methodologies themselves.

5.2 Unit Step Function

The unit step function (Fig. 5.1), also known as the Heaviside step function, is a mathematical function that is zero for negative arguments and one for positive arguments. It is often denoted by the symbol $u(t)$ or $H(t)$. Here is the formal definition: $H(t) = \begin{cases} 1 & t \geq 0 \\ 0 & t < 0 \end{cases}$.

F. Asadi, *A Problem-Solving Approach to Electric Circuits*,
https://doi.org/10.1007/978-3-031-95493-1_5

Fig. 5.1 Unit step function

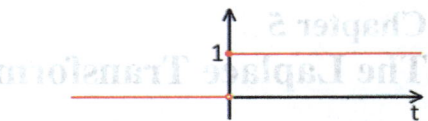

Example 5.1

When $0 < a < b$, the function shown in Fig. 5.2 can be written as $f(t) = k(H(t-a) - H(t-b))$.

Fig. 5.2 Function $f(t)$

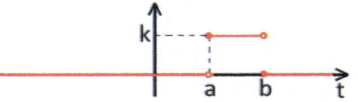

5.3   Unit Impulse Function

The unit impulse function, also known as the Dirac delta function, is an idealized mathematical function that is zero everywhere except at zero, where it is infinitely high, and its integral over the entire real line is equal to one. The Dirac delta function is typically denoted by the symbol $\delta(t)$.

The functions shown in Figs. 5.3 and 5.4 can be considered as approximations of the Dirac delta function. While the true Dirac delta function is an idealized concept with infinite height and zero width, these functions represent practical approximations with finite height and width.

Fig. 5.3 Rectangular pulse approximation of Dirac delta function ($\Delta \to 0$)

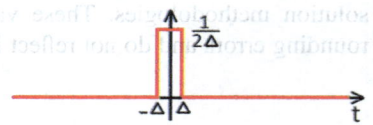

Fig. 5.4 Triangular pulse approximation of Dirac delta function ($\Delta \to 0$)

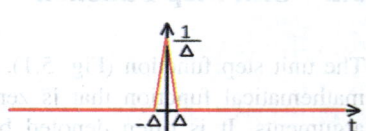

Sifting property is the most important property of unit impulse function. When the unit impulse function is multiplied by another function and integrated, it effectively "sifts out" the value of the other function at the point where the impulse is located. The sifting property of the Dirac delta function is mathematically expressed as

$$\int_{-\infty}^{+\infty} f(t)\delta(t-a)dt = f(a)$$

Sifting property is crucial in many applications. While the unit impulse function is an idealized concept, it is a powerful tool for analyzing systems and signals, particularly in the context of linear systems theory.

5.4 Laplace Transform

The Laplace transform is a mathematical tool that converts a function of a real variable (often representing time) into a function of a complex variable (often representing frequency). This transformation can be useful for solving differential equations, particularly those that arise in engineering and physics.

The Laplace transform of a function $f(t)$ is typically denoted as $F(s)$ or $\mathcal{L}(f(t))$. The Laplace transform of a function $f(t)$ is defined as (note that s is a complex variable, i.e., $s = \sigma + j\omega$):

$$F(s) = \mathcal{L}(f(t)) = \int_0^\infty f(t)e^{-st}dt$$

The inverse Laplace transform is the reverse operation of the Laplace transform. It takes a function in the s-domain (the complex frequency domain) and transforms it back into a function in the time domain. Fortunately, a Laplace transform table can be employed to determine the inverse Laplace transform of most functions.

Example 5.2
Determine the Laplace transform of the Heaviside step function.

Solution:

$$H(t) = \begin{cases} 1 & t \geq 0 \\ 0 & t < 0 \end{cases}$$

$$H(s) = \mathcal{L}(H(t)) = \int_0^\infty H(t)e^{-st}dt = \int_0^\infty e^{-st}dt = \frac{e^{-st}}{-s}\bigg|_0^\infty = \frac{e^{-(\sigma+j\omega)t}}{-(\sigma+j\omega)}\bigg|_0^\infty$$

When $\sigma > 0$, $\lim\limits_{t \to \infty} e^{-(\sigma+j\omega)t} = 0 \Rightarrow \lim\limits_{t \to \infty} \frac{e^{-(\sigma+j\omega)t}}{-(\sigma+j\omega)} = 0$. Therefore,

$$\frac{e^{-(\sigma+j\omega)t}}{-(\sigma+j\omega)}\bigg|_0^\infty = 0 - \frac{e^{-(\sigma+j\omega)0}}{-(\sigma+j\omega)} = \frac{1}{(\sigma+j\omega)} = \frac{1}{s}$$

Example 5.3
Determine the Laplace transform of $f(t) = tH(t)$. Note that $H(t)$ shows the Heaviside step function.

Solution:

$$F(s) = \mathcal{L}(f(t)) = \int_0^\infty tH(t)e^{-st}dt = \int_0^\infty te^{-st}dt$$

Tabular integration (Fig. 5.5) can be used to calculate $\int_0^\infty te^{-st}dt$.

Fig. 5.5 Calculation of $\int te^{-st}dt$

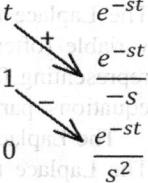

According to Fig. 5.5, $\int_0^\infty te^{-st}dt = \left(\frac{t}{-s} - \frac{1}{s^2}\right)e^{-st}$. Therefore,

$$\int_0^\infty te^{-st}dt = \left(\frac{t}{-s} - \frac{1}{s^2}\right)e^{-st}\Bigg|_0^\infty = \lim_{t \to \infty}\left(\frac{t}{-s} - \frac{1}{s^2}\right)e^{-st} - \left(\frac{0}{-s} - \frac{1}{s^2}\right)e^{-s0}$$

$$= \lim_{t \to \infty}\left(\frac{t}{-s} - \frac{1}{s^2}\right)e^{-st} + \frac{1}{s^2} = \lim_{t \to \infty}\left(\frac{t}{-se^{st}} - \frac{1}{s^2e^{st}}\right) + \frac{1}{s^2}$$

$$= \lim_{t \to \infty}\left(\frac{t}{-se^{st}}\right) - \lim_{t \to \infty}\left(\frac{1}{s^2e^{st}}\right) + \frac{1}{s^2}$$

Remember that s is a complex variable, that is, $s = \sigma + j\omega$. When $Re(s) = \sigma > 0$, $\lim_{t \to \infty}\left(\frac{t}{-se^{st}}\right) - \lim_{t \to \infty}\left(\frac{1}{s^2e^{st}}\right) = 0 - 0 = 0$. Note that $\lim_{t \to \infty}\left(\frac{t}{-se^{st}}\right) = 0$ since exponential function grows much faster than linear functions. $\lim_{t \to \infty}\left(\frac{1}{s^2e^{st}}\right) = 0$ since numerator is a constant divided by infinity. Therefore,

$$F(s) = \mathcal{L}(f(t)) = \int_0^\infty te^{-st}dt = \frac{1}{s^2}$$

5.5   Commonly Used Laplace Transforms

Table 5.1 provides a compendium of common Laplace transform pairs, essential for solving differential equations and analyzing linear systems.

Table 5.1 Laplace transform of commonly used functions

Signal	$f(t)$	$F(s)$
Impulse	$\delta(t)$	1
Step function	$H(t)$	$\frac{1}{s}$
Ramp	$tH(t)$	$\frac{1}{s^2}$
Exponential	$e^{-at}H(t)$	$\frac{1}{s+a}$
Damped ramp	$te^{-at}H(t)$	$\frac{1}{(s+a)^2}$
Sine	$\sin(\omega t)H(t)$	$\frac{\omega}{s^2+\omega^2}$
Cosine	$\cos(\omega t)H(t)$	$\frac{s}{s^2+\omega^2}$
Damped sine	$e^{-at}\sin(\omega t)H(t)$	$\frac{\omega}{(s+a)^2+\omega^2}$
Damped cosine	$e^{-at}\cos(\omega t)H(t)$	$\frac{s+a}{(s+a)^2+\omega^2}$
Damped polynomial	$e^{-at}t^nH(t)$	$\frac{n!}{(s+a)^{n+1}}$

5.6   Inverse Laplace Transform

Partial fraction decomposition and a table of Laplace transform pairs (Table 5.1) can help us to calculate the inverse Laplace transform of most functions that appear in engineering. Let us study some numeric examples.

Example 5.4

Determine the inverse Laplace transform of $\frac{5s^2+s+4}{(s+4)(s^2+4)}$.

Solution:

$$\frac{5s^2+s+4}{(s+4)(s^2+4)} = \frac{4}{s+4} + \frac{s-3}{s^2+4} = \frac{4}{s+4} + \frac{s}{s^2+4} + \frac{-3}{s^2+4}$$

$$\mathcal{L}^{-1}\left(\frac{5s^2+s+4}{(s+4)(s^2+4)}\right) = \mathcal{L}^{-1}\left(\frac{4}{s+4}\right) + \mathcal{L}^{-1}\left(\frac{s}{s^2+4}\right) + \mathcal{L}^{-1}\left(\frac{-3}{s^2+4}\right)$$

$$= 4\mathcal{L}^{-1}\left(\frac{1}{s+4}\right) + \mathcal{L}^{-1}\left(\frac{s}{s^2+4}\right) + \frac{-3}{2}\mathcal{L}^{-1}\left(\frac{2}{s^2+4}\right)$$

$$= \left(4e^{-4t} + \cos(2t) - \frac{3}{2}\sin(2t)\right).H(t)$$

Example 5.5

Determine the inverse Laplace transform of $\frac{10(s+2)}{s(s^2+2s+2)}$.

Solution:

$$\frac{10(s+2)}{s(s^2+2s+2)} = \frac{A}{s} + \frac{Bs+C}{s^2+2s+2}$$

$$A = \frac{10(s+2)}{s^2+2s+2}\bigg|_{s=0} = \frac{20}{2} = 10$$

$$\frac{10(s+2)}{s(s^2+2s+2)} = \frac{10}{s} + \frac{Bs+C}{s^2+2s+2} \Rightarrow \frac{10(s+2)}{s(s^2+2s+2)} - \frac{10}{s} = \frac{Bs+C}{s^2+2s+2}$$

$$\frac{10(s+2)}{s(s^2+2s+2)} - \frac{10(s^2+2s+2)}{s(s^2+2s+2)} = \frac{-10s^2-10s}{s(s^2+2s+2)} = \frac{-10s-10}{s^2+2s+2} = \frac{Bs+C}{s^2+2s+2}$$

$$-10s - 10 = Bs + C \Rightarrow B = -10, C = -10$$

$$\frac{10(s+2)}{s(s^2+2s+2)} = \frac{10}{s} + \frac{-10s-10}{s^2+2s+2} = \frac{10}{s} - \frac{10(s+1)}{(s+1)^2+1}$$

$$\mathcal{L}^{-1}\left(\frac{10(s+2)}{s(s^2+2s+2)}\right) = 10\left(\mathcal{L}^{-1}\left(\frac{1}{s}\right) - \mathcal{L}^{-1}\left(\frac{(s+1)}{(s+1)^2+1}\right)\right)$$

$$= 10(1 - e^{-t}\cos(t))H(t)$$

Example 5.6

Determine the inverse Laplace transform of $\frac{s+1}{s^2-6s+13}$.

Solution:

$$\mathcal{L}^{-1}\left(\frac{s+1}{s^2-6s+13}\right) = \mathcal{L}^{-1}\left(\frac{s-3+4}{(s-3)^2+4}\right)$$

$$= \mathcal{L}^{-1}\left(\frac{s-3}{(s-3)^2+4}\right) + 2\mathcal{L}^{-1}\left(\frac{2}{(s-3)^2+4}\right)$$

$$= e^{3t}\cos(2t)H(t) + 2e^{3t}\sin(2t).H(t)$$

$$= e^{3t}(\cos(2t) + 2\sin(2t)).H(t)$$

Example 5.7

Determine the inverse Laplace transform of $\frac{10(s+2)}{s(s^2+2s+2)}$.

Solution:

$$\mathcal{L}^{-1}\left(\frac{10(s+2)}{s(s^2+2s+2)}\right) = \mathcal{L}^{-1}\left(\frac{10(s+2)}{s\left((s+1)^2+1\right)}\right) = \mathcal{L}^{-1}\left(\frac{A}{s}\right)$$

$$+ \mathcal{L}^{-1}\left(\frac{B(s+1)+C}{(s+1)^2+1}\right)$$

$$A = \frac{10(s+2)}{(s+1)^2+1}\bigg|_{s=0} = 10$$

$$\frac{10(s+2)}{s(s^2+2s+2)} - \frac{10}{s} = \frac{10(s+2)}{s(s^2+2s+2)} - \frac{10(s^2+2s+2)}{s(s^2+2s+2)}$$

$$= \frac{10s+20-10s^2-20s-20}{s(s^2+2s+2)} = \frac{-10s-10s^2}{s(s^2+2s+2)}$$

$$= \frac{-10-10s}{s^2+2s+2} = \frac{-10(s+1)+0}{(s+1)^2+1} \Rightarrow B = -10, C = 0$$

$$\mathcal{L}^{-1}\left(\frac{10(s+2)}{s(s^2+2s+2)}\right) = \mathcal{L}^{-1}\left(\frac{10}{s}\right) - \mathcal{L}^{-1}\left(\frac{10(s+1)}{(s+1)^2+1}\right)$$

$$= (10 - 10e^{-t}\cos(t))H(t)$$

5.7   Properties of Laplace Transform

The followings are the important properties of Laplace transform. Note that $f(t) \longleftrightarrow F(s)$, $f_1(t) \longleftrightarrow F_1(s)$, $f_2(t) \longleftrightarrow F_2(s)$, $H(t)$ shows the Heaviside step function, and $k_1, k_2 \in \mathbb{R}$.

1. $k_1 f(t) \longleftrightarrow k_1 F(s)$
2. $k_1 f_1(t) + k_2 f_2(t) \longleftrightarrow k_1 F_1(s) + k_2 F_2(s)$
3. for $a > 0, f(at) \longleftrightarrow \frac{1}{a}F\left(\frac{s}{a}\right)$
4. $f(t-t_0)H(t-t_0) \longleftrightarrow e^{-t_0 s}F(s)$
5. $e^{-at}f(t) \longleftrightarrow F(s+a)$
6. $\dot{f}(t) = \frac{df}{dt} \longleftrightarrow sF(s) - f(0)$
7. $\ddot{f}(t) = \frac{d^2f}{dt^2} \longleftrightarrow s^2F(s) - sf(0) - \dot{f}(0)$
8. $\dots f(t) = \frac{d^3f}{dt^3} \longleftrightarrow s^3F(s) - s^2f(0) - s\dot{f}(0) - \ddot{f}(0)$
9. $\int_0^t f(\tau)d\tau \longleftrightarrow \frac{F(s)}{s}$
10. $t.f(t) \longleftrightarrow -\frac{d}{ds}(F(s)) = -F'(s)$
11. $\frac{f(t)}{t} \longleftrightarrow \int_s^\infty F(\sigma)d\sigma$
12. $\lim_{t \to 0^+} f(t) = \lim_{s \to \infty} s.F(s)$
13. $\lim_{t \to \infty} f(t) = \lim_{s \to 0} s.F(s)$
14. $f_1(t) * f_2(t) = \int_{-\infty}^{+\infty} f_1(\tau)f_2(t-\tau)d\tau \longleftrightarrow F_1(s).F_2(s)$
15. $if \forall t \geq 0 : f(t+T) = f(t) \Rightarrow \mathcal{L}(f(t)) = \frac{\int_0^T e^{-st}f(t)dt}{1-e^{-Ts}}$

Example 5.8

Determine the Laplace transform of $f(t) = \begin{cases} +1 & 0 \le t < 1 \\ -1 & 1 \le t < 2 \end{cases}$, $f(t+2) = f(t)$. The graph of $f(t)$ is shown in Fig. 5.6.

Fig. 5.6 Periodic function $f(t)$

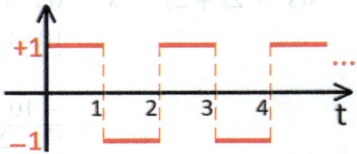

Solution:

$$if \; \forall t \ge 0 : f(t+T) = f(t) \Longrightarrow L(f(t)) = \frac{\int_0^T e^{-st}f(t)dt}{1 - e^{-Ts}}$$

$$T = 2 \Longrightarrow L(f(t)) = \frac{\int_0^2 e^{-st}f(t)dt}{1 - e^{-2s}} = \frac{\int_0^1 e^{-st}dt + \int_1^2 -e^{-st}dt}{1 - e^{-2s}} = \frac{\left.\frac{e^{-st}}{-s}\right|_0^1 + \left.\frac{e^{-st}}{s}\right|_1^2}{1 - e^{-2s}}$$

$$= \frac{\frac{e^{-s}}{-s} + \frac{1}{s} + \frac{e^{-2s}}{s} - \frac{e^{-s}}{s}}{1 - e^{-2s}} = \frac{1 - 2e^{-s} + e^{-2s}}{(1 - e^{-2s})s}$$

$$L(f(t)) = \frac{1 - 2e^{-s} + e^{-2s}}{(1 - e^{-2s})s} = \frac{(1 - e^{-s})^2}{(1 - e^{-s})(1 + e^{-s})s} = \frac{(1 - e^{-s})}{(1 + e^{-s})s}$$

Example 5.9

Determine the Laplace transform of $f(t) = \sin(t)$.

Solution:

$$if \forall t \ge 0 : f(t+T) = f(t) \Longrightarrow L(f(t)) = \frac{\int_0^T e^{-st}f(t)dt}{1 - e^{-Ts}}$$

$$f(t) = \sin(t) \Longrightarrow T = 2\pi \Longrightarrow L(f(t)) = \frac{\int_0^{2\pi} e^{-st}\sin(t)dt}{1 - e^{-2\pi s}} = \frac{\frac{1 - e^{-2\pi s}}{s^2 + 1}}{1 - e^{-2\pi s}} = \frac{1}{s^2 + 1}$$

5.8   Laplace Transform for Solving Differential Equations

The Laplace transform converts differential equations into algebraic equations. Initial conditions are directly included in the transformed equation, making it easier to handle. Memorizing the following two formulas is highly recommended, as they are instrumental in solving a majority of problems.

$$\dot{f}(t) = \frac{df}{dt} \longleftrightarrow sF(s) - f(0)$$

$$\ddot{f}(t) = \frac{d^2 f}{dt^2} \longleftrightarrow s^2 F(s) - sf(0) - \dot{f}(0)$$

Let us study some numeric examples.

Example 5.10

Calculate the inverse Laplace transform of $X(s) = \frac{s-13}{s^2-2s+5}$ and $Y(s) = \frac{7s-17}{s^2-2s+5}$.

Solution:

We use the partial fraction decomposition and Table 5.1 to calculate the given inverse Laplace transforms.

$$\begin{cases} x(t) = L^{-1}(X(s)) = L^{-1}\left(\frac{s-13}{s^2-2s+5}\right) = L^{-1}\left(\frac{s-1}{(s-1)^2+2^2} - 6\frac{2}{(s-1)^2+2^2}\right) = e^t\cos(2t) - 6e^t\sin(2t) \\[4mm] y(t) = L^{-1}(Y(s)) = L^{-1}\left(\frac{7s-17}{s^2-2s+5}\right) = L^{-1}\left(\frac{7(s-1)}{(s-1)^2+2^2} - 5\frac{2}{(s-1)^2+2^2}\right) = 7e^t\cos(2t) - 5e^t\sin(2t) \end{cases}$$

Example 5.11

Solve the $\ddot{y}(t) + 2\dot{y}(t) + y(t) = 6H(t), y(0) = 1, \dot{y}(0) = 2$.

Solution:

Taking the Laplace transform of both sides:

$$L(\ddot{y}(t)) + 2L(\dot{y}(t)) + L(y(t)) = 6L(H(t))$$

$$L(\ddot{y}(t)) = s^2 Y(s) - sy(0) - \dot{y}(0) = s^2 Y(s) - s - 2$$

$$L(\dot{y}(t)) = sY(s) - y(0) = sY(s) - 1$$

$$s^2 Y(s) - s - 2 + 2(sY(s) - 1) + Y(s) = \frac{6}{s} \Longrightarrow Y(s) = \frac{s + 4 + \frac{6}{s}}{s^2 + 2s + 1}$$

$$= \frac{s + 4}{(s + 1)^2} + \frac{6}{s(s + 1)^2}$$

$$\frac{s + 4}{(s + 1)^2} = \frac{A}{s + 1} + \frac{B}{(s + 1)^2}$$

$$B = (s + 4)|_{s = -1} = -1 + 4 = 3$$

$$A = \frac{d}{ds}(s + 4)\Big|_{s = -1} = 1|_{s = -1} = 1$$

Therefore, $\frac{s+4}{(s+1)^2} = \frac{1}{s+1} + \frac{3}{(s+1)^2}$.

$$\frac{6}{s(s+1)^2} = \frac{A}{s} + \frac{B}{s+1} + \frac{C}{(s+1)^2}$$

$$A = \frac{6}{(s+1)^2}\bigg|_{s=0} = \frac{6}{1} = 6$$

$$C = \frac{6}{s}\bigg|_{s=-1} = \frac{6}{-1} = -6$$

$$B = \frac{d}{ds}\left(\frac{6}{s}\right)\bigg|_{s=-1} = -\frac{6}{s^2}\bigg|_{s=-1} = -6$$

Therefore, $\frac{6}{s(s+1)^2} = \frac{6}{s} - \frac{6}{s+1} - \frac{6}{(s+1)^2}$.

$$Y(s) = \frac{s+4}{(s+1)^2} + \frac{6}{s(s+1)^2} = \frac{s+1+3}{(s+1)^2} + \frac{6}{s} - \frac{6}{s+1} - \frac{6}{(s+1)^2}$$

$$Y(s) = \frac{s+1+3}{(s+1)^2} + \frac{6}{s} - \frac{6}{s+1} - \frac{6}{(s+1)^2}$$

$$= \frac{1}{s+1} + \frac{3}{(s+1)^2} + \frac{6}{s} - \frac{6}{s+1} - \frac{6}{(s+1)^2}$$

$$Y(s) = \frac{-5}{s+1} - \frac{3}{(s+1)^2} + \frac{6}{s}$$

Using Table 5.1, $y(t) = (-5e^{-t} - 3te^{-t} + 6)H(t)$.

Example 5.12

Solve the $\ddot{y}(t) + 4y(t) = H(t) - H(t-1), y(0) = \dot{y}(0) = 0$.

Solution:

Taking the Laplace transform of both sides:

$$L(\ddot{y}(t)) + L(4y(t)) = L(H(t)) - L(H(t-1))$$

$$L(\ddot{y}(t)) + 4L(y(t)) = L(H(t)) - L(H(t-1))$$

$$s^2 Y(s) - sy(0) - \dot{y}(0) + 4Y(s) - 4y(0) = \frac{1}{s} - \frac{1}{s}e^{-s}$$

$$Y(s) = \frac{1}{s(s^2+4)} - \frac{1}{s(s^2+4)}e^{-s}$$

$$\frac{1}{s(s^2+4)} = \frac{A}{s} + \frac{Bs+C}{s^2+4}$$

$$A = \frac{1}{s^2 + 4}\Big|_{s=0} = \frac{1}{4}$$

$$\frac{1}{s(s^2+4)} = \frac{\frac{1}{4}}{s} + \frac{Bs+C}{s^2+4} \implies \frac{1}{s(s^2+4)} - \frac{\frac{1}{4}}{s} = \frac{Bs+C}{s^2+4} \implies \frac{1 - \frac{1}{4}(s^2+4)}{s(s^2+4)} = \frac{Bs+C}{s^2+4}$$

$$\frac{1 - \frac{1}{4}s^2 - 1}{s(s^2+4)} = \frac{Bs+C}{s^2+4} \implies B = \frac{-1}{4}, C = 0 \implies \frac{1}{s(s^2+4)} = \frac{\frac{1}{4}}{s} + \frac{-\frac{1}{4}s}{s^2+4}$$

$$Y(s) = \frac{1}{s(s^2+4)} - \frac{1}{s(s^2+4)}e^{-s} = \frac{\frac{1}{4}}{s} - \frac{\frac{1}{4}s}{s^2+4} - \frac{\frac{1}{4}}{s}e^{-s} + \frac{\frac{1}{4}s}{s^2+4}e^{-s}$$

$$y(t) = L^{-1}\left(\frac{\frac{1}{4}}{s}\right) - L^{-1}\left(\frac{\frac{1}{4}s}{s^2+4}\right) - L^{-1}\left(\frac{\frac{1}{4}}{s}e^{-s}\right) + L^{-1}\left(\frac{\frac{1}{4}s}{s^2+4}e^{-s}\right)$$

$$y(t) = \frac{1}{4}H(t) - \frac{1}{4}\cos(2t)H(t) - \frac{1}{4}H(t-1) + \frac{1}{4}\cos(2(t-1))H(t-1)$$

$$y(t) = \frac{1}{4}(1 - \cos(2t))H(t) - \frac{1}{4}(1 - \cos(2(t-1)))H(t-1)$$

$$y(t) = \begin{cases} \frac{1}{4}(1 - \cos(2t)) & 0 \le t < 1 \\ \frac{1}{4}\cos(2t-2) - \frac{1}{4}\cos(2t) & t \ge 1 \end{cases}$$

Example 5.13
Solve the $\ddot{y}(t) + \dot{y}(t) + y(t) = \delta(t), y(0) = \dot{y}(0) = 0$.

Solution:
Taking the Laplace transform of both sides:

$$L(\ddot{y}(t)) + L(\dot{y}(t)) + L(y(t)) = L(\delta(t))$$

$$(s^2 + s + 1)Y(s) = 1 \implies Y(s) = \frac{1}{(s^2 + s + 1)}$$

$$y(t) = L^{-1}\left(\frac{1}{s^2+s+1}\right) = L^{-1}\left(\frac{1}{(s+\frac{1}{2})^2 + \frac{3}{4}}\right) = L^{-1}\left(\frac{1}{(s+\frac{1}{2})^2 + \left(\frac{\sqrt{3}}{2}\right)^2}\right)$$

$$= \frac{1}{\frac{\sqrt{3}}{2}}e^{-\frac{1}{2}t}\sin\left(\frac{\sqrt{3}}{2}t\right)H(t)$$

Therefore, $y(t) = \frac{2}{\sqrt{3}}e^{-\frac{1}{2}t}\sin\left(\frac{\sqrt{3}}{2}t\right)H(t)$ is the solution of this problem.

Example 5.14
Find the general solution of $\ddot{y}(t) + 3\dot{y}(t) + 2y(t) = \sin(3t)$.

Solution:
Homogenous solution can be found easily:

$$s^2 + 3s + 2 = 0 \Rightarrow s_1 = -1, s_2 = -2 \Rightarrow y_h(t) = C_1 e^{-t} + C_2 e^{-2t}$$

Let us find the particular solution with undetermined coefficient method. According to the method of undetermined coefficients, a particular solution to the differential equation $\ddot{y}(t) + 3\dot{y}(t) + 2y(t) = \sin(3t)$ can be assumed to have the form $y_p(t) = A \sin(3t) + B \cos(3t)$. Therefore,

$$y_p(t) = A\sin(3t) + B\cos(3t)$$

$$\dot{y}_p(t) = 3A\cos(3t) - 3B\sin(3t)$$

$$\ddot{y}_p(t) = -9A\sin(3t) - 9B\cos(3t)$$

Substitution in the $\ddot{y}(t) + 3\dot{y}(t) + 2y(t) = \sin(3t)$ leads to

$$- 9A\sin(3t) - 9B\cos(3t) + 3(3A\cos(3t) - 3B\sin(3t))$$
$$+ 2(A\sin(3t) + B\cos(3t)) = \sin(t)$$

which can be simplified to $(-7A - 9B) \sin(3t) + (9A - 7B) \cos(3t) = \sin(3t)$. Therefore,

$$\begin{cases} -7A - 9B = 1 \\ 9A - 7B = 0 \end{cases} \Rightarrow A = \frac{-7}{130}, B = \frac{-9}{130} \Rightarrow y_p(t) = A\sin(3t) + B\cos(3t)$$

$$= \frac{-7}{130}\sin(3t) + \frac{-9}{130}\cos(3t)$$

Note that $a \times \sin(\omega t) + b \times \cos(\omega t) = \text{sign}(a).\sqrt{a^2 + b^2} \sin\left(\omega t + \tan^{-1}\left(\frac{b}{a}\right)\right)$. Therefore,

$$y_p(t) = \frac{-7}{130}\sin(3t) + \frac{-9}{130}\cos(3t) = -\sqrt{\frac{1}{130}}\sin\left(3t + \tan^{-1}\left(\frac{\frac{-9}{130}}{\frac{-7}{130}}\right)\right)$$

$$= -\sqrt{\frac{1}{130}}\sin\left(3t + \tan^{-1}\left(\frac{9}{7}\right)\right) = -0.0877\sin(3t + 0.9098)$$

$$= 0.0877\sin(3t + 0.9098 - \pi) = 0.0877\sin(3t - 2.2318)$$

The general solution of this equation is

$$y(t) = y_h(t) + y_p(t) = C_1 e^{-t} + C_2 e^{-2t} + 0.0877\sin(3t - 2.2318)$$

Steady-state response ($y_{ss}(t)$) is the behavior of a dynamic system after a long enough time, when all transient effects (solutions of homogenous equation) have died out. In simpler terms, it is the long-term behavior of the system. Steady-state response is commonly studied with sinusoidal inputs. In this example, the first two terms go toward zero as t increases. The steady-state response in this example is $y_{ss}(t) = 0.877 \sin(3t - 2.2318)$ since $\lim_{t \to \infty} C_1 e^{-t} + C_2 e^{-2t} = 0$.

Sometimes, the transient behavior of the system (solutions of the homogenous equation) is not important and we just look for the steady-state response of the system. The Laplace transform can be used to find the steady-state response. Let us find the steady-state response of this example using the Laplace transform.

Given differential equation $\ddot{y}(t) + 3\dot{y}(t) + 2y(t) = \sin(3t)$ can be written as of $\ddot{y}(t) + 3\dot{y}(t) + 2y(t) = u(t)$ where of $u(t)$ shows the system input. In this example input is sinusoidal, that is, $u(t) = \sin(3t)$. Transfer function of this system is

$$L(\ddot{y}(t)) + L(3\dot{y}(t)) + L(2y(t)) = L(u(t)) \Rightarrow L(\ddot{y}(t)) + 3L(\dot{y}(t)) + 2L(y(t))$$
$$= L(u(t)) \Rightarrow s^2 Y(s) + 3sY(s) + 2Y(s)$$
$$= U(s) \Rightarrow (s^2 + 3s + 2)Y(s) = U(s) \Rightarrow T(s)$$
$$= \frac{Y(s)}{U(s)} = \frac{1}{s^2 + 3s + 2}$$

Note that the initial conditions are assumed to be zero when calculating the transfer function. The input is $u(t) = \sin(3t)$, so we need to calculate $T(3j)$:

$$T(s) = \frac{1}{s^2 + 3s + 2} \Rightarrow T(3j) = \frac{1}{(3j)^2 + 3 \times 3j + 2} = \frac{1}{-7 + 9j} = 0.0877 e^{-j2.2318}$$

Therefore, steady state solution (particular solution) of given differential equation is

$$y_{ss}(t) = y_p(t) = 0.0877 \sin(3t - 2.2318)$$

which is identical to the result obtained from the previous method.

Example 5.15
Find the general solution of $\ddot{y}(t) + 3\dot{y}(t) + 2y(t) = \sin\left(3t - \frac{\pi}{4}\right)$.

Solution:
Homogenous solution can be found easily:

$$s^2 + 3s + 2 = 0 \Rightarrow s_1 = -1, s_2 = -2 \Rightarrow y_h(t) = C_1 e^{-t} + C_2 e^{-2t}$$

In this example, the input is $u_2(t) = \sin\left(3t - \frac{\pi}{4}\right) = \sin\left(3\left(t - \frac{\pi}{12}\right)\right)$. As seen in the previous example, for the input $u_1(t) = \sin(3t)$, a particular solution is $y_{p1}(t) = 0.0877 \sin(3t - 2.2318)$. Therefore, particular solution for

$$u_2(t) = u_1\left(t - \tfrac{\pi}{12}\right) = \sin\left(3\left(t - \tfrac{\pi}{12}\right)\right) = \sin\left(3t - \tfrac{\pi}{4}\right) \quad \text{must be} \quad H(t) = \begin{cases} 1 & t \ge 0 \\ 0 & t < 0 \end{cases}.$$

Therefore, the general solution of $\ddot{y}(t) + 3\dot{y}(t) + 2y(t) = \sin\left(3t - \tfrac{\pi}{4}\right)$ is

$$y(t) = y_h(t) + y_{p2}(t) = C_1 e^{-t} + C_2 e^{-2t} + 0.0877 \sin(3t - 3.0172)$$

Example 5.16

Find the general solution of $\ddot{y}(t) + 3\dot{y}(t) + 2y(t) = 6\sin\left(5t + \tfrac{\pi}{6}\right)$.

Solution:

Homogenous solution can be found easily:

$$s^2 + 3s + 2 = 0 \Rightarrow s_1 = -1, s_2 = -2 \Rightarrow y_h(t) = C_1 e^{-t} + C_2 e^{-2t}$$

We found the transfer function in the previous example: $T(s) = \frac{Y(s)}{U(s)} = \frac{1}{s^2+3s+2}$. Input is $6\sin\left(5t + \tfrac{\pi}{6}\right)$ therefore, we need to calculate $T(5j)$.

$$T(s) = \frac{1}{s^2 + 3s + 2} \Rightarrow T(5j) = \frac{1}{(5j)^2 + 3 \times 5j + 2} = \frac{1}{-23 + 15j} = 0.0364 e^{-j2.5637}$$

$$y_{ss}(t) = y_p(t) = 0.0364 \times 6 \times \sin\left(5t + \frac{\pi}{6} - 2.5637\right) = 0.2184 \sin(5t - 2.0401)$$

General solution of this equation is

$$y(t) = y_h(t) + y_p(t) = C_1 e^{-t} + C_2 e^{-2t} + 0.2184 \sin(5t - 2.0401)$$

Example 5.17

Find the general solution of $\ddot{y}(t) + 3\dot{y}(t) + 2y(t) = 9\cos\left(7t + \tfrac{\pi}{3}\right)$.

Solution:

Homogenous solution can be found easily:

$$s^2 + 3s + 2 = 0 \Rightarrow s_1 = -1, s_2 = -2 \Rightarrow y_h(t) = C_1 e^{-t} + C_2 e^{-2t}$$

The transfer function for $\ddot{y}(t) + 3\dot{y}(t) + 2y(t) = u(t)$ is: $T(s) = \frac{Y(s)}{U(s)} = \frac{1}{s^2+3s+2}$. Input is $9\cos\left(7t + \tfrac{\pi}{3}\right)$ therefore, we need to calculate $T(7j)$.

$$T(s) = \frac{1}{s^2 + 3s + 2} \Rightarrow T(7j) = \frac{1}{(7j)^2 + 3 \times 7j + 2} = \frac{1}{-47 + 21j} = 0.0194 e^{-j2.7214}$$

$$y_{ss}(t) = y_p(t) = 0.0194 \times 9 \times \cos\left(7t + \frac{\pi}{3} - 2.7214\right) = 0.1746 \sin(7t - 1.6742)$$

General solution of this equation is

$$y(t) = y_h(t) + y_p(t) = C_1 e^{-t} + C_2 e^{-2t} + 0.1746 \sin(7t - 1.6742)$$

Example 5.18

Find the general solution of $\ddot{y}(t) + 3\dot{y}(t) + 2y(t) = 9\cos\left(7t + \frac{\pi}{3}\right) + 6\sin\left(5t + \frac{\pi}{6}\right)$.

Solution:

Homogenous solution can be found easily:

$$s^2 + 3s + 2 = 0 \Rightarrow s_1 = -1, s_2 = -2 \Rightarrow y_h(t) = C_1 e^{-t} + C_2 e^{-2t}$$

We use the superposition principle to find the particular solution. According to previous examples, particular solution of $\ddot{y}(t) + 3\dot{y}(t) + 2y(t) = 9\cos\left(7t + \frac{\pi}{3}\right)$ is $y_{p1}(t) = 0.1746 \sin (7t - 1.6742)$. Particular solution of $\ddot{y}(t) + 3\dot{y}(t) + 2y(t) = 6\sin\left(5t + \frac{\pi}{6}\right)$ is $y_{p2}(t) = 0.2184 \sin (5t - 2.0401)$. Therefore, particular solution of $\ddot{y}(t) + 3\dot{y}(t) + 2y(t) = 9\cos\left(7t + \frac{\pi}{3}\right) + 6\sin\left(5t + \frac{\pi}{6}\right)$ is

$$y_p(t) = y_{p1}(t) + y_{p2}(t) = 0.1746 \sin(7t - 1.6742) + 0.2184 \sin(5t - 2.0401)$$

General solution of this equation is

$$y(t) = y_h(t) + y_p(t) = C_1 e^{-t} + C_2 e^{-2t} + 0.1746 \sin(7t - 1.6742)$$
$$+ 0.2184 \sin(5t - 2.0401)$$

Example 5.19

Find the general solution of $\ddot{y}(t) + 3\dot{y}(t) + 2y(t) = 7e^{-4t}$.

Solution:

Homogenous solution can be found easily:

$$s^2 + 3s + 2 = 0 \Rightarrow s_1 = -1, s_2 = -2 \Rightarrow y_h(t) = C_1 e^{-t} + C_2 e^{-2t}$$

The transfer function for $\ddot{y}(t) + 3\dot{y}(t) + 2y(t) = u(t)$ is: $T(s) = \frac{Y(s)}{U(s)} = \frac{1}{s^2+3s+2}$. Input is $7e^{-4t}$. Therefore, we need to calculate $T(-4)$.

$$T(s) = \frac{1}{s^2 + 3s + 2} \Rightarrow T(-4) = \frac{1}{(-4)^2 + 3 \times -4 + 2} = \frac{1}{6}$$

$$y_p(t) = \frac{1}{6} \times 7e^{-4t} = \frac{7}{6} e^{-4t}$$

General solution of this equation is

$$y(t) = y_h(t) + y_p(t) = C_1 e^{-t} + C_2 e^{-2t} + \frac{7}{6} e^{-4t}$$

Example 5.20
Find the general solution of $\ddot{y}(t) + 3\dot{y}(t) + 2y(t) = 5e^{4t}$.

Solution:
Homogenous solution can be found easily:

$$s^2 + 3s + 2 = 0 \Longrightarrow s_1 = -1, s_2 = -2 \Longrightarrow y_h(t) = C_1 e^{-t} + C_2 e^{-2t}$$

The transfer function for $\ddot{y}(t) + 3\dot{y}(t) + 2y(t) = u(t)$ is: $T(s) = \frac{Y(s)}{U(s)} = \frac{1}{s^2+3s+2}$. Input
is $5e^{4t}$. Therefore, we need to calculate $T(+4)$.

$$T(s) = \frac{1}{s^2 + 3s + 2} \Longrightarrow T(4) = \frac{1}{(4)^2 + 3 \times 4 + 2} = \frac{1}{30}$$

$$y_p(t) = \frac{1}{30} \times 5e^{4t} = \frac{1}{6} e^{4t}$$

General solution of this equation is

$$y(t) = y_h(t) + y_p(t) = C_1 e^{-t} + C_2 e^{-2t} + \frac{1}{6} e^{4t}$$

Example 5.21
Find the general solution of $\ddot{y}(t) + 3\dot{y}(t) + 2y(t) = 7e^{j4t}$.

Solution:
Homogenous solution can be found easily:

$$s^2 + 3s + 2 = 0 \Longrightarrow s_1 = -1, s_2 = -2 \Longrightarrow y_h(t) = C_1 e^{-t} + C_2 e^{-2t}$$

The transfer function for $\ddot{y}(t) + 3\dot{y}(t) + 2y(t) = u(t)$ is: $T(s) = \frac{Y(s)}{U(s)} = \frac{1}{s^2+3s+2}$. Input
is $7e^{j4t}$. Therefore, we need to calculate $T(4j)$.

$$T(s) = \frac{1}{s^2 + 3s + 2} \Longrightarrow T(4) = \frac{1}{(4j)^2 + 3 \times 4j + 2} = \frac{1}{-14 + 12j}$$

$$y_p(t) = \frac{1}{-14 + 12j} \times 7e^{j4t} = \frac{7}{-14 + 12j} \times e^{j4t} = (-0.2882 - j0.2471)e^{j4t}$$
$$= 0.3796 e^{-j2.433} e^{j4t} = 0.3796 e^{j(4t-2.433)}$$

Therefore, the general solution of this equation is

$$y(t) = y_h(t) + y_p(t) = C_1 e^{-t} + C_2 e^{-2t} + 0.3796 e^{j(4t-2.433)}$$

Let us employ the undetermined coefficient method to see whether the obtained result is correct.

$$y_p(t) = Ae^{j4t}$$

$$\dot{y}_p(t) = j4Ae^{j4t}$$

$$\ddot{y}_p(t) = -16Ae^{j4t}$$

Note that $A \in \mathbb{C}$, that is, $A = a + bj$.

$$\ddot{y}_p(t) + 3\dot{y}_p(t) + 2y_p(t) = 7e^{j4t} \Rightarrow -16Ae^{j4t} + 3 \times j4Ae^{j4t} + 2 \times Ae^{j4t}$$
$$= 7e^{j4t} \Rightarrow (-14 + 12j)A = 7 \Rightarrow (-14 + 12j)(a + bj)$$
$$= 7 + 0j \Rightarrow -14a - 12b + j(12a - 14b) = 7 + 0j$$

Remember that two complex numbers are equal if and only if their real parts are equal and their imaginary parts are equal. Therefore,

$$\begin{cases} -14a - 12b = 7 \\ 12a - 14b = 0 \end{cases} \Rightarrow a = -0.2882, b = -0.2471$$

Particular solution is

$$y_p(t) = (-0.2882 - j0.2471)e^{j4t} = 0.3796e^{-j2.433}e^{j4t} = 0.3796e^{j(4t-2.433)}$$

Therefore, the general solution is

$$y(t) = y_h(t) + y_p(t) = C_1e^{-t} + C_2e^{-2t} + 0.3796e^{j(4t-2.433)}$$

The result obtained is identical to the one from the previous method.

Example 5.22
Find the general solution of $\ddot{y}(t) + 3\dot{y}(t) + 2y(t) = 3e^{j4t}$.

Solution:
Homogenous solution can be found easily:

$$s^2 + 3s + 2 = 0 \Rightarrow s_1 = -1, s_2 = -2 \Rightarrow y_h(t) = C_1e^{-t} + C_2e^{-2t}$$

The transfer function for $\ddot{y}(t) + 3\dot{y}(t) + 2y(t) = u(t)$ is $T(s) = \frac{Y(s)}{U(s)} = \frac{1}{s^2+3s+2}$. Input is $3e^{j4t}$. Therefore, we need to calculate $T(j4)$.

$$T(s) = \frac{1}{s^2 + 3s + 2} \Rightarrow T(j4) = \frac{1}{(j4)^2 + 3 \times j4 + 2} = \frac{1}{-14 + 12j}$$

$$y_p(t) = \frac{1}{-14 + 12j} \times 3e^{j4t} = 0.0542e^{-j2.433} \times 3e^{j4t} = 0.1626e^{j(4t - 2.433)}$$

Therefore, the general solution of this equation is

$$y(t) = y_h(t) + y_p(t) = C_1 e^{-t} + C_2 e^{-2t} + 0.1626e^{j(4t - 2.433)}$$

Example 5.23

Find the general solution of (A) $\ddot{y}(t) + 3\dot{y}(t) + 2y(t) = 3\cos(4t)$ (B) $\ddot{y}(t) + 3\dot{y}(t) + 2y(t) = 3\cos(4t)$.

Solution:

Homogenous solution can be found easily:

$$s^2 + 3s + 2 = 0 \Rightarrow s_1 = -1, s_2 = -2 \Rightarrow y_h(t) = C_1 e^{-t} + C_2 e^{-2t}$$

As demonstrated in the preceding example, a particular solution to the input $u = 3e^{j4t}$ is given by $y_p(t) = 0.1626e^{j(4t - 2.433)}$. The input in part (A) is $u_A(t) = 3\cos(4t) = Re\{3e^{j4t}\}$. Therefore, particular solution of $\ddot{y}(t) + 3\dot{y}(t) + 2y(t) = 3\cos(4t)$ is $y_p(t) = Re\{0.1626e^{j(4t - 2.433)}\} = 0.1626\cos(4t - 2.433)$. The general solution of part (A) is: $y(t) = y_h(t) + y_p(t) = C_1 e^{-t} + C_2 e^{-2t} + 0.1626\cos(4t - 2.433)$.

The input in part (B) is $u_B(t) = 3\sin(4t) = Im\{3e^{j4t}\}$. Therefore, particular solution of $\ddot{y}(t) + 3\dot{y}(t) + 2y(t) = 3\sin(4t)$ is $y_p(t) = Im\{0.1626e^{j(4t - 2.433)}\} = 0.1626\sin(4t - 2.433)$. The general solution of part (B) is $y(t) = y_h(t) + y_p(t) = C_1 e^{-t} + C_2 e^{-2t} + 0.1626\sin(4t - 2.433)$.

Example 5.24

Find the general solution of $\ddot{y}(t) + 3\dot{y}(t) + 2y(t) = 6$.

Solution:

Homogenous solution can be found easily:

$$s^2 + 3s + 2 = 0 \Rightarrow s_1 = -1, s_2 = -2 \Rightarrow y_h(t) = C_1 e^{-t} + C_2 e^{-2t}$$

The transfer function for $\ddot{y}(t) + 3\dot{y}(t) + 2y(t) = u(t)$ is: $T(s) = \frac{Y(s)}{U(s)} = \frac{1}{s^2 + 3s + 2}$. Input is $6e^{j0t}$. Therefore, we need to calculate $T(0)$.

$$T(s) = \frac{1}{s^2 + 3s + 2} \Rightarrow T(0) = \frac{1}{(0)^2 + 3 \times 0 + 2} = \frac{1}{2}$$

$$y_p(t) = \frac{1}{2} \times 6e^{j0t} = 3$$

Therefore, the general solution of this equation is

$$y(t) = y_h(t) + y_p(t) = C_1 e^{-t} + C_2 e^{-2t} + 3$$

Example 5.25

Find the general solution of $\ddot{y}(t) + 3\dot{y}(t) + 2y(t) = 4 + 7\cos\left(3t + \frac{\pi}{12}\right)$.

Solution:

Homogenous solution can be found easily:

$$s^2 + 3s + 2 = 0 \Rightarrow s_1 = -1, s_2 = -2 \Rightarrow y_h(t) = C_1 e^{-t} + C_2 e^{-2t}$$

The transfer function for $\ddot{y}(t) + 3\dot{y}(t) + 2y(t) = u(t)$ is: $T(s) = \frac{Y(s)}{U(s)} = \frac{1}{s^2+3s+2}$. Input is $4 + 7\cos\left(3t + \frac{\pi}{12}\right)$. Let us use superposition to find the particular solution. For $u_1(t) = 4$, output is

$$y_{p1}(t) = T(0) \times u_1(t) = \frac{1}{(0)^2 + 3 \times 0 + 2} \times 4 = \frac{1}{2} \times 4 = 2$$

For $u_2(t) = 7\cos\left(3t + \frac{\pi}{12}\right)$, output is

$$T(j3) = \frac{1}{(j3)^2 + 3 \times j3 + 2} = \frac{1}{-7 + j9} = 0.0877 e^{-j2.2318}$$

$$y_{p2}(t) = 0.0877 \times 7 \cos\left(3t + \frac{\pi}{12} - 2.2318\right) = 0.6139 \cos(3t - 1.9700)$$

Therefore, the particular solution is

$$y_p(t) = y_{p1}(t) + y_{p2}(t) = 2 + 0.6139 \cos(3t - 1.9700)$$

Therefore, the general solution of this equation is

$$y(t) = y_h(t) + y_p(t) = C_1 e^{-t} + C_2 e^{-2t} + 2 + 0.6139 \cos(3t - 1.9700)$$

Exercise 5.1

Use the superposition principle to fine the particular solution of $\ddot{y}(t) + 3\dot{y}(t) + 2y(t) = 3e^{j4t} = 3\cos(4t) + j3\sin(4t)$.

Example 5.26

Solve the differential equation $\begin{cases} \dfrac{dx(t)}{dt} = 2x(t) + 3y(t) \\ \dfrac{dy(t)}{dt} = x(t) + 4y(t) \end{cases}$ with initial conditions

$x(0) = 0$ and $y(0) = 1$.

Solution:

Applying the Laplace transform to the given equations:

$$\begin{cases} sX(s) = 2X(s) + 3Y(s) \\ sY(s) - 1 = X(s) + 4Y(s) \end{cases}$$

To simplify notation, let $X(s)$ be denoted by X and $Y(s)$ by Y:

$$\begin{cases} sX = 2X + 3Y \\ sY - 1 = X + 4Y \end{cases} \Rightarrow \begin{cases} (2-s)X + 3Y = 0 \\ X + (4-s) = -1 \end{cases} \Rightarrow \begin{bmatrix} 2-s & 3 \\ 1 & 4-s \end{bmatrix} \begin{bmatrix} X \\ Y \end{bmatrix}$$

$$= \begin{bmatrix} 0 \\ -1 \end{bmatrix} \Rightarrow \begin{bmatrix} X \\ Y \end{bmatrix} = \begin{bmatrix} 2-s & 3 \\ 1 & 4-s \end{bmatrix}^{-1} \begin{bmatrix} 0 \\ -1 \end{bmatrix}$$

$$= \frac{1}{(2-s)(4-s) - 3} \begin{bmatrix} 4-s & -3 \\ -1 & 2-s \end{bmatrix} \begin{bmatrix} 0 \\ -1 \end{bmatrix} = \frac{1}{s^2 - 6s + 5} \begin{bmatrix} 3 \\ s-2 \end{bmatrix}$$

$$= \frac{1}{(s-1)(s-5)} \begin{bmatrix} 3 \\ s-2 \end{bmatrix} \Rightarrow \begin{cases} X(s) = \dfrac{3}{(s-1)(s-5)} \\ Y(s) = \dfrac{s-2}{(s-1)(s-5)} \end{cases}$$

$$x(t) = \mathcal{L}^{-1}(X(s)) = \mathcal{L}^{-1}\left(\frac{3}{(s-1)(s-5)} \right) = \mathcal{L}^{-1}\left(\frac{-\frac{3}{4}}{s-1} + \frac{\frac{3}{4}}{s-5} \right) = \frac{3}{4}\left(e^{5t} - e^{t} \right)$$

$$y(t) = \mathcal{L}^{-1}(Y(s)) = \mathcal{L}^{-1}\left(\frac{s-2}{(s-1)(s-5)} \right) = \mathcal{L}^{-1}\left(\frac{\frac{1}{4}}{s-1} + \frac{\frac{3}{4}}{s-5} \right) = \frac{1}{4}\left(3e^{5t} - e^{t} \right)$$

Exercise 5.2

Solve $\begin{cases} \dfrac{dx(t)}{dt} = 3x(t) - 2y(t) \\ \dfrac{dy(t)}{dt} = 4x(t) - y(t) \end{cases}$ with the initial condition of $x(0) = 1$ and $y(0) = 1$.

5.9   Problem Solving with MATLAB®

This section provides examples of how MATLAB can be employed to solve the types of problems encountered in this chapter.

Example 5.27

The code in Fig. 5.7 calculates $\int te^{-st}dt$.

Fig. 5.7 Calculation of $\int te^{-st}dt$

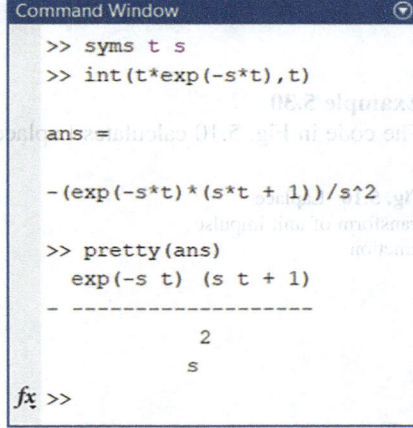

```
Command Window
>> syms t s
>> int(t*exp(-s*t),t)

ans =

-(exp(-s*t)*(s*t + 1))/s^2

>> pretty(ans)
  exp(-s t) (s t + 1)
- -------------------
           2
          s
fx >>
```

Example 5.28

The code in Fig. 5.8 calculates $\int e^{-j\omega_0 t}e^{-st}dt$.

Fig. 5.8 Calculation of $\int e^{-j\omega_0 t}e^{-st}dt$

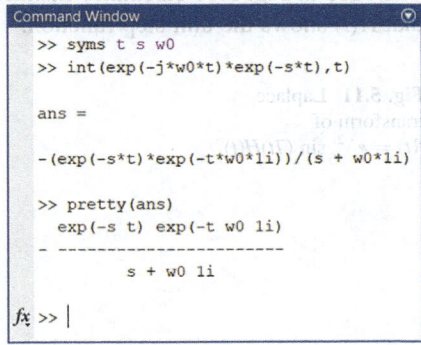

```
Command Window
>> syms t s w0
>> int(exp(-j*w0*t)*exp(-s*t),t)

ans =

-(exp(-s*t)*exp(-t*w0*1i))/(s + w0*1i)

>> pretty(ans)
  exp(-s t) exp(-t w0 1i)
- -----------------------
        s + w0 1i
fx >> |
```

Example 5.29

The code in Fig. 5.9 calculates Laplace transform of the unit step function.

Fig. 5.9 Laplace transform of unit step function

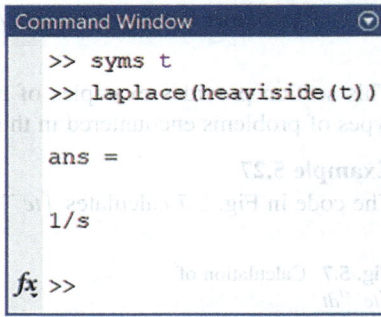

Example 5.30
The code in Fig. 5.10 calculates Laplace transform of the unit impulse function.

Fig. 5.10 Laplace transform of unit impulse function

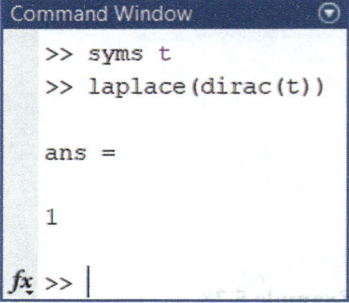

Example 5.31
The code in Fig. 5.11 calculates Laplace transform of $f(t) = e^{-2t} \sin (7t)H(t)$. Note that $H(t)$ shows the unit step function.

Fig. 5.11 Laplace transform of $f(t) = e^{-2t} \sin (7t)H(t)$

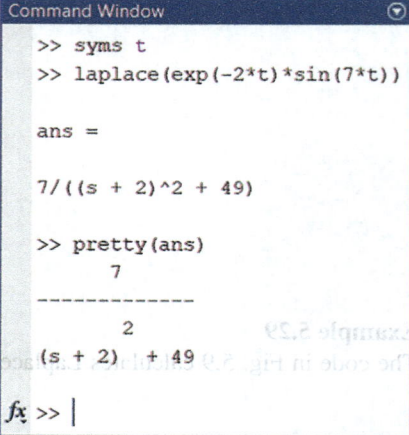

Example 5.32

The code in Fig. 5.12 calculates Laplace transform of $f(t) = \sin\left(5t + \frac{\pi}{4}\right)H(t)$.

Fig. 5.12 Laplace transform of $f(t) = \sin\left(5t + \frac{\pi}{4}\right)H(t)$

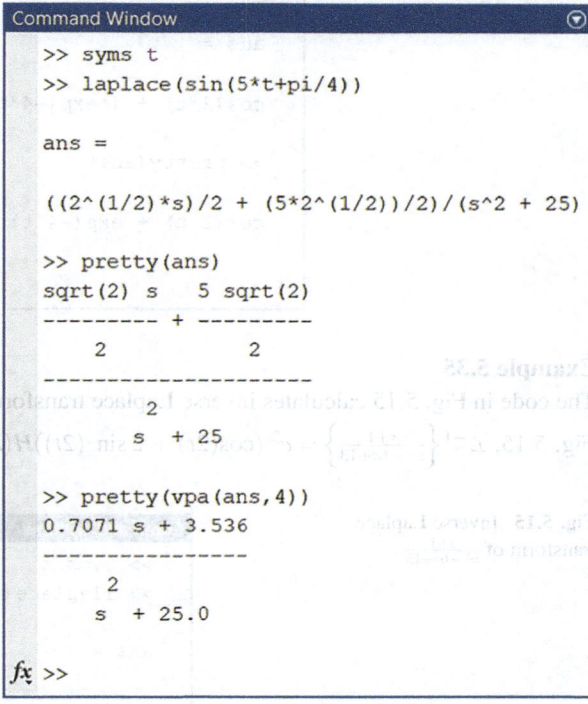

```
Command Window

>> syms t
>> laplace(sin(5*t+pi/4))


ans =


((2^(1/2)*s)/2 + (5*2^(1/2))/2)/(s^2 + 25)


>> pretty(ans)
sqrt(2) s     5 sqrt(2)
--------- + ---------
   2            2
---------------------
       2
      s  + 25

>> pretty(vpa(ans,4))
0.7071 s + 3.536
----------------
     2
    s  + 25.0

fx >>
```

Example 5.33

The code in Fig. 5.13 calculates Laplace transform of $f(t) = te^{-6t}H(t)$.

Fig. 5.13 Laplace transform of $f(t) = te^{-6t}H(t)$

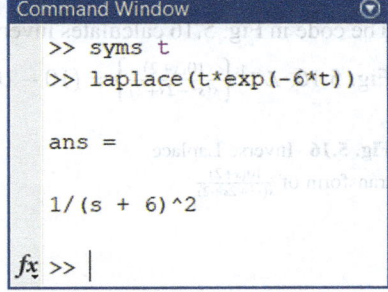

```
Command Window

>> syms t
>> laplace(t*exp(-6*t))


ans =

1/(s + 6)^2

fx >> |
```

Example 5.34

The code in Fig. 5.14 calculates inverse Laplace transform of $\frac{5s^2+s+4}{(s+4)(s^2+4)}$. According to Fig. 5.14, $\mathcal{L}^{-1}\left\{\frac{5s^2+s+4}{(s+4)(s^2+4)}\right\} = \left(4e^{-4t} + \cos(2t) - \frac{3}{2}\sin(2t)\right)H(t)$.

Fig. 5.14 Inverse Laplace
transform of $\frac{5s^2+s+4}{(s+4)(s^2+4)}$

```
Command Window
>> syms s
>> ilaplace((5*s^2+s+4)/((s+4)*(s^2+4)))

ans =

cos(2*t) + 4*exp(-4*t) - (3*sin(2*t))/2

>> pretty(ans)
                                  sin(2 t) 3
cos(2 t) + exp(-4 t) 4 - -----------
                                      2
fx >>
```

Example 5.35

The code in Fig. 5.15 calculates inverse Laplace transform of $\frac{s+1}{s^2-6s+13}$. According to
Fig. 5.15, $\mathcal{L}^{-1}\left\{\frac{s+1}{s^2-6s+13}\right\} = e^{3t}(\cos(2t) + 2\sin\ (2t))H(t)$.

Fig. 5.15 Inverse Laplace
transform of $\frac{s+1}{s^2-6s+13}$

```
Command Window
>> syms s
>> ilaplace((s+1)/(s^2-6*s+13))

ans =

exp(3*t)*(cos(2*t) + 2*sin(2*t))

fx >>
```

Example 5.36

The code in Fig. 5.16 calculates inverse Laplace transform of $\frac{10(s+2)}{s(s^2+2s+2)}$. According to
Fig. 5.16, $\mathcal{L}^{-1}\left\{\frac{10(s+2)}{s(s^2+2s+2)}\right\} = (10 - 10e^{-t}\cos(t))H(t)$.

Fig. 5.16 Inverse Laplace
transform of $\frac{10(s+2)}{s(s^2+2s+2)}$

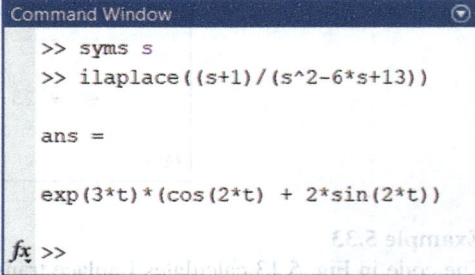

```
Command Window
>> syms s
>> ilaplace(10*(s+2)/s/(s^2+2*s+2))

ans =

10 - 10*exp(-t)*cos(t)

fx >>
```

Example 5.37

The code in Fig. 5.17 calculates the partial fraction decomposition of $\frac{5s^2+s+4}{(s+4)(s^2+4)}$.

According to Fig. 5.17, $\frac{5s^2+s+4}{(s+4)(s^2+4)} = \frac{4}{s+4} + \frac{0.5+0.75j}{s-2j} + \frac{0.5-0.75j}{s+2j}$.

Fig. 5.17 Partial fraction decomposition of $\frac{5s^2+s+4}{(s+4)(s^2+4)}$

```
Command Window

>> [r,p,k]=residue([5 1 4],conv([1 4],[1 0 4]))

r =

    4.0000 + 0.0000i
    0.5000 + 0.7500i
    0.5000 - 0.7500i


p =

   -4.0000 + 0.0000i
    0.0000 + 2.0000i
    0.0000 - 2.0000i


k =

    []

fx >> |
```

The code shown in Fig. 5.18 simplifies $\frac{0.5+0.75j}{s-2j} + \frac{0.5-0.75j}{s+2j}$. According to

Fig. 5.18, $\frac{0.5+0.75j}{s-2j} + \frac{0.5-0.75j}{s+2j} = \frac{s-3}{s^2+4}$. Therefore, $\frac{5s^2+s+4}{(s+4)(s^2+4)} = \frac{4}{s+4} + \frac{0.5+0.75j}{s-2j}$

$+ \frac{0.5-0.75j}{s+2j} = \frac{4}{s+4} + \frac{s-3}{s^2+4}$.

```
Command Window

>> simplify((.5+.75*j)/(s-2*j)+(.5-.75*j)/(s+2*j))

ans =

(s - 3)/(s^2 + 4)

fx >>
```

Fig. 5.18 Simplifying $\frac{0.5+0.75j}{s-2j} + \frac{0.5-0.75j}{s+2j}$

You can use the following technique to simplify the obtained expression as well (remember that $z + \bar{z} = 2\,Re\{z\}$):

$$\frac{5s^2+s+4}{(s+4)(s^2+4)} = \frac{4}{s+4} + \frac{0.5+0.75j}{s-2j} + \frac{0.5-0.75j}{s+2j} = \frac{4}{s+4} + 2\,Re\left\{\frac{0.5+0.75j}{s-2j}\right\}$$

$$= \frac{4}{s+4} + 2\,Re\left\{\frac{(0.5+0.75j)(s+2j)}{s^2+4}\right\}$$

$$= \frac{4}{s+4} + 2\,Re\left\{\frac{0.5s+j+0.75js-1.5}{s^2+4}\right\} = \frac{4}{s+4} + 2\times\frac{0.5s-1.5}{s^2+4}$$

$$= \frac{4}{s+4} + \frac{s-3}{s^2+4}$$

Example 5.38

The code in Fig. 5.19 calculates the partial fraction decomposition of $\frac{10(s+2)}{s(s^2+2s+2)}$.

According to Fig. 5.19, $\frac{10(s+2)}{s(s^2+2s+2)} = \frac{-5}{s-(-1+j)} + \frac{-5}{s-(-1-j)} + \frac{10}{s}$.

Fig. 5.19 Partial fraction decomposition of $\frac{10(s+2)}{s(s^2+2s+2)}$

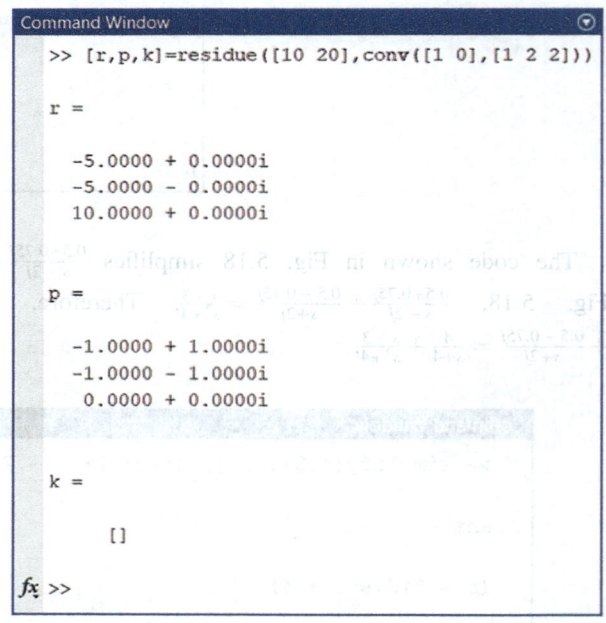

```
Command Window

>> [r,p,k]=residue([10 20],conv([1 0],[1 2 2]))

r =

   -5.0000 + 0.0000i
   -5.0000 - 0.0000i
   10.0000 + 0.0000i


p =

   -1.0000 + 1.0000i
   -1.0000 - 1.0000i
    0.0000 + 0.0000i


k =

   []

fx >>
```

The code shown in Fig. 5.20 simplifies $\frac{-5}{s-(-1+j)} + \frac{-5}{s-(-1-j)}$. According to Fig. 5.20, $\frac{-5}{s-(-1+j)} + \frac{-5}{s-(-1-j)} = \frac{-(10s+10)}{s^2+2s+2}$. Therefore, $\frac{10(s+2)}{s(s^2+2s+2)} = \frac{10}{s} + \frac{-5}{s-(-1+j)} + \frac{-5}{s-(-1-j)} = \frac{10}{s} - \frac{(10s+10)}{s^2+2s+2}$.

Fig. 5.20 Simplifying $\frac{-5}{s-(-1+j)} + \frac{-5}{s-(-1-j)}$

```
Command Window
>> simplify(-5/(s+1-j)-5/(s+1+j))

ans =

-(10*s + 10)/(s^2 + 2*s + 2)
fx >> |
```

Example 5.39

The code in Fig. 5.21 solves $\ddot{y}(t) + 5\dot{y}(t) + 6y(t) = 4\sin\left(t + \frac{\pi}{3}\right)$.

```
Command Window
>> syms y(t)
>> ode=diff(y,t,2)+5*diff(y,t)+6*y==4*sin(t+pi/3);
>> dsolve(ode)

ans =

C1*exp(-3*t) - (2*2^(1/2)*cos(t + (7*pi)/12))/5 + C2*exp(-2*t)
fx >>
```

Fig. 5.21 The code for solving $\ddot{y}(t) + 5\dot{y}(t) + 6y(t) = 4\sin\left(t + \frac{\pi}{3}\right)$

The vpa function takes the symbolic expression as input and the desired number of decimal places as the second argument. It then evaluates the expression numerically and displays the result with the specified precision. By using vpa, you can ensure that the numerical values in the solution are displayed with the desired level of accuracy.

To obtain a more precise numerical approximation, let us apply the vpa function to the calculated result (Fig. 5.22).

```
Command Window
>> syms y(t)
>> ode=diff(y,t,2)+5*diff(y,t)+6*y==4*sin(t+pi/3);
>> dsolve(ode)

ans =

C1*exp(-3*t) - (2*2^(1/2)*cos(t + (7*pi)/12))/5 + C2*exp(-2*t)

>> vpa(ans,5)

ans =

C2*exp(-2.0*t) - 0.56569*cos(t + 1.8326) + C1*exp(-3.0*t)
fx >>
```

Fig. 5.22 Utilize the vpa command to numerically evaluate the symbolic expression

Example 5.40

The code in Fig. 5.23 solves $\ddot{y}(t) + 5\dot{y}(t) + 6y(t) = 4\sin\left(t + \frac{\pi}{3}\right), y(0) = 0, \dot{y}(0) = 1$.

```
Command Window
>> syms y(t)
>> ode=diff(y,t,2)+5*diff(y,t)+6*y==4*sin(t+pi/3);
>> ic=[y(0)==0,subs(diff(y,t),0)==1];
>> dsolve(ode,ic)

ans =

exp(-2*t)*((6*2^(1/2)*(2^(1/2)/4 - 6^(1/2)/4))/5 - (2*2^(1/2)*(2^(1/2)/4 +

>> vpa(ans,5)

ans =

0.014359*exp(-2.0*t) - 0.16077*exp(-3.0*t) - 0.56569*cos(t + 1.8326)

fx >>
```

Fig. 5.23 The code for solving $\ddot{y}(t) + 5\dot{y}(t) + 6y(t) = 4\sin\left(t + \frac{\pi}{3}\right), y(0) = 0, \dot{y}(0) = 1$

Example 5.41

The code in Fig. 5.24 solves $\ddot{y}(t) + 2\dot{y}(t) + y(t) = \delta(t), y(0) = \dot{y}(0) = 0$. Note that $\delta(t)$ shows the unit impulse function.

Fig. 5.24 The code for solving $\ddot{y}(t) + 2\dot{y}(t) + y(t) = \delta(t), y(0) = \dot{y}(0) = 0$

```
Command Window
>> syms y(t)
>> ode=diff(y,t,2)+2*diff(y,t)+y==dirac(t);
>> ic=[y(0)==0,subs(diff(y,t),0)==0];
>> dsolve(ode,ic)

ans =

(t*exp(-t)*sign(t))/2

fx >> |
```

According to Fig. 5.24, the solution is $y(t) = \frac{te^{-t}}{2}\text{sign}(t)$. Remember that $\text{sign}(t) = \begin{cases} +1 & t > 0 \\ 0 & t = 0 \\ -1 & t < 0 \end{cases}$. Therefore, the solution of $\ddot{y}(t) + 2\dot{y}(t) + y(t) = \delta(t), y(0) = \dot{y}(0) = 0$ for $t > 0$ is $y(t) = \frac{te^{-t}}{2}$.

Example 5.42

The code in Fig. 5.25 solves $\ddot{y}(t) + 2\dot{y}(t) + y(t) = 6H(t), y(0) = \dot{y}(0) = 0$. Note that $H(t)$ shows the unit step function.

```
Command Window

>> syms y(t)
>> ode=diff(y,t,2)+2*diff(y,t)+y==6*heaviside(t);
>> ic=[y(0)==1,subs(diff(y,t),0)==2];
>> dsolve(ode,ic)

ans =

-exp(-t)*(3*sign(t) - 3*exp(t) - 3*exp(t)*sign(t) + 3*t*sign(t) + 2)

>> expand(ans)

ans =

3*sign(t) - 2*exp(-t) - 3*exp(-t)*sign(t) - 3*t*exp(-t)*sign(t) + 3

fx >> |
```

Fig. 5.25 The code for solving $\ddot{y}(t) + 2\dot{y}(t) + y(t) = 6H(t), y(0) = \dot{y}(0) = 0$

According to Fig. 5.25, the solution for $t > 0$ is $y(t) = 6 - 5e^{-t} - 3te^{-t}$.

Example 5.43

The code in Fig. 5.26 solves $\begin{cases} \dfrac{dx(t)}{dt} = 3x(t) + 4y(t) \\ \dfrac{dy(t)}{dt} = -4x(t) + 3y(t) \end{cases}$ $\begin{bmatrix} x(0) \\ y(0) \end{bmatrix} = \begin{bmatrix} 0 \\ 1 \end{bmatrix}$.

```
Command Window                                    ⊙
    >> syms x(t) y(t)
    >> eq1=3*x+4*y==diff(x,t);
    >> eq2=-4*x+3*y==diff(y,t);
    >> ic=[x(0)==0,y(0)==1];
    >> [xSol,ySol]=dsolve([eq1,eq2],ic)


    xSol =


    sin(4*t)*exp(3*t)


    ySol =


    cos(4*t)*exp(3*t)

fx >>
```

Fig. 5.26 The code for solving $\begin{cases} \dfrac{dx(t)}{dt} = 3x(t) + 4y(t) \\ \dfrac{dy(t)}{dt} = -4x(t) + 3y(t) \end{cases}$ $\begin{bmatrix} x(0) \\ y(0) \end{bmatrix} = \begin{bmatrix} 0 \\ 1 \end{bmatrix}$

Example 5.44

The code in Fig. 5.27 plots the impulse response, that is, the output for the unit impulse input with zero initial conditions, of $\ddot{y}(t) + 3\dot{y}(t) + 4y(t) = 8\dot{u}(t) + u(t)$ on $[0, 6]$ interval. Note that $y(t)$ and $u(t)$ show the output and input, respectively. The transfer function of given differential equation is (the initial conditions are zero):

$$L\{\ddot{y}(t)\} + L\{3\dot{y}(t)\} + L\{4y(t)\} = L\{8\dot{u}(t)\} + L\{u(t)\} \Longrightarrow (s^2 + 3s + 4)Y(s)$$

$$= (8s + 1)U(s) \Longrightarrow T(s) = \frac{Y(s)}{U(s)} = \frac{8s + 1}{s^2 + 3s + 4}$$

Fig. 5.27 Plotting $\ddot{y}(t) + 3\dot{y}(t) + 4y(t) = 8\dot{u}(t) + u(t)$ on $[0, 6]$ interval

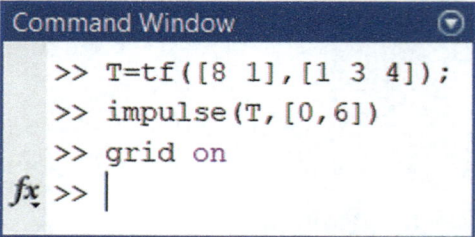

```
Command Window                                    ⊙
    >> T=tf([8 1],[1 3 4]);
    >> impulse(T,[0,6])
    >> grid on
fx >> |
```

Output of the code shown in Fig. 5.27 is shown in Fig. 5.28.

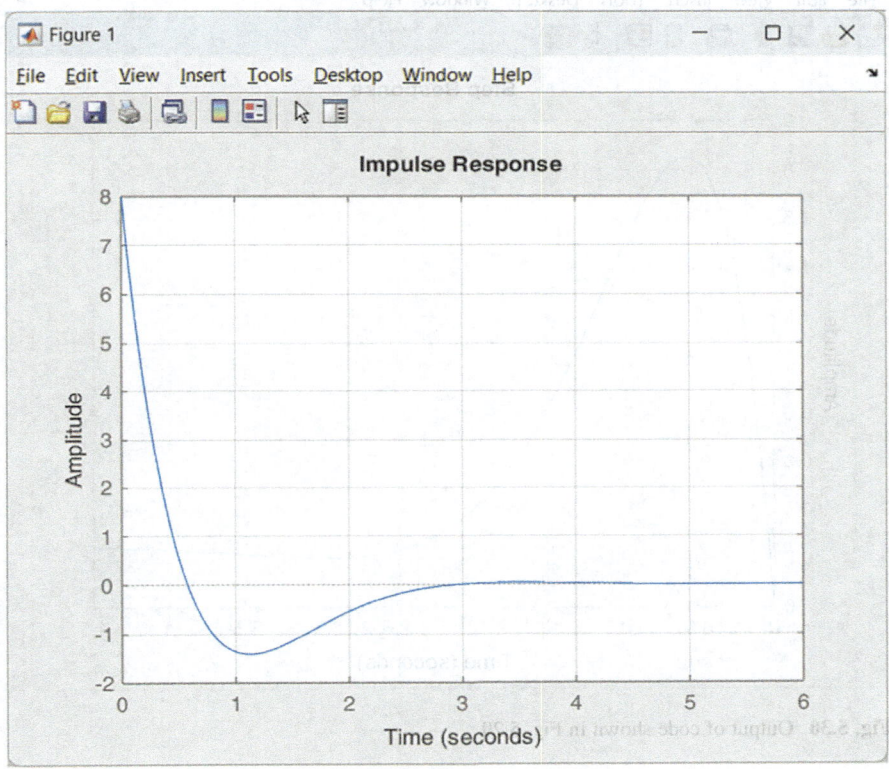

Fig. 5.28 Output of the code shown in Fig. 5.27

Example 5.45
The code in Fig. 5.29 plots the step response, that is, the output for the unit step input with zero initial conditions, of $\ddot{y}(t) + 3\dot{y}(t) + 4y(t) = 8\dot{u}(t) + u(t)$ on $[0, 4.5]$ interval. Note that $y(t)$ and $u(t)$ show the output and input, respectively. Output of the code shown in Fig. 5.29 is shown in Fig. 5.30.

Fig. 5.29 Plotting step
response of $\ddot{y}(t) + 3\dot{y}(t) +$
$4y(t) = 8\dot{u}(t) + u(t)$ on
$[0, 4.5]$ interval

```
Command Window
>> T=tf([8 1],[1 3 4]);
>> step(T,[0,4.5])
>> grid on
fx >>
```

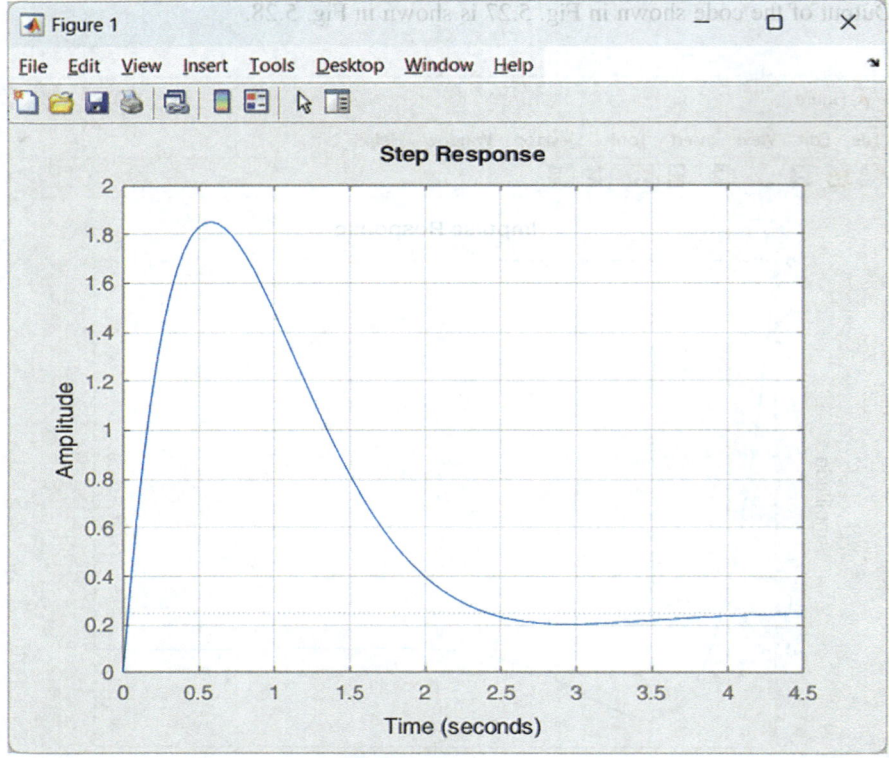

Fig. 5.30 Output of code shown in Fig. 5.29

Further Reading

1. Advanced Engineering Mathematics (10th edition), Erwin Kreyszig, Wiley, 2011.
2. Calculus (8th edition), James Stewart, Cengage Learning, 2015.
3. Applied Numerical Analysis with MATLAB®/Simulink®: For Engineers and Scientists, Farzin Asadi, Springer, 2023.

Index